11000 Years old cave painting probably illustrates the landing skills of a cormorant (see page 3)

Oxford Ornithology Series
Edited by T. R. Birkhead FRS

1. *Bird Population Studies: Relevance to Conservation and Management* (1991)
   Edited by C. M. Perrins, J.-D. Lebreton, and G. J. M. Hirons
2. *Bird-Parasite Interactions: Ecology, Evolution, and Behaviour* (1991)
   Edited by J. E. Loye and M. Zuk
3. *Bird Migration: A General Survey* (1993)
   Peter Berthold
4. *The Snow Geese of La Pérouse Bay: Natural Selection in the Wild* (1995)
   Fred Cooke, Robert F. Rockwell, and David B. Lank
5. *The Zebra Finch: A Synthesis of Field and Laboratory Studies* (1996)
   Richard A. Zann
6. *Partnerships in Birds: The Study of Monogomy* (1996)
   Edited by Jeffery M. Black
7. *The Oystercatcher: From Individuals to Populations* (1996)
   Edited by John D. Goss-Custard
8. *Avian Growth and Development: Evolution within the Altricial-Precocial Spectrum* (1997)
   Edited by J. M. Starck and R. E. Ricklefs
9. *Parasitic Birds and Their Hosts* (1998)
   Edited by Stephen I. Rothstein and Scott K. Robinson
10. *The Evolution of Avian Breeding Systems* (1999)
    J. David Ligon
11. *Harriers of the World: Their Behaviour and Ecology* (2000)
    Robert E. Simmons
12. *Bird Migration* 2e (2001)
    Peter Berthold
13. *Avian Incubation: Behaviour, Environment, and Evolution* (2002)
    Edited by D. C. Deeming
14. *Avian Flight* (2005)
    John J. Videler

Avian Flight

JOHN J. VIDELER
Marine Zoology, Groningen University
Evolutionary Mechanics, Leiden University
The Netherlands

OXFORD
UNIVERSITY PRESS

OXFORD
UNIVERSITY PRESS

Great Clarendon Street, Oxford OX2 6DP

Oxford University Press is a department of the University of Oxford.
It furthers the University's objective of excellence in research, scholarship,
and education by publishing worldwide in

Oxford New York

Auckland Cape Town Dar es Salaam Hong Kong Karachi
Kuala Lumpur Madrid Melbourne Mexico City Nairobi
New Delhi Shanghai Taipei Toronto

With offices in

Argentina Austria Brazil Chile Czech Republic France Greece
Guatemala Hungary Italy Japan Poland Portugal Singapore
South Korea Switzerland Thailand Turkey Ukraine Vietnam

Published in the United States
by Oxford University Press Inc., New York

First published 2005
First published in paperback 2006

British Library Cataloguing in Publication Data
Data available

Library of Congress Cataloging in Publication Data
Data available

Typeset by Newgen Imaging Systems (P) Ltd., Chennai, India
Printed in Great Britain
on acid-free paper by
Biddles Ltd., King's Lynn

ISBN 0–19–856603–4 978–0–19–856603–8
ISBN 0–19–929992–7 (Pbk.) 978–0–19–929992–8 (Pbk.)

1 3 5 7 9 10 8 6 4 2

To my seven muses (in order of appearance)

Hanneke
Tennie
Hetteke
Sasha
Elise
Sanne
Hanna

Preface

Birds are feathered quadruped vertebrates who use their front legs as wings. The number of recognized species approaches 10,000. Not all birds fly. There are about 8500 extant volant species and each one flies in its own particular way. The variation is tremendous although birds also share common characteristics directly related to flight. Both the diversity and the basic requirements are fascinating. The fact that vertebrate structure and movement, adapted to satisfy aerodynamic principles, can defeat gravity is fascinating and we want to know how it is achieved. A variety of disciplines is involved in the subject. The demands of the physical laws governing the conquest of the air are clearly reflected in details of morphological, physiological and behavioural adaptations. A large body of knowledge is dispersed in specialized literature and is not easily accessible. Aerodynamic principles are often described in mathematical terms in papers where there is no room for explanations aimed at those primarily interested in birds. On the other hand, the complexity of the biological phenomena is difficult to grasp for scholars who are physically and mathematically orientated. I would like to display the interrelationships among the variety of aspects around avian flight in an attempt to make the subject comprehensible for interested readers from all disciplines.

Structure and function of bird wings differ fundamentally from aircraft wings in many ways. Bird wings consist of an arm and a hand part and can be stretched and folded. During gliding and flapping flight the wing shape can change dramatically. Flapping wings provide both lift and thrust and not just lift forces. It will become clear that aerodynamic theories explaining how conventional aeroplanes fly, are not sufficient to clarify flapping and gliding flight in birds. Most birds can land on a branch. To understand how they do that is an objective of this book.

'Avian flight' covers main aspects of aerial locomotion by birds including sections on history of science, aerodynamics, functional morphology, evolution, kinematics, physiology, energetics and costs of flight. The reader is expected to have some basic understanding of biology, physics and mathematics, but a university degree in any of these directions is not required. This book aims at a wide range of readers interested in the relationships between form and function. It should attract the attention of both professional ornithologists and amateur bird watchers. It is also aimed to be of interest to anyone preoccupied with the miracle of flight in general or with man made flying machines in particular. Equations, models and complex technical details are presented in separate boxes, studying the contents of these is not required to follow the main flow of arguments and information.

Common and scientific names follow Dickinson's (2003) 'The Howard & Moore complete checklist of the birds of the world'. Only common names are used in the text. Appendix 1 lists these in alphabetical order with the accompanying scientific names; Appendix 2 provides the scientific names alphabetically with the associated common names. Biological terms are explained (mainly using the eleventh edition of Henderson's dictionary, 1996) in Appendix 3.

My personal interest in this topic was raised in the early 1960s by a book entitled 'De vliegkunst in het dierenrijk' (The art of flight in the animal kingdom) written by E. J. Slijper (who was my professor in zoology at the university of Amsterdam) in collaboration with the aero- and hydrodynamicist J. M. Burgers of Delft Technical University. More than twenty years later studies on the flight performance and energetics of common kestrels, in collaboration with the ecologist Serge Daan and his group (Groningen University, the Netherlands) and the aerodynamicist Daniel Weihs (Technion Haifa, Israel), offered the opportunity to actually contribute to the field. Kestrels and rough-legged buzzards kept as falconry birds for scientific studies made me an addict to the subject for the rest of my life.

There are many reasons why bird flight is fascinating. An important one is the fact that, as I hope to show, we don't fully understand how birds actually do it. Many features specific to the way birds behave during take-off, in flapping or gliding flight and during landing are still enigmatic. The movements are usually so fast that these must be studied in slow motion to appreciate what is actually happening. The study of the events inside flying birds is even more difficult. Another important reason for our lack of understanding is the problem that we are unable to see or measure the reaction of the air as a bird moves through it. So, there is still a veil of mystery over the interaction between bird, air and gravity, let us find out how far it can be lifted.

Watching a great variety of birds with special attention for the movements, helps to learn more about the flight mechanisms but it also emphasizes the feeling that we have a long way to go before we will fully understand avian flight. The background knowledge offered in this book is intended to make such observations even more rewarding. The following section gives an example of the kind of experiences I am aiming at.

Bird flight watching

The beauty, the variety and fascinating easiness of bird flight can be observed almost everywhere, however some places are better than others. My work as a marine biologist brings me every year to my favourite spot while teaching a field course in the marine life of Mediterranean rocky shores. I am privileged to be lodged at La Revellata lighthouse high on a cliff at the end of a rocky peninsula, overlooking the Bay of Calvi on Corsica, the island of beauty. It is a focus point for migrating birds, but there are also many residents breeding on the steep cliffs or hiding in the rich maquis vegetation. There is a large variety in species and sizes, from the tiny Marmora's warbler to the osprey, but it is not so much this type of variety that strikes me most. Every species shows different flight techniques, even those that seem to match closely in size, shape, and behaviour. This is demonstrated by the following personal impression of the variety of flight behaviours observed one morning from that vantage-point.

The fishing ospreys use flapping flight predominantly. The amplitude of the wing beats is shallow and it sometimes looks as if there is a snap in the middle between the hand and the arm part. They came earlier this morning soaring, using the upward movement of the sea wind obstructed by the high rocks of the peninsula. Ospreys are among the largest birds that are able to windhover. Windhovering is flying against the wind at the speed of the wind. They use it while fishing, remaining briefly at one spot above the water before diving into the water with the wings half closed. Ospreys make a peculiar shaking wing beat just after they take-off from the surface obviously to get rid of the water trapped between the feathers. It usually takes several strikes before a plunge is successful. The struggle in the water can be heavy. A bird with a fish in its claws wildly flaps its swept back hand wings while attempting to get out of the water. Once airborne, the fish is kept in an aerodynamically favourable head on position. Even with a fish, the ospreys can circle in up-draught without a single wing beat, the wings kept stretched with separated primary feathers, quickly gaining height up to well above the lighthouse before gliding down in the direction of the nest.

There are three falcons on the peninsula. The kestrels typically windhover, flapping the wings while keeping the head in a remarkably fixed position. They also make good use of up-draughts, enabling them to hang motionless on one spot. I am not sure if Eleonora's falcon breeds here, but they pass regularly causing havoc among the songbirds. The flight is impressive. The tips of the hand wings point backwards, while the dark grey birds shoot like slender arrows past cliffs and rocky outcrops. They never windhover. The peregrines that are nesting on the steep cliffs are more robust and bigger than the other two species of falcon. Hunting starts with circling high up in the air where they await a chance to stoop down on their prey at tremendous speed. The flight of the three falcon species could hardly be more different.

Two species of gull can be spotted on the small island rock just offshore from the lighthouse. The Audouin's gulls only occasionally show up. Apart from the red beak with black band and the dark green feet they look similar to the abundant yellow-legged gulls. These have yellow beaks with a red dot. The Audouin's wings appear slightly more slender than those of the yellow-legged gulls. Both species make good use of favourable wind conditions for soaring and gliding around the rocks. They spend hours effortlessly sailing the wind. From my high position I can observe the flight from above, the side and below. Gulls can glide absolutely motionless with stretched arm wings and swept-back hand wings. Their body is beautifully streamlined and looks small with respect to the wings. Sometimes they change to slow flapping flight with stretched wings, easily rowing. The Audouin's gulls forage in a typical way by flying against the wind very closely over the water surface. They reduce speed until they stall and than grab small fish from the water with a quick snap of the beak. They also can do it without wind by slowly gliding over the water at a height of half their wing span. Stalling is induced by quickly rotating the wings a few times around the span-wise axis. They grab the food and have to flap the wings briefly to gain altitude again. The Audouin's obviously make use of ground effect which increases the efficiency of lift generation by the wings. The phenomenon is explained in Chapter 4.

The European shags here use this trick all the time when they fly from the roosting places to the feeding area, very low over the water surface. Oceanic birds near the lighthouse, mainly European storm petrels and Manx shearwaters, also keep close to the surface, probably for the same reason.

One pair of common ravens nests on the steep cliffs next to the lighthouse. The Corsican raven is smaller than its relative from the mainland and is probably an endemic race. Ravens know all the tricks of the flight trade. The wing beats make a strong impression; they also soar in up-draught. Courtship flight is extremely versatile, turning, wheeling and flying upside-down during breathtaking aerobatics.

There was one hooded crow this morning. It did not seem to be in good shape, its feathers were a bit in a mess and a few primaries missing. It took off in north-western direction straight towards the mainland coast of France. The wing beats were continuous but irregular. The animal was loosing height over the sea and it sometimes flapped vigorously to gain some altitude. It did not look as if this bird was going to make the 180 km to the French coast. Flight skills can obviously be seriously affected by a reduced condition of the flight apparatus.

On days without much wind the lighthouse attracts many insects providing food for aerial plankton feeders belonging to two unrelated groups of bird species, the swifts and the swallows. Due to convergent evolution driven by the extreme common feeding habit these swirling insectivores look alike at first sight. Three species of Apodids (common swifts, pallid swifts and alpine swifts) and three species of hirundines, (barn swallows, house martins and crag martins) circle around the lighthouse. The swifts are very similar in their body shape, gliding, flapping, turning and wheeling behaviour. There are small differences in flight techniques among the three species but it is difficult to detect the kinematic details that are actually causing this impression. The body shapes of the swallows and both martins differ. The tail, for example, of the crag martin is cut off straight, the house martin has a short V-shaped tail and that of the barn swallow is deeply forked with extremely extended outer feathers. Also the wings are different but it requires more than casual observation to find out how these differences result in different flight patterns during aerial feeding.

Sitting on the roof is the blue rock thrush. It jumps down from this high perch singing a short song as loud as the call of a mistle thrush. Its downward flight consists of short steep glides interrupted by spells of wing fluttering. The loud song and the striking flight pattern are obviously meant to advertise its position and its claim on the area. The way back up goes often in stages and in silence.

A spotted flycatcher perches in the top of last year's 4–5 m high dry flower stem of an agave. Its attitude is very alert and it dashes away on a quick flight excursion only a few metres away from the perch. This snap and return feeding strategy is typical for flycatchers all over the world. There will be differences in details of the flight kinematics but the general pattern is remarkably similar.

Marmora's warbler is very abundant here; it perches and sings in the highest branches of the small shrubs and bushes of the maquis. From there it makes nervous short flights or actually jumps with a short flutter of the wings from one perch to the next probably to establish and advertise its territory. It seems to forage by creeping over the vegetation looking for invisibly small insects. When alarmed these warblers fly over distances of up to a few hundred metres in a very condense type of bounding flight. This consists of a few high frequency wing beats followed by a short flight upward with the wings closed. It looks less regular and stable than the bounding flight of the European goldfinches. These migrants fly in a steady cadence clearly gaining speed during fast wing beats followed by a distinct bounce upwards. It is the steady pattern of a galloping horse. The closely related Eurasian siskin again flies similar but different. I could sit here for ages watching the variety in flight skills.

Bird flight watching brings strong impressions of the huge diversity in flight techniques. Each of these keeps the bird aloft but how? It seems hard to believe that there is one common principle. This book tries to summarize the knowledge about the various aspects of avian flight. Its purpose is to increase the interest in this aspect of bird life and to unveil our ignorance.

Acknowledgements

About 25 years ago I had my first hands on experience with experimental bird flight research. A generous sum of money granted by the Hasselblad foundation to Serge Daan and myself made it possible to buy the equipment required to measure various aspects of bird flight in the laboratory and in the field. An expedition to Kenya organised to make high speed films of windhovering birds made me an addict to bird flight for the rest of my life.

Common kestrels and rough-legged buzzards, taken out of the wild for a while to be trained for scientific falconry, deserve to be acknowledged. These birds hopefully enjoyed a prolific life after their release (or escape).

I wish to thank the colleagues and students who were authors and co-authors of bird flight related papers bearing my name as well. These are in alphabetical order: Serge Daan, Marc Gnodde, Alex Groenewegen, Dirkjan Masman, David Povel, Eize Stamhuis, Gerrit Vossebelt and Danny Weihs.

My research group has been heavily involved in discussions about many flight related issues and I am grateful for the stimulating atmosphere. Honorary appointments first in Leiden and later in Groningen have been very motivating.

Several friends and colleagues helped by commenting on various parts in successive stages of the making of this book:

The input of Eize Stamhuis improved most chapters and helped to clarify numerous aerodynamic issues. In the early stages Felix Hess taught me basic aerodynamics for birds. Charlie Ellington was involved in the aerodynamic parts in Chapters 4 and 5 and commented on the ideas regarding the energetics of flight. Richard Bonser helped with Chapter 3 and contributed to the story about the mechanical properties of feathers during discussions by e-mail and at SEB conferences. Jeremy Rayner supported the idea to write this book from the beginning and his list of bird flight references has been an invaluable tool. I decided not to use Ulrike Müller's in depth explanation of the physics of vortex rings in 'Avian Flight' but still use it as a valuable teaching tool and thank her for preparing it. Jeroen Nienhuis, Serge Daan and Theunis Piersma made numerous suggestions improving the last two chapters. Henk Visser corrected the contents of the box on the doubly labelled water technique in chapter 8. At Stareso, the director Pierre Lejeune and visitors Francis Neat and Rudolf Zurmühlen contributed with data and suggestions for improvement. The expert knowledge on proteins of my daughter Hortense Videler greatly enhanced the clarity of the complex story on how muscles work in Box 7.1.

The mail box bearing the name of the title of the book contains various messages of contributing colleagues: Douglas Altshuler, Herman Berkhoudt, Andy Biewener, Charles Bishop, Sophia Engel, Ted

Goslow, Marcel Klaassen, Werner Nachtigall, David Povel, Geoff Spedding, Henk Tennekes, Brett Tobalske, and York Winter. I am grateful for the fruitful and encouraging interactions also in those cases where opinions differed.

The palaeontologists Paul Lambers, David Unwin, Sandra Chapman, Angela Milner, Winfried Werner and Martina Kölbl-Ebert facilitated access to the original *Archaeopteryx* fossils.

The X-ray pictures of the wandering albatross wing used in Fig. 2.6b were made by Michael Bougaardt, with Leonard Compagno of the Shark Research Centre (Iziko Museums Cape Town, South Africa) in strong support. My PhD student Henk Jan Hoving assisted in many ways with the albatross study. Comments of my brother in law Jo Spijkers improved Box 1.1.

Pat Butler and Thomas Alerstam commented as external reviewers in detail on the general structure; their constructive remarks made me change the style of the book making it hopefully more accessible to a wider audience.

My wife Hanneke acted as jack of all trades by training birds, analyzing films and by reading each chapter several times. I thank her most of all for the way she encouraged me to carry on in spite of the negative impact the writing process sometimes had on our family life.

It took far too long to write this book. Luckily, generations of editors at Oxford University Press were extremely patient and (surprisingly) never seemed to loose faith in the project.

Avian Flight reflects my personal opinion about the subject and I take full responsibility for the contents no matter how controversial. Criticism will be greatly appreciated.

Contents

1 *Acquisition of knowledge*

1.1 *Introduction*

It would be unwise to start thinking about any scientific problem without knowing what previous generations discovered. Bird flight has a long history of widespread interest witnessed by prehistoric cave drawings. A great deal of that interest arose from the desire to fly like birds. This chapter concentrates on the development of the understanding of the principles of flight in relation to birds.

In historical times, the ideas of the Greek philosopher and natural scientist Aristotle (384–322 BC) serve as a starting point. His way of thinking is not familiar to us and it is not easy to find out what he actually conceived. Some of his outspoken opinions stood firm for more than 20 centuries. Approximately half way the second millennium AD, with the start of the Renaissance, more experimentally inclined science starts to emerge. Freethinkers like Leonardo da Vinci, Galileo Galilei, and later Giovanni Borelli challenge the ancient ideas. Physics explodes in the seventeenth and eighteenth centuries with the discovery of its basic laws that still govern modern thinking. Contributions of Christiaan Huygens, Isaac Newton, Gottfried Leibniz, Daniel Bernoulli, and Leonhard Euler constitute the firm base for the principles of the mechanics of fluids (liquids and gases). The apotheosis of these principles is laid down in the Navier–Stokes equations, describing the distribution of pressures and velocities of fluid in a three-dimensional space.

Experimental biology, studying bird flight directly rather than its theoretical principles, starts with George Cayley around the turn of the eighteenth century. About a century later Otto Lilienthal flies the machines he designed using bird wings as examples, Étienne-Jules Marey makes the first high-speed films of flying birds, offering the possibility to study flapping wings in slow motion and Osborne Reynolds provides insight in the scaling principles determining various patterns of fluid flow.

During the first half of the twentieth century, steady-state aerodynamic theory for fixed wings is fully established and the aircraft industry emerges. Consequently, explanations of how birds fly are based on the attached flow aerodynamics of conventional aircraft wings although no direct experimental evidence exists showing that this is the case in flying birds. Nevertheless, the phenomenon is considered understood.

After the Second World War, the aircraft industry discovers that delta wings can obtain lift from separating flow at the sharp sweptback leading edges. This leading edge vortex lift turns out to be used by insects during gliding and flapping flight. Further discoveries of the unsteady effects in insect flight mechanics reveal that the established feeling of having understood bird flight was probably premature.

A general picture of the functional anatomy of the flight apparatus of birds gradually emerges during the twentieth century, but is based on an extremely limited number of species. During the last decades, novel techniques have measured crucial aspects of avian flight directly, either in the laboratory or in the field. Most of the remaining part of this book deals with results obtained during the last 40 years or so. I use examples from my own work whenever possible since these are most familiar to me and form the basis of my thinking about the subject.

Five boxes complement this chapter. Box 1.1 provides the Standard International (SI) units, definitions of fundamental and derived quantities. Box 1.2 has Newton's laws of motion. Box 1.3 introduces basic kinematic equations, dealing with movements without reference to forces, and applies these to freely falling objects. The equations for potential and kinetic energy are also included in this box. Box 1.4 gives the *Re* number equation and Box 1.5 summarizes important properties of air in the atmosphere.

1.2 *Ancient thoughts*

The question 'how do birds fly' is probably as old as humanity and we are still struggling to find an answer. The frontispiece of this book shows a remarkable bird painted in black high against the ceiling of a half-open cave, the '*Cueva de la Soledad*', in the Sierra de San Francisco in the central desert of Baja California, Mexico. The painting is pre-Indian and has been estimated by carbon dating to be about 11,000 years old. Prehistoric cave drawings tell us about the skill and knowledge of the artist without further written information. In this case, the drawing shows some unusual details, suggesting accurate observation of the use of the wings during landing. The picture is remarkable because the artist emphasized the strongly swept-back hand wings of this bird. The arm wings are very short, ending close to the body. There is an impression of extended bastard wings or alulae. The feet are drawn spread out as if the bird, probably Brandt's cormorant, is preparing for landing. Is this painting telling us something about the function of the hand wings during landing?

About 24 centuries ago Aristotle made accurate observations and developed theories regarding bird flight *The complete works of Aristotle* (Barnes, 1991) contain several notes related to bird flight but no formal explanation of how birds do it. Aristotle tried to explain phenomena, such as flight through air and movement in water in general terms, on the basis of casual observations and a theoretical model approach. Unfortunately, he believed his theories more than his observations.

In his way of thinking, force was proportional to mass and velocity. The notion of the concept of acceleration or deceleration, or change of velocity existed but not in a useful physical form. Aristotle thought that if in a given time, a force moves a body over a certain distance, that force can move half the body over twice the distance in the same time. In other words, a force that can move a body at a given speed can move half that body at twice the speed. Aristotle's general concept of motion was that it could only exist as long as a force acts on a body and hence motion stops as soon as the force stops acting on it. Motion furthermore requires three things: the mover, the subject moved, and time. The subject can only move as long as the mover is acting on it. The mover has to be immovable or must support itself on something immovable. These principles make it difficult to understand movement in air or water. In air, a stone thrown or a flying arrow (the subject moved) is no longer in contact with the mover. Aristotle's solution for this problem was as follows: the original mover gives its power of being a mover to air, the mover throws the stone and simultaneously moves a volume of air, as soon as the hand of the thrower stops, the air volume will stop too but now has the property of a mover and causes the stone and a new volume of air to move on, the new volume of air stops and becomes the next mover and so on.

So much for Aristotle's philosophical approach of movement in a fluid medium. We shall now try to see the progression of his thoughts on how birds actually fly.

In *Parts of Animals*, Book IV, he says

> In birds the arms or forelegs are replaced by a pair of wings, and this is their distinctive character. For it is part of the substance of a bird that it shall be able to fly; and it is by the extension of wings that this is made possible. Moreover, birds cannot as a fact fly if their legs be removed, nor walk without their wings.

To find out how Aristotle thinks that wings enable the bird to fly, we have to read Chapter 2 of the book on the *Movement of Animals*. There he tries to match the observations with the theory:

> For just as there must be something immovable within the animal, if it is to be moved, so even more must there be without it something immovable, by supporting itself upon which that which is moved moves. For were that something always to give way (as it does for tortoises walking on mud or persons walking in sand) advance would be impossible, and neither would there be any walking unless the ground were to remain still, nor any flying or swimming were not the air and the sea to resist.

This notion of resistance or drag approaches Newtonian mechanics very closely. However, in the next statement Aristotle ruins it all by returning to his philosophical concept on the mover and the moved: 'And this which resists must be different from what is moved, the whole of it from the whole of that, and what is thus immovable must be no part of what is moved: otherwise there will be no movement'. Evidence for this statement comes from the fact that 'one only can move a boat by pushing it while standing on the shore, not by pushing the mast while standing on the ship'.

Another point of view that was accepted as the truth for many ages was Aristotle's idea about the function of the tail: 'In winged creatures the tail serves like a ships rudder, to keep the flying thing in its course'. Although it is not the right explanation, it indicates that Aristotle recognized the problem of dynamic stability inherent to moving in a fluid medium.

A strange description of the function of the crest on the breastbone in birds and of the large pectoral muscles can be found in *Parts of Animals*, Book IV:

> The breast in all birds is sharp-edged, and fleshy. The sharp edge is to minister to flight, for broad surfaces move with considerable difficulty, owing to the large quantity of air which they have to displace; while the fleshy character acts as a protection, for the breast, owing to its form, would be weak, were it not amply covered.

Box 1.1 ***SI units and definitions of fundamental and derived quantities***

Fundamental quantity	Dimension	SI unit
Length	L	Metre (m)
Mass	M	Kilogramme (kg)
Time	T	Second (s)

Derived quantity	Descriptions	SI unit
Velocity or speed	The rate of change of displacement	$\mathrm{m\,s^{-1}}$
Acceleration	The rate of change of velocity	$\mathrm{m\,s^{-2}}$
Force	Mass times acceleration One unit accelerates 1 kg with 1 $\mathrm{m\,s^{-2}}$	Newton (N) = $\mathrm{kg\,m\,s^{-2}}$
Momentum or impetus	Property of a moving body equal to mass times velocity (the rate of change of momentum is force)	$\mathrm{kg\,m\,s^{-1}} = \mathrm{Ns}$
Impulse	Force times duration	Ns
Work	Force times distance is an amount of energy	Joule (J) = Nm
Power	Energy consumption: work per unit time; force times velocity; momentum times acceleration	Watt (W) = $\mathrm{J\,s^{-1}} = \mathrm{N\,m\,s^{-1}}$

This set of contradictions is highly confusing. It might be that the carina on the sternum is indicated as the sharp edge. However, even if that is the case the remarks remain enigmatic.

Aristotle was very close to the concepts of inertia (the property of matter to stay in its state of rest or of uniform motion in a straight line) and of momentum or impetus, (the force of motion of a body gained in movement, equal to the product of mass and velocity). (See Box 1.1). Unfortunately, he looked for the cause of movement outside the moved body because he thought that the medium continued to generate the force required to keep the body moving. Two centuries later Hipparchus generated the idea that earth thrown upwards contained the force of throwing (Sambursky 1987). However, it was Johannes Philoponus from Alexandria in the sixth century AD who clearly reached the conclusion that the medium does not help. He describes how kinetic power is transferred at the instant of throwing from the thrower to the object thrown, by virtue of which it is kept moving in its forced motion, that is, the object is given momentum. A full explanation, however, was not possible until the finding of acceleration by Galileo Galilei, a millennium later.

1.3 *A chronicle of cognition*

It takes us to around 1500 AD before new contributions to the knowledge of flight emerge. Leonardo da Vinci's (1452–1519) notebook '*Sul volo degli Uccelli*' was lost for centuries. Only at the turn of the nineteenth century the mirror image handwriting was transcripted and subsequently translated into English (Hart 1963). Leonardo da Vinci's interest was focussed on the interaction between birds and air because he dreamt about man-powered flight. Several sketches in his notebook show designs of the essential parts of artificial wings. The structure of these wings is obviously not based on the anatomy of bird wings but inspired by the interpretation of the forces on the wings of birds in flight. Most of the bird drawings in the notebook show tracings of the presumed air movements (Fig. 1.1). Leonardo had a very quick eye; his sketches resemble tracings from high-speed film pictures of birds in fast flapping flight. He translated his observations into laws governing the flight of birds. These sound like instructions for birds. For example, on page 6 *verso* he writes: 'If the wing tip is beaten by the wind from below, the bird could be overturned if it does not use one of two remedies. It must either immediately lower the beaten wing tip under the wind or it should beat the distal half of the opposite wing down'. It is important to note that the codex on the flight of birds describes for the first time but repeatedly that in flying birds the centre of gravity does not coincide with the centre of lift.

Da Vinci, intrigued by flow in water and air, made sketches of flow patterns with photographic precision. He discovered that rapids occur in

Fig. 1.1 Drawings of flying birds and the direction of the wind in Leonardo da Vinci's notebook *Sul volo degli Uccelli* (page 8, *verso*). The lowest drawing is used to explain the stabilizing function of the tail.

shallow or narrow parts of a river and that the flow in wider, deeper parts is slow. His conclusion was that the product of the cross-sectional area and the flow velocity had to be constant in order to keep the passing mass of water per unit time constant. This is the first mention in history of the continuity equation for an incompressible fluid. Da Vinci rediscovered that air induces resistance, and initially reached the false conclusion that the wings in flight compress the air to generate lift forces. (Velocities approaching the speed of sound are required to compress air in open space). Six years before his death he had obviously changed his mind when he wrote in *Codex E*:

> What quality of air surrounds birds in flying? The air surrounding birds is above thinner than the usual thinness of the other air, as below it is thicker than the same, and it is thinner behind than above in proportion to the velocity of the bird in its motion forwards, in comparison with the motion of its wings towards the ground; and in the same way the thickness of the air is thicker in front of the bird than below, in proportion to the said thinness of the two said airs.

Anderson (1997) indicates that if the words *'thinness'* and *'thickness'* are replaced by 'pressure' and *'thinner'* and *'thicker'* by 'lower and higher pressure' we obtain a clear explanation of the pressure distribution around a conventional wing. That would imply that Da Vinci understood what causes lift and pressure drag forces on a wing. He also generated the idea, still applied in water- and wind-tunnels today that it makes no difference aerodynamically whether the flow passes a stationary body or a body moves through a stationary flow (we will use this concept in Chapter 4 to show what happens with the fluid around a wing section tested in a water-tunnel). In the *Codex Atlanticus* he wrote: 'As it is to move the object against the motionless air so it is to move the air against the motionless object' and 'The same force as is made by the thing against air, is made by air against the thing'. He believed that this force was proportional to the surface area and the velocity of the body. As will become clear later on, he was right about the surface area but not about velocity. In birds and aircraft, drag forces are proportional to the square of the velocity (Anderson 1997). That not only the surface area but also the shape of the body determines resistance to fluid flow was also clear to Da Vinci, witnessed by various drawings of streamlined bodies and projectiles based on the shapes of fish.

The problem with Da Vinci's scientific production is that little of it was accessible for the following generations. He hardly published his ideas and tried to keep his notes secret by mirrored writing. Furthermore, Da Vinci used the Italian language because he lacked the ability to write in Latin, the lingua franca of science in those days.

Falling bodies and the laws of gravity were certainly within the broad range of Da Vinci's interests. However, Galileo Galilei (1564–1642) used experimental results to calculate the acceleration of falling bodies first. He falsified Aristotle's idea that a heavy spherical body uses less time to fall from a given height than a lighter one of the same size. He also found that differences in the resistance of objects varying in shape caused differences in falling periods. Galileo discovered that aerodynamic resistance is proportional to the density of the air but considered it small enough to ignore it.

Giovanni Alphonso Borelli (1608–1679), a mathematician with a strong inclination towards biomechanics, lived in Southern Italy. Towards the end of his life, he wrote his masterpiece '*De motu animalium*'. The first part of it appeared a year after his death in 1680. The book consists of propositions or statements, given one after the other each supported by an explanation (Borelli 1680; Maquet 1989). The propositions 182 to 204 in Chapter 22 are about flight. Borelli describes the structure and function of the flight apparatus (his translated ideas are given below):

> The wings have a stiff skeleton on the front side and are covered by flexible windtight feathers. The body has heavy pectoral muscles, in strength comparable with the heart muscle, but it is otherwise lightly built containing hollow bones and air sacs and it is covered with light feathers. The ribs, shoulders, and wings contain little flesh. The muscles of the hind legs are weakly developed. The pectoral muscles are 4 times as strong. Birds fly by beating the air with their wings. They

jump as it were through the air just as a person can jump on the ground. The wings compress the air making it to react to the wing beat as solid ground reacts to the push off of feet. The air offers resistance because it does not want to be displaced and mixed with stationary air. The air particles rub against each other and that causes resistance. Apart from that is air elastic. Wing beats compress the air and the air bounces back. The jumps occur during the downstrokes. During the upstrokes, the wings move with the stiff leading edge forward, followed by the flexible feathers. They then do not meet resistance similar to a sword that moves with the sharp edge forward. Repeated jumping through the air costs a tremendous amount of force.

Borelli estimated this force to be 10,000 times as large as the weight of the bird. He thinks that is the reason why birds must be lighter than quadrupeds. *The downstrokes keep the bird in the air and can even cause the bird to gain altitude as long as the wing beat frequency is high enough.* The question is how the wing movement can generate propulsive forces as well. Borelli developed the plum stone theory to answer it. The explanation is illustrated in frames 2 and 3 of his table XIII (Fig. 1.2).

The stiff leading edges of the wings with the flexible planes of feathers behind them form a wedge that is driven through the air during the downstroke. The oblique trailing edges of the wedge formed by the flexible feathers push the air backward and the bird forward. The action can be compared with shooting a slippery plum stone from between the thumb and the index finger. The stone shoots off in a direction perpendicular to the direction of compression due to its wedge shape.

Aristotle's idea that the tail of a bird functioned as a ship's rudder had still many followers but Borelli fiercely opposed it. According to him the function of the tail is solely to steer the bird up and downward and not to

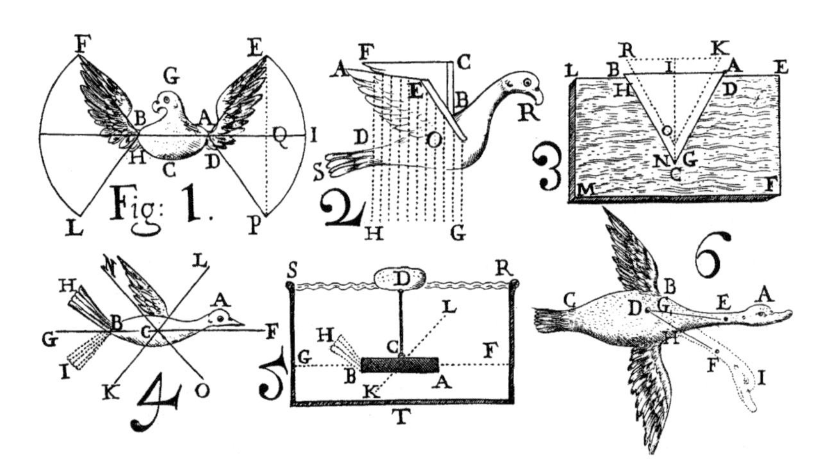

Fig. 1.2 Tabel XIII of Borelli's *De motu animalium* (1680). See text for the explanation of the individual figures.

the right or the left. An experiment with a model in water proved his view (Fig. 1.2, pictures 4 and 5):

> To function as the rudder of a boat, the tail would have to be implanted vertically. Birds change their horizontal direction by beating the left and right wing at different speeds. The action can be compared with the way a rower alters coarse by pulling harder on one oar than on the other. Birds with long necks can use these to steer up and downwards but not to the left or the right. Moving the head to the left or the right would change the position of the centre of gravity to a position next to the flight direction and that would cause serious imbalance.

Borelli believed that such behaviour would be useless and stupid, unworthy of the cleverness of nature (‘*indigna naturae solertia*’).

> Birds can glide without beating their wings because they have impetus (momentum). They fly just as missiles such as the plum stone along parabolic trajectories. Large birds of prey are blown upwards by the action of the wind on their excessively large wings, similar to the flight of clouds.

Borelli also paid attention to landing.

> The impetus must disappear during landing otherwise accidents will happen. Birds can avoid these accidents in different ways. Wings and tail can be spread and kept perpendicular to the flight direction. Just prior to landing, the wings can beat actively against the flight direction and the bending legs may absorb the remaining bit of impetus.

Borelli is often regarded the father of biomechanics. His fascinating ideas were inspiring and surely influenced contemporary scientists in Europe. However, he missed fundamental insight in fluid mechanical principles and that made his contribution to the knowledge of how birds fly anecdotal and his designs of scuba diving equipment, published in the same book, lethal. The tragedy is that the lacking principles were discovered during his lifetime.

1.4 *The rise of aerodynamics*

The air in which birds fly is a gas, physicists regard both gases and liquids as fluids. A fluid is a substance without a shape of its own, consisting of freely moving particles. The substance deforms easily under the slightest pressure and fills any space completely. Normally, fluids are continuous without any holes or empty gaps. Gases are distinct from liquids by the fact that they easily compress and expand. Despite this difference, common laws and rules describe and predict the behaviour of fluids and the related forces.

Near the end of the seventeenth century, time was ripe for the discovery of what we now call the classical mechanics (Box 1.2). Isaac Newton (1642–1727) described the laws of motion in *Philosophiae Naturalis Principia Mathematica* in 1686. The text of the first law as it appeared in the first English translation (Motte 1729) reads: ‘Every body perseveres in its state of rest or of uniform motion in a right line, unless it is compelled to

Box 1.2 ***Newton's basic laws of motion***

First	An object will remain at rest or in uniform motion in a straight line in the absence of external forces. This property of the object is its *inertia*. (For a unit of mass, m, of an object rotating at a distance, r, around an axis, the *moment of inertia*, I, is mr^2.)
Second	The net force on an object is equal to the mass of the object times the acceleration. (In a rotating system the net force is equal to the moment of inertia times the angular acceleration.)
Third	Every force exerted on an object will meet an equal force in opposite direction.

change that state by forces impress'd thereon'. This implies that there are no net resulting forces acting on a bird flying at a uniform speed, at one altitude along a straight track. The thrust force equals drag and the lift equals the force of gravity. Changes of velocity and direction require force.

The second law (force equals mass times acceleration) tells us, in a bird flight related context, that the rate at which the velocity of a bird changes is equal to the resultant of all the forces on the bird divided by its body mass. The direction of the rate of change of velocity is the direction of the resultant force. This law indicates that acceleration or deceleration is caused by an uncompensated force in one direction. If the thrust is bigger than the drag, the bird accelerates. The magnitude of the acceleration is inversely related to the accelerated mass.

Newton's third law reveals that every action of a bird in the air will be opposed by an equal reaction in the opposite direction. This means that birds push off against the air and generate reaction forces that make them fly.

There were more bright minds active in Newton's time. My fellow compatriot Christiaan Huygens (1629–1695), while working on the pendulum clock, realized that the resisting force on a moving body in a fluid medium was not proportional to the velocity but to the velocity squared and was the first to prove that experimentally. He also started to think in terms of conservation of energy during collisions of moving bodies. Gottfried Leibniz (1646–1716) was Huygens's pupil when they were both in Paris. He expanded on the idea that energy is conserved by transferring it from one form to another. A body can have the potential to do work due to its height above the ground. This is equal to mhg, which is the product of its mass (m) the height (h) above the ground and the acceleration due to gravity (g). When a body is moving, an amount of work is required to stop it. The pendulum of a clock exchanges the potential to do work due to its height with that due to its motion. Leibniz termed the energy due to motion *vis viva* and considered it to be equal to the moving mass and the velocity squared. Much later towards the end of the nineteenth century, Lord Kelvin (W. Thompson) introduced the terms potential and kinetic energy for the two exchangeable possibilities to do work (Box 1.3). The kinetic energy is the work a moving body is capable of doing when it is stopped and it equals half the *vis viva*.

Box 1.3 ***Basic kinematics and energy equations***

Flight is movement, defined as a continuous change of position in three-dimensional space. In an earth-bound frame of reference a rectangular (orthogonal) x-y-z coordinate system can be defined where x-y is the plane of the earth and z is the perpendicular axis into the air. In a flying bird different points of the body move along different paths. The position of each point can be specified by its projections on the three axes.

Consider a simple case of one point in rectilinear horizontal motion along the x-axis. It displaces over a certain distance Δx (m) in a certain time Δt (s). Over that time span, it has an average velocity of Δx divided by Δt (m s^{-1}). When Δt is small and approaching 0 we have the instantaneous velocity v or the rate of change of displacement which is the derivative of x with respect to t. In mathematical notation,

$$v = \lim_{\Delta t \to 0} \frac{\Delta x}{\Delta t} = \frac{dx}{dt} \quad (\mathrm{m\,s^{-1}}) \qquad (1.3.1)$$

If the motion is uniform, this differential equation will be constant because the displacement with respect to time and hence the velocity v is constant. Velocity changes require accelerations or decelerations. Acceleration (a) is the rate of change of velocity and is, analogous to equation (1.3.1), described in mathematical shorthand as another differential equation:

$$a = \lim_{\Delta t \to 0} \frac{\Delta v}{\Delta t} = \frac{dv}{dt} \quad (\mathrm{m\,s^{-2}}) \qquad (1.3.2)$$

The velocity of an object in uniformly accelerated motion can now be defined as,

$$v = v_0 + at \quad (\mathrm{m\,s^{-1}}) \qquad (1.3.3)$$

where v_0 is the speed at an initial position x_0 at time $t = 0$.

The displacement with respect to the initial position x_0 of a moving object at time t is,

$$\Delta x = x - x_0 = v_0 t + \tfrac{1}{2}at^2 \quad (\mathrm{m}) \qquad (1.3.4)$$

Falling is a special case of displacement. In our frame of reference it is a motion with nearly constant acceleration along the z-axis in the direction of the centre of the earth. Free falling bodies are attracted by the earth with the acceleration caused by gravitation g of about 9.81 m s^{-2}. If we ignore the drag of the air flow on a falling object we can show that the duration of a fall and the final velocity are constant and not dependent on body mass. Assume a fall from a height h. Equation (1.3.4) tells us that,

$$h = v_0 t + \tfrac{1}{2} g t^2 \quad \text{(m)} \tag{1.3.5}$$

The fall starts with zero initial velocity, so the first term, $v_0 t$, is zero. The duration of the fall t depends only on the height h and the gravitational acceleration g:

$$t = \sqrt{\frac{2h}{g}} \quad \text{(s)} \tag{1.3.6}$$

The final velocity at the impact on the ground after falling from height h (v_h) is according to equation (1.3.3) and assuming v_0 to be zero,

$$v_h = gt \quad (\text{m s}^{-1}) \tag{1.3.7}$$

For example, a tennis ball has a mass m of about 58 g. Its weight is mass times g which is approximately 0.57 N. The radius of the ball is 3.2 cm giving it a volume of slightly more than 137 cm^3. If that volume would be filled with lead instead of air the mass would be more than 1.5 kg. If we drop these balls from 1 m height they would both reach the ground after 0.45 s and at a speed of 0.45 m s^{-1}.

However, the potential energy before the drop and the kinetic energy lost during the impact on the ground, are different.

Potential energy

The potential energy (E_p) is proportional to the mass (m), the acceleration due to gravity (g), and the height above the ground (h):

$$E_\text{p} = mgh \quad (\text{kg m s}^{-2}\text{m} = \text{Nm} = \text{J}) \tag{1.3.8}$$

At an elevation of 1 m the lead filled ball has about 15 J and the normal one 0.57 J potential energy.

Kinetic energy

The kinetic energy (E_k) is

$$E_\text{k} = \tfrac{1}{2} m v^2 \quad (\text{kg}\,(\text{m s}^{-1})^2 = \text{Nm} = \text{J}) \tag{1.3.9}$$

The kinetic energy at impact of the tennis ball is 0.006 J and that of the leaden ball 0.151 J, which is 25 times as much.

The potential energy equation is completely transparent but the formula for kinetic energy is less clear. The kinetic energy concept is important for the understanding of bird flight. It is intuitively easy to imagine that mass and velocity are the contributing factors to kinetic energy, but it is not obvious why the velocity is squared and where the factor $\frac{1}{2}$ comes from. The analysis of an example of total loss of kinetic energy answers these questions. Imagine what happens when a bird hits a window at full speed because it did not see it. The work done on the window equals that on the unfortunate bird. It is the force times the distance it decelerated. The force is the mass of the bird m times its deceleration (Newton's second law). Although the deceleration probably changes in the time t between the instant the beak of the bird hits the window and the end of the crash, it is convenient to choose an average value. That will be equal to the flight velocity v just before the crash divided by the time t it took to decrease to zero. The same time t provides the distance over which the force is applied by multiplying it with the average velocity during the crash. Again, for convenience, a linear velocity decrease is assumed which makes $\frac{1}{2}v$ the average velocity between v at the beginning of the crash and velocity $= 0$ at the end. The distance is hence equal to $\frac{1}{2}vt$. The term t is lost when multiplying the distance with the deceleration, vt^{-1}, and leaves the equation describing the work done equal to $\frac{1}{2}mv^2$.

Potential and kinetic energy are physical properties of a flying bird and it is good to know these properties. However, in order to understand the total phenomenon of flight we also need to know more about the behaviour of the air in static, but more importantly in dynamic interaction with the bird.

More than 2200 years ago, Archimedes discovered that any static body submerged in fluid is subjected to a force exerted by the fluid on that body. The magnitude of the force equals the weight of the displaced fluid. The density of air is only about 1.23 $\mathrm{kg\,m^{-3}}$ and this static force is negligibly small for birds. Forces related to relative movement between bodies and fluid can be much larger and are far more complex.

In early 1700, Daniel Bernoulli (1700–1782) was born in Groningen where his father Johann was the professor of mathematics. He studied philosophy, mathematics, physics, and medicine at various European universities. One of the many problems he tried to solve was that of the relationship between pressure and flow velocity of blood. Blood flows through the veins and the arteries varying in diameter. Bernoulli showed (as Leonardo da Vinci did before him) that fluid flows faster where a vessel is narrow and slower when it is wide. The pressures he could measure were higher in the slow flow and lower in the fast flow. He knew Leibniz's work and realized that conservation and exchange of energy could explain the phenomenon he had discovered as well. The pressure in a fluid is expressed in Pa (Pascal) or $\mathrm{N\,m^{-2}}$. In other words pressure is energy per unit volume (in SI units: $\mathrm{J\,m^{-3}} = \mathrm{N\,m\,m^{-3}} = \mathrm{N\,m^{-2}}$). So again, energy is conserved by exchanging two forms of it: static pressure and dynamic pressure. The dynamic pressure being the dynamic energy per unit volume equals $\frac{1}{2}\rho v^2$ (with density ρ instead of mass m as in Leibnitz's equation because mass equals volume times density; v is velocity as before).

The static pressure is the sum of the ambient pressure plus the excess pressure due to the elevation of the fluid. This potential pressure is the potential energy per unit volume or mgh divided by volume, which is ρgh (g is the gravitational acceleration and h the elevation). Bernoulli published a treatise entitled *Hydrodynamica* in 1738. It describes the relationship between velocity and pressure in a fluid of constant density and negligible viscosity, which is flowing steadily without rotation. This is the definition of laminar flow. Under these conditions, the sum of the static pressure and the dynamic pressure is constant. Velocity increase will enlarge the dynamic pressure and this decreases the static pressure. Bernoulli's law, as it has been termed later, proved to be rather robust even when the underlying assumptions were not exactly met. For bird flight, the constant density condition is met. Outside a very thin layer adjacent to the bird, viscosity has little or no effect. Laminar flow conditions around flapping wings are of more concern.

Leonhard Euler (1707–1783), who joined Daniel Bernoulli in St Petersburg, developed differential equations relating pressures and velocities in a fluid in three dimensions. The equations are based on Newton's second law, the law of conservation of mass, and on the principle of continuity. Euler's model can be explained as follows. Imagine an arbitrary small cubic volume within a fluid. The ribs of the cube run in three perpendicular X-, Y-, and Z-directions. The fluid moves and passes through the cube in an arbitrary direction. That means that there are components of movement in each of the three directions X, Y, and Z. The velocity of the mass of fluid is allowed to change while passing through the imaginary cube. This implies that there are velocity gradients in the three orthogonal directions. By assuming that compression effects are negligibly small, Euler was able to show that the sum of the velocity changes in each direction was zero, satisfying the requirement that the volume and mass of the fluid in the cube had to be constant. Under the conditions indicated, Euler's equations can describe fluid flow quantitatively. The equations can also predict the forces on an object in a flow. However, they are not extremely realistic because viscous forces close to the object are important too. About one century later Jean-Claude Barré de Saint-Venant (1797–1886) in a paper based on earlier work of Claude-Louis-Marie-Henri Navier (1785–1836), added the impact of viscosity to the Euler equations, grossly increasing their complexity. Two years later, in 1845, George Gabriel Stokes (1819–1903), Lucasian professor at Cambridge, independently derived and published the same equations. They were termed the Navier–Stokes equations. Fast computers can easily deal with the complexity of the calculations and the equations are still widely used to model flow phenomena.

Although the basic principles governing fluid flow were developed and understood during the eighteenth century, little thought had been given to the consequences for the understanding of the interaction between animals and the fluid environment, whether water or air. Practically inclined scientists started to fill that gap during the next century.

1.5 Application of principles

The contribution of Sir George Cayley (1773–1857) is a milestone in the development of thinking about flight (Gibbs-Smith 1962). This English baronet continued along Leonardo da Vinci's approach by carefully observing nature. He recognized that thrust and lift are two separate forces working on a bird. However, he made a false start by assuming in his first notebook of 1801 that birds produce thrust and lift on alternate downstrokes. He must have realized that this was wrong because there is no mention of this idea in his official papers of 1809 and 1810 (Cayley 1809, 1810). The last paper contains the first quantitative kinematic data regarding the flapping flight of a bird: the rook flies at 34.5 ft per second (10.5 $m s^{-1}$) and covers 12.9 ft (3.9 m) during one wing beat cycle. The vertical wing excursion was estimated at 0.75 ft (0.23 m). From these figures, Cayley calculated the vertical speed of the wing at 4 $ft s^{-1}$ (1.2 $m s^{-1}$). It is not at all clear how he did these measurements. He discovered that an oblique airflow on feathers and wings generates lift force, which varies as the square of the relative airspeed multiplied by the density. Cayley designed *solids of least resistance* using the body shapes of trout, dolphin, and the Eurasian woodcock as examples. These resembled ideal streamlined bodies with a round front part, a pointed end, and the largest thickness at about one-third of the length from the front. Cayley did not indicate in his work that he knew that these bodies, with a thickness over length ratio close to one-quarter, provide the lowest resistance for the largest volume.

Cayley designed and built the first manned aeroplanes. At least two of those were tested. An unnamed groom, a young lad, operated the first flapping machine. It crashed and he got hurt after flying over a short distance. Cayley is said to have accused the boy of causing the failure. The machine came down because he did not keep the wings flapping fast enough, was too fat and got frightened. Cayley's coachman made the second attempt in an improved version in 1853. After his crash landing, he allegedly complained that he was hired to drive and not to fly. Cayley was aware of the dangers because he predicted in 1846: 'A hundred necks have to be broken before all the sources of accident can be ascertained and guarded against'. Unlike the next student of flight in this survey, he made sure that it was not his own neck.

Otto Lilienthal (1848–1896) was an engineer who owned an engine factory in Berlin. He desperately wanted to learn the art of flight from birds and studied their flight apparatus in detail (Fig. 1.3). He rediscovered and concentrated on the lift forces generated by cambered wings in airflow. Lilienthal designed and built a simple force balance to measure lifting forces on wing-shaped objects. In his opinion, the flight of birds is based on the fact that wings, due to air resistance, generate a force during the downstroke, which is equal and opposite to the gravitational force. In still air and during flight at uniform velocity the wings generate thrust forces equal and opposite to the drag at that speed. Lilienthal described how the arm part of the wings of a

Fig. 1.3 Lilienthal's (1889) studies of the outlines of flying birds taken from *Der Vogelflug als Grundlage der Fliegekunst*, comparing species with slotted and closed hand wing surfaces.

seagull generates lift force and how the thrust comes from the movements of the hands. He considers the aerodynamic forces proportional to the velocity squared and the surface area of the wing. Lilienthal indicates that the hand wing is constructed very lightly to keep the moments of inertia (Box 1.2) about the axis of rotation in the shoulder joint small. The arm can be heavier because even the most distal part of it moves up and down over a short distance only. Lilienthal distinguished three types of flight: flapping flight on one spot with regard to the surrounding air, rowing flight during displacements relative to the earth, and gliding flight. He was obsessed with the flight abilities of European white storks. He was convinced that these animals wanted to live in villages and towns close to people because God sent them to show humans how to fly. Lilienthal reached his goal and flew but paid a high price for it. He crashed in one of his self-constructed gliding planes and died from the injuries.

Marey (1830–1904) made the first three-dimensional high-speed films of flying birds and analysed the kinematics of the wing beats. He was also the first to do physiological experiments on flying birds. His major contributions are in '*Le Vol des Oiseaux*' (Marey 1890). Marey was a physiologist who designed experiments to monitor the activity of flight muscles and wing beat parameters of a flying bird (Fig. 1.4). He also made mechanical models of birds to measure lift forces during flapping movements. In his opinion, the resistance on a flying object in air should be equal to the resultant of all negative and positive pressure forces exerted on it. He designed and built ingenious experimental equipment with numerous manometers to measure pressure changes and drag. The experiments supported the direct relationship between the drag and the surface area of an object but also revealed that a coefficient is required. This drag coefficient is the ratio of the drag of an object and that of a flat plate of one square metre that moves at $1\ \text{m s}^{-1}$. It has to be empirically determined and is commonly based on force measurements of solid bodies in a physical experimental set-up, for example, wing sections in a wind-tunnel. It is not easy to relate the relevance of the resulting coefficients with those required for birds. One of the problems is

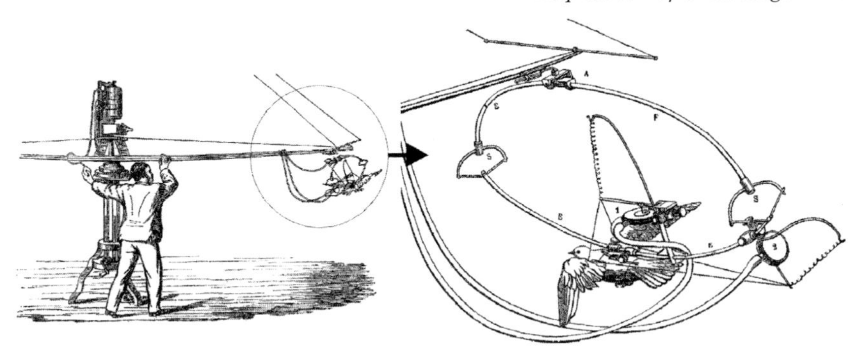

Fig. 1.4 Experimental equipment used by Marey to study bird flight empirically. Illustration taken from *Le Vol des Oiseaux* (1890).

that of scale. The results of measurements on a streamlined body in a flume must be translated to predict the flow around the fuselage of an aircraft. Major scaling problems were solved in principle by the laws of similarity discovered by Osborne Reynolds (1842–1912). The title of the key paper published in 1883 was: 'An experimental investigation of the circumstances which determine whether the motion of water shall be direct or sinuous, and of the law of resistance in parallel channels'. It shows that the determining circumstances are limited to the magnitude of four parameters only: the length dimension in the direction of the flow, the flow velocity, the density, and the viscosity of the fluid. Resistance or drag depends in a complex way on the behaviour of the fluid. If the flow is steady and smooth and flow particles are following one direction, the flow regime is direct or laminar. With length and/or velocity increase, the flow (sometimes rather abruptly) becomes sinuous or turbulent. In that case, the flow pattern is highly irregular and full of rotation. Reynolds showed that for each flow situation a dimensionless number can be calculated. The magnitude of the number gives an indication of the flow regime. The number has been called the *Re* number after the discoverer. (The equation is given in Box 1.4). It is the ratio of density over viscosity multiplied by the velocity of the flow and the length. Both the numerator and the denominator of this fraction have the SI unit $\mathrm{N\,s\,m^{-2}}$ (momentum per unit area), which makes it a dimensionless number. Density ($\mathrm{kg\,m^{-3}}$) divided by viscosity ($\mathrm{kg\,m^{-1}\,s^{-1}}$) is the inverse of the kinematic viscosity ($\mathrm{m^2\,s^{-1}}$). It is a measure of the tendency of the fluid to spread; its unit is time per square metre. At low *Re* numbers ($\ll 1$) viscous forces dominate and the flow is laminar. At higher values, inertia becomes more important and the transition to a turbulent regime will occur.

An example of the use of the *Re* number in one of my own studies on bird flight provides some feeling of its relevance. I wanted to visualize the flow pattern around the wing of a common swift in fast gliding flight at about 15 $\mathrm{m\,s^{-1}}$. The experimental set-up to do the measurement in air was not available. However, a closed circuit water-tunnel filled with particle-seeded water to visualize the flow was at my disposal. To obtain the same flow

Box 1.4 ***The Re number equation***

The Reynolds number (*Re*) conveniently expresses the relative importance of inertial over viscous forces in a dimensionless way:

$$Re = vl\,\nu^{-1} \quad (-) \qquad (1.4.1)$$

v is the velocity in $\mathrm{m\,s^{-1}}$; l a relevant measure of length in m. The kinematic viscosity ν is expressed in $\mathrm{m^2\,s^{-1}}$. It is the ratio of viscosity μ (momentum per unit area: $\mathrm{N\,s\,m^{-2}}$) to density ρ (mass per unit volume: $\mathrm{kg\,m^{-3}}$).

patterns in air and water the same *Re* number is required. At 20°C, the ratio of density over viscosity in water is 10^6 ($\mathrm{s\,m^{-2}}$) and that in air 15 times less. If the water in the tunnel could run safely at 1 $\mathrm{m\,s^{-1}}$ (15 times slower than the airspeed), the dimension of the wing could be the same as that of a real swift. The *Re* number of the swift wing in air at 15 $\mathrm{m\,s^{-1}}$ was 5×10^4 if the cord length of 5 cm is taken as the relevant length measurement in the direction of the flow. In fresh water, the same number is reached at a flow speed of 1 $\mathrm{m\,s^{-1}}$. However, a swift wing is not suitable to be used in water where it gets wet and where it meets forces, which will be higher than those in air. To overcome these problems, a polyester swift wing model was used in water to show the flow patterns occurring during glides in the air.

Re numbers for flying birds (based on body length) range from approximately 20,000 (2×10^4) to 200,000 (2×10^5) from gold crest to mute swan, commercial aircraft fly at numbers around 10^8. Viscous forces are relatively unimportant within the range relevant for birds, but in a thin layer close to the bird, where the airspeed is reduced to almost zero, viscosity may influence the flow pattern and the drag. The importance of this so-called boundary layer will become apparent in Section 1.6.

1.6 *Accumulation of knowledge in the twentieth century*

Calculations using the Navier–Stokes equations were originally only applicable assuming laminar flow conditions, and often provided results that deviated from measured values. At a conference in Heidelberg in 1904, Ludwig Prandtl suggested a practical solution for these problems by dividing the flow near a solid object into two regions: a layer of fluid close to the object where viscosity is a dominant factor and the flow outside that boundary layer where Euler's equations could be applied. (Properties of the fluid air are summarized in Box 1.5.) This approach focuses the attention to what happens in the boundary layer. Very close to a solid surface in a flow, the fluid particles do not move because they stick to that surface. In the direction away from the surface, there is a gradient of increased flow velocities until the

Box 1.5 ***Important properties of air for birds***

The International Standard Atmosphere at sea level

Quantity	Symbol	Value	Unit
Temperature	t	15	°C
Pressure	p	101,325	Pa ($N\,m^{-1}$)
Density	ρ	1.23	$kg\,m^{-3}$
(Dynamic) viscosity	μ	1.79×10^{-5}	$kg\,m^{-1}\,s^{-1}$ ($N s\,m^{-2}$)
Kinematic viscosity	ν	1.46×10^{-5}	$m^2\,s^{-1}$
Gravitational acceleration	g	9.81	$m\,s^{-2}$

The air temperature

Lapse rate is the meteorological term to indicate the rate of fall in temperature with increasing height. In standard atmosphere this rate is $0.0065\ K\,m^{-1}$ (Temperature in Kelvin (K) minus 273 = °C).

The density of air in standard atmosphere decreases with increasing temperature and depends on the relative pressure at the flying altitude:

$$\rho = 1.23 p p_0^{-1} 288 (t + 273)^{-1} \quad kg\,m^{-3} \tag{1.5.1}$$

where p is the pressure at altitude, p_0 the pressure at sea level, and t is the temperature in °C. At sea level extreme humidity values and temperatures approaching 40°C can decrease the density by approximately 3% maximally. Densities decrease at sea level ($pp_0^{-1} = 1$) with increasing temperature.

The viscosity of air increases with temperature following Sutherland's law:

$$\mu = 0.1458 \times 10^{-5} (t + 273)^{0.5} (1 + 110 (t + 273)^{-1})^{-1} \quad N s\,m^{-2} \tag{1.5.2}$$

The kinematic viscosity increases from about 1×10^{-5} at −40°C to $1.7 \times 10^{-5}\ m^2\,s^{-1}$ at +40°C.

fluid moves at the speed of the flow far away from the solid surface reaching the free stream velocity. Fluid particles in the velocity gradient are subjected to shear stresses because the neighbouring particles are moving at different speeds. Shear stretches and rotates the particles and is proportional to the viscosity and to the steepness of the velocity gradient (the resulting force per square metre is termed Newton's shear stress). Due to viscosity causing shear, the flow in boundary layers can easily become unstable. With increasing *Re* numbers fluid particles no longer follow straight laminar pathways but start to move along wavy tracks, make rotating movements; finally, the flow will become turbulent. A rough estimate of the thickness of the boundary layer can be obtained from the ratio of the length over the square root of the *Re* number (Lighthill 1990). It is less than 1 mm for the flow over the body of a 9-cm long gold crest and a few millimetres for a 1.5-m long mute swan with *Re* numbers of 20,000 and 200,000 respectively. With regard to bird flight, we should anticipate that most of the airflow is turbulent and energy is transferred through rotating masses of air. At best, the scale of the turbulence can be small with respect to the size of the bird. Vastly increased computer power and clever modelling make it currently possible to calculate viscous and unsteady flow fields near the wings of animals in flapping flight. Solutions of the full three-dimensional Navier–Stokes equations for incompressible unsteady fluid flow have been reached in studies on insect flight (i.e. Liu *et al.* 1998; Sun and Tang 2002). The results closely match empirically obtained data. Insect wings however are relatively simple compared to bird wings and operate at lower *Re* numbers. Dynamic changes in shape during bird wing flapping cycles are an extra challenge for computational fluid dynamicists. So far, the flow related to flapping flight in birds has not been modelled successfully. Empirical measurements of the three-dimensional airflow around a flying bird are required to reach full understanding of avian flight.

Empirical investigations of the lift and drag performance of cambered wings, originally based on cross-sectional shapes of bird arm wings, were of fundamental importance at the start of the development of manned flying machines. After powered flight became established however, aeroplane wing design was no longer based on studies of bird wings. Generations of reliable aircraft evolved rapidly during the first decennia of the twentieth century no longer inspired directly by nature.

Aeroplanes were already abundantly flying about when Herbert Wagner published the theory behind the development of upward force by accelerating wings in 1925. Importantly, Wagner indicates that it takes time before a wing, accelerated in still air, generates its maximum lift. Figure 1.5 shows the development of lift generating flow relative to a wing cross section in three successive time steps from top to bottom. In the upper figure, the flow has just started to approach the wing at a small angle of attack. The air stagnates below the wing and starts to flow at high velocity around the round leading edge over the top part (why that happens will be discussed in Chapter 4). Two areas of high and two areas of low pressure develop as indicated. There are high velocities and low pressures below the sharp trailing edge and on top of the rounded leading edge. Positive pressure develops due to stagnation of the flow below the wing, where the flow hits the lower part of the round leading edge, and on top of the wing close to the trailing edge. Due to this pressure distribution, there will be on average some obliquely backward directed upward force on the wing during this first stage. The pressure differences close to each other near the sharp trailing edge of the cross section create an unstable situation and the air will start to roll up into a vortex, which at first remains attached to the trailing edge as shown in Fig. 1.5(b).

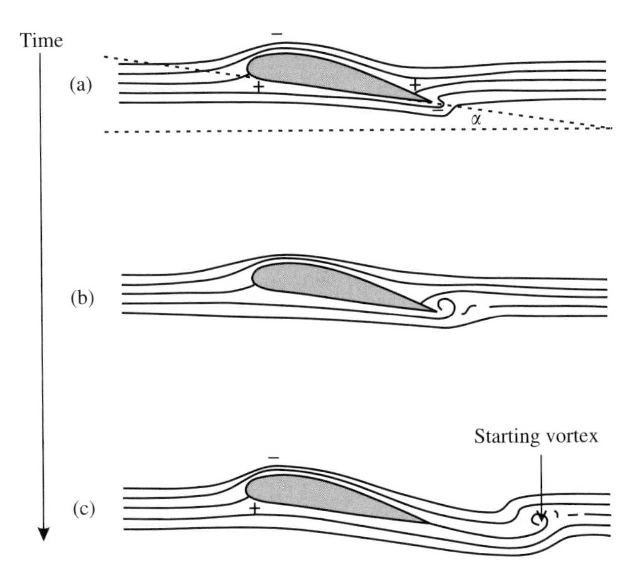

Fig. 1.5 The development in time of pressure differences around a fixed wing accelerated from rest (Wagner 1925). The pictures (a)–(c) represent three time steps. The angle of attack, $\alpha°$, between the bottom of the wing and the horizontal airflow is indicated in (a).

A vortex is a rotating motion of fluid easily caused by pressure and velocity differences. The anticlockwise rotating vortex at the trailing edge does not remain attached but is shed soon (and named 'the starting vortex'). From that moment on, the pressure difference between high pressure below the front part of the wing and the low pressure over the upper side, and hence the lift force, is fully established and stable (Fig. 1.5(c)). It is possible to witness the instant when the starting vortex is shed while sitting in a seat overlooking the wing in a large aircraft such as a Boeing 747. Just prior to take-off, while the aircraft increases speed over the first part of the runway, the wing tips droop down a little bit. At the instant when the wings shed the starting vortex, the wing tips abruptly swing up over a considerable distance and the aircraft takes off. Aeronautical engineers know about the time delay before maximum lift develops and call it 'the Wagner effect'. After take-off the pressure difference between the lower and upper side of a lift-generating wing exists all along the wing. Only at a wingtip air can escape from the high-pressure region below the wing to the upper part with low pressure. The escaping air rotates up and inwards and forms a wing tip vortex. Wingtip vortices emerge at the instant when the starting vortex is released and lift develops. They are in fact continuous with the starting vortex. Vortices have a centre where the rotational velocity is minimal; away from that centre the rotational velocity increases to a maximum and gradually dies out beyond that. Fluid dynamic theory tells us that, in the direction along its centre, a vortex cannot end in the fluid. It either can end at the boundary of the fluid or must form a closed loop with itself or another vortex (Lighthill 1986). The starting vortex of an aircraft forms a closed loop with the wing tip or trailing vortices and the bound vortex on the lifting wings. The bound vortex around the wings becomes visible by subtracting the average flow velocity from the local velocities around the wing. The bound vortex is shed during landing when the speed of the aircraft drops below the value needed to maintain the pressure differences.

Between the wing tip vortices, the air rushes down from the trailing edges of the wings. This downwash is the mass of air diverted downward per unit time by the presence of the wing (Anderson and Eberhardt 2001). The bound vortex, the trailing vortices, and the starting vortex form one giant closed vortex ring (Fig. 1.6). The trailing vortices of many commercial aircraft are visualized on sunny days with blue skies because engine exhaust fumes are sucked into them. The parallel lines formed by the two vortices can be followed across the sky when the wind is calm at the flying altitudes. It is fascinating to imagine that a departing plane leaves a starting vortex somewhere halfway down the runway and produces a double trail connected to this starting vortex to end with a stopping vortex during the landing, producing a giant extremely elongated ring.

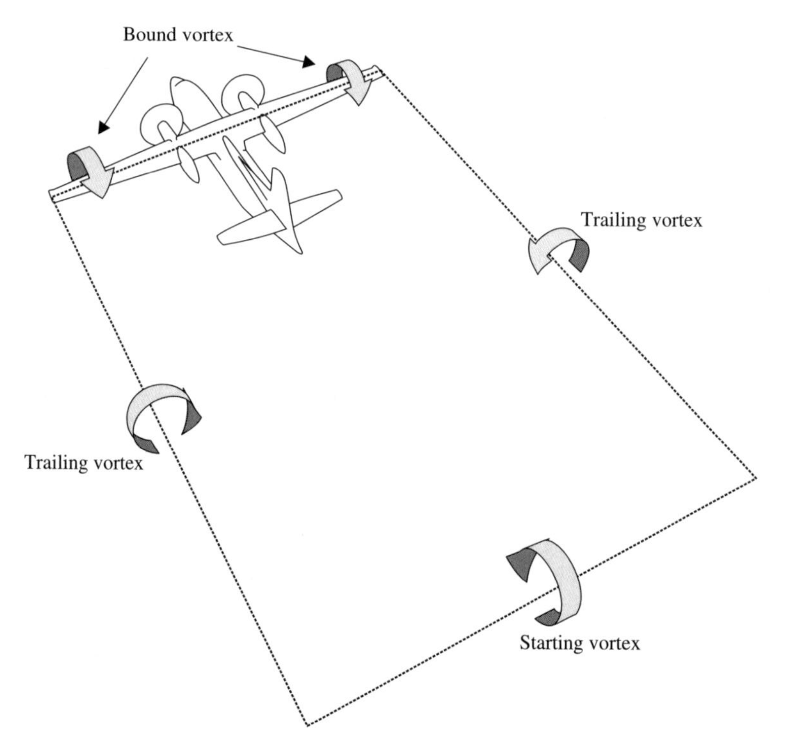

Fig. 1.6 A simplified vortex system of a conventional aircraft (Fokker 50).

Over the past century, bird flight aerodynamics was considered explained by fixed wing aircraft theory. Although no direct evidence is available, that point of view is probably correct as far as fast steady gliding of large birds is concerned. There is also no experimental proof to show that attached flow is present on bird wings during low-speed gliding, that is, prior to landing on a perch, or during most forms of flapping flight. Can we expect built-up of attached circulation on a flapping wing? Several studies show the structure of the wake behind flying birds (treated in Chapter 4). The total disturbance caused by the bird and hence the energy exchange between bird and air can be measured by analysing the wake, but information about what happens at the wings cannot be conclusively derived from wake structures. This means that our present knowledge on how birds fly ends here.

In an attempt to overcome some of the negative aspects related to conventional wings, engineers developed aircraft based on a fundamentally different aerodynamic principle after the Second World War. The lift generating flow of conventional wings can easily be disturbed because it needs to stay attached to the wing surface to avoid dramatic loss of lift. Flow separation from the upper part of the wing, for example, caused by an overly large angle of attack, causes the wing to stall, which implies loss of lift and strong increase of drag. Contrastingly, the principle used in delta wing aircraft, for example, the Concorde, is based on flow separation right at the leading edge of sweptback wings. Delta wings with sharp leading edges readily stall even at small angles of attack in a controlled way. The separating flow forms an attached vortex along the entire leading edge: a LEV. It increases in size towards the rear and the forward velocity of the aircraft is reflected in the spiral pathways of the air in the LEV. The conical shape of such a spiral vortex on top of the Concorde wings is shown in Fig. 1.7. The principle was already well known from empirical studies before Polhamus (1971) developed analytical methods to predict the lift and drag

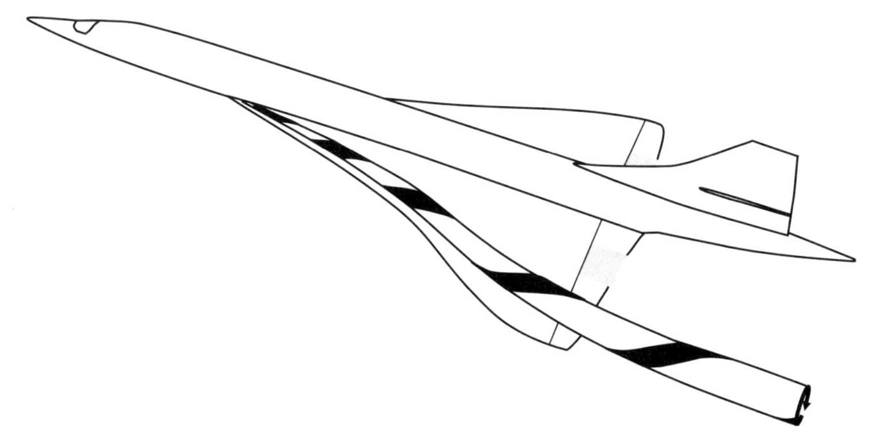

Fig. 1.7 The conical spiral leading edge vortex on the wings of the Concorde.

characteristics of leading edge vortices. LEVs can produce high lift and high drag almost instantaneously and the physical conditions required are not at all critical. The velocities at which LEVs function can be very low.

Recent research indicates that LEVs and other novel lift generating mechanisms play an important role in insect flight aerodynamics during gliding and flapping (Sane 2003). Circumstantial evidence that birds widely use LEVs will be given in Chapter 4. Further empirical research should concentrate on the interactions between wings and air during flight to find out if birds and insects have more tricks in common.

1.7 *Novel initiatives*

With most attention going to aerodynamic developments related to aircraft design, little was added to the knowledge of bird specific flight mechanics during the first half of the twentieth century. In Germany, Erich von Holst built artificial birds that could perform a kind of flapping flight (Holst 1943). However, this 'Spielerei' did not bring new insights.

In the same period the functional anatomy of the flight apparatus became well established (i.e. Herzog 1968; Sy 1936), although some important details of wing (Vazquez 1994) and feather (Ennos *et al.* 1995) structure were only recently discovered as is shown in the following chapters.

We see many new developments during the last four decades of the twentieth century. During the late 1960s and early 1970s mathematical models based on aerodynamic principles were developed by several specialists to calculate the mechanical costs of flight in birds. Wind tunnels were especially designed and built for bird flight studies under controlled conditions. X-ray film techniques combined with electromyograms of the most important flight muscles show how the movements of the skeleton and the timing of muscle contractions are related. The forces exerted by the main flight muscles on the wings of starlings and magpies in full flight have been measured directly (i.e. Dial *et al.* 1997). Energy spent in flight has been measured in various ways. The interaction between a flying bird and air results in rotating masses of air (vortices) behind the bird. Vortex theory describing these forces and the energy involved in such rotational flows is developed for flying animals by Ellington (1984) and Rayner (1979*a,b*).

In the field, the total energy budgets over longer periods are determined from the exchange of respiratory gases by measuring the drop in concentration of heavy isotopes of hydrogen and oxygen injected in the blood of experimental birds. Heart rate logging and satellite transmitters or satellite operated global positioning systems (GPS) attached to big birds offer a wealth of information on flight performance (Butler *et al.* 1998). Advanced radar techniques also contribute significantly to our knowledge of flight behaviour and strategies (Alerstam *et al.* 1993). Several investigators flew in small aircraft with flocks of birds to collect data on the flight conditions as experienced by birds.

Recent biological research is changing our insight in the art of bird flight although conventional aerodynamic theory is still widely used to explain most aspects of it. Steinbeck's (1958) statement probably applies here: 'When a hypothesis is deeply accepted it becomes a growth which only a kind of surgery can amputate'. Birds are not miniature aircraft, clumsily fluttering the wings to stay aloft. Birds travel over large distances and fly in dynamic interaction with the medium air. They can start and land anywhere, hover on the spot and make use of favourable wind conditions; we still have to discover how they do that.

1.8 *Summary and conclusions*

The unfortunate conflict between Aristotle's philosophy and his accurate observations probably precluded the discovery of the basic principles of bird flight 24 centuries ago. His complex thoughts dominated the way of thinking about the subject until well into the Renaissance. From that period onwards, two more or less separate scientific approaches developed, each contributing to what we presently know about bird flight. One focussed on birds and the way they remain airborne and a second approach was concerned with distantly flight related fundamental physical processes. Until the second part of the twentieth century, few scientists actually concentrated on the flight of birds for the sake of the subject itself. Da Vinci, Borelli, Cayley, and Lilienthal studied bird flight because they were dreaming of manned flight and wanted to learn the art from birds. The scholars representing the physical approach were not at all interested in birds but tried to explain the principles of movement in a fluid medium such as air. The discovery of the laws of gravity by Galilei started an era of increasing fundamental insight. The development of the concepts of potential and kinetic energy provided insight in the energetics of moving objects in fluid and Bernoulli's law was an important step in understanding the relationships between pressures and velocities in that fluid. Newton's laws explain the basic principles of flight.

Insight in the aerodynamic properties of curved plates derived from Lilienthal's studies of the arm wings of large gliding birds started the development of manned aircraft in the beginning of the twentieth century. From that period onwards, explanations of bird flight are entirely based on lift generation due to attached flow over a curved wing. Recently, the leading edge vortex, the dominant lift generating mechanism of delta-winged aircraft is found to be used by insects. It could well be an important additional mechanism used by other flying animals including birds.

Details of bird wings provided in Chapters 2 and 3 indicate which aerodynamic principles are most likely used. These are also discussed in Chapter 4.

2 *The flight apparatus*

2.1 *Introduction*

Birds are unique because they use feathered front legs as wings. The basic quadruped organization is modified to accommodate the needs required by the special function. Natural selection sacrificed other functions of the arms and hands during the evolutionary process. The whole body plan was reshaped resulting in a bipedal organism with completely decoupled anterior extremities.

Ornithological textbooks usually provide overviews of the anatomy of birds in general and of the flight apparatus in particular. An elaboration of the same here would be superfluous. Instead, brief summaries of flight-related anatomical aspects will be given together with some insight into the relations between form and function and in the variation among groups. The aim is to find out what the apparatus tells us about how birds fly.

Wings will receive most of the attention. There are as many different wings as there are bird species and any classification into a limited number of functional groups only reduces the interesting complexity. Understanding the relation between form and function demands detailed attention for the specific wing structure of each species. On the other hand, wings of all flying birds have important common features: they all consist of an arm and a hand part. The basic construction of the internal and external wing will be described first. Insight into the internal and external anatomy of wings is needed to understand the principle movements allowed by the structural constraints. Wings must be folded and stretched as well as be moved up and down and rotated fore and aft. Bird wings have been the subject of many scaling exercises and it is good to know the main conclusions from these studies without attempting to go into the details of the numerical analyses. Aristotle had already found out that a bird without wings cannot fly, but which parts of the wings are absolutely necessary to allow take-off and flight? Experiments with mutilated wings of live birds have a long history and we will see if the results obtained increase our insight in the function of bird wings.

Hummingbirds and swifts deviate so much from that basic pattern that their wing design receives a separate paragraph. For the same reason the long slender wings of albatrosses and giant petrels require special treatment.

Tails are the next dominant intrinsic parts of the flight apparatus. The internal anatomy and the outer shape have flight-related features and we must try to find out how much of it is understood.

Not only the wings and the tail but also the head, the neck, the body, and the hind limbs have features directly related to flight in many species. Some obvious ones will be discussed.

Feathers are crucial in relation to flight since they determine to a large extend the shape of the bird including the main parts of its flight apparatus. Detailed treatment of the knowledge about the structure, function, and mechanical properties of the main flight feathers requires a separate chapter that will follow this one.

2.2 *Wing morphology*

The basic structure of wings in general must be treated first before deviating patterns can be distinguished. Internally bird wings have a modified quadruped arm skeleton; on the outside the shape is uniquely determined by feathers, the hallmark structure of birds. Variation on the general pattern can be large and is illustrated by groups containing the smallest and the largest extant birds.

2.2.1 Internal wing design

Wings have to be both strong and light. Strong because they have to exchange forces with the air. During flapping flight the oncoming air flow is deflected and air is accelerated by the wing action. Wings must also be able to receive the reactive forces from the air and transmit these to the body. They must be light to reduce inertial forces during various phases of the wing beat cycle, but in particular during the acceleration phase of the downstroke when the wings are fully stretched. Therefore, the amount of heavy tissue such as muscles and bones decreases towards the tip.

The schematic drawing of Fig. 2.1 shows the approximate position and the names of the main internal parts of a wing. (Appendix 3 explains the meaning of the scientific names and jargon.) The design enables the wings to flap up and down and to fold and extend. Wings are connected to the body at the shoulder joint where the proximal condyle of the humerus, the caput humeri, articulates in the glenoid cavity formed by the scapula and the coracoid. The coracoid is firmly connected to the sternum, and its length determines the distance between the shoulder joint and the sternum. Paired clavicles are fused to form the wishbone (the furcula), which is attached to the dorsal parts of the left and right coracoids. The sternum of flying birds has a central bony keel, the carina. Ribs, vertebral column, and the sternum form a closed cage. The main flight muscles, the pectoralis and the supracoracoideus, have their origin on the sternum, on the carina and on the coracoid. The pectoralis inserts from below on the anterior crest of the humerus. It pulls the wing down and causes forward rotation (pronation) of the wing during the downstroke. The supracoracoideus is situated underneath the pectoralis. It forms a tendon which passes through the triosseal

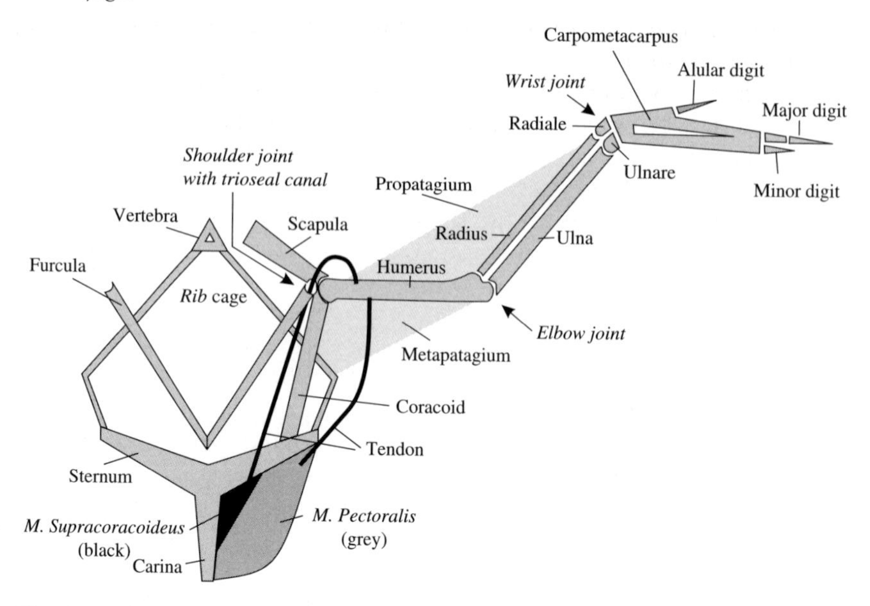

Fig. 2.1 Schematic overview of the flight-related internal anatomy of the left wing and rib cage of a generalized bird (see text for further explanation).

canal in the shoulder joint to insert on the dorsal tubercle on the upper part of the humerus, approaching the insertion from above. The triosseal canal is formed either by the scapula, the coracoid and the furcula, by the scapula and the coracoid, or even by the coracoid alone. The canal forms an important anatomical feature because it acts as a pulley to make it possible that the supracoracoideus contributes to lifting the wing. Together with a few other small muscles the supracoracoideus is responsible for the upstroke and for the rearward rotation (supination) of the wing during the upstroke or prior to landing.

The shoulder joint faces laterally and allows the humerus a great deal of freedom. It can move up and down and fore and backward over large angles in most birds. The wings are supported by the humerus, the radius and ulna, and by the hand skeleton. The propatagium is a fold of skin inside the front part of the arm wing. The metapatagium connects the elbow with the trunk. The radius and ulna articulate with the humerus at the elbow and with two carpal bones (the radiale and the ulnare) in the wrist joint. The wrist is a double joint because the carpal bones articulate also with the carpometacarpus of the hand skeleton. The elbow and wrist joints in avian wings extend and flex in synchrony due to a special configuration of skeletal and muscular elements. A separate paragraph is devoted below to the detailed anatomy and freedom of movement of wings. The hand skeleton consists of the carpometacarpus (fused carpals and metacarpals) and some digits. There are usually only three digits with one or two phalanges each.

The first digit is the skeleton of the alula or bastard wing; the others support primary feathers.

The actual and relative dimensions of the 10 skeletal elements of the wing differ among birds. The differences provide insight into specific functions of the arm and hand parts of the wing. Figure 2.2 shows the skeletons of the forelimb of five species scaled in such a way that the skeletons of the hand wing are of the same length. The relative importance of the role of the hand wing is given away by the bowing of radius and ulna. The wider the gap between these bones the more room there is for forelimb musculature inserting on the hand wing. The Laysan albatross has obviously less dynamic control over its hand wing than the others and is less coordinated in unsteady flight situations during flapping, starting, and landing. The blue grouse shows the opposite features. Hummingbirds and their close relatives the swifts have extremely long hand skeletons. In flight the arm part of the wing appears shorter than it actually is because the wrist is kept close to the body. X-ray pictures in Fig. 2.3 compare the skeleton of the common swift with that of a song bird (the European goldfinch) to show the substantial difference between the extremely aerial swift and a generalist. The angle at the elbow of the swift is virtually fixed. Blood vessels and nerves run straight across from the humerus to the radius and ulna. The song bird can stretch and fold the arm almost completely.

Bird wings contain 45 muscles, 11 of these are subdivided into two or three parts. Eight muscles have more than one insertion point (Vanden Berge

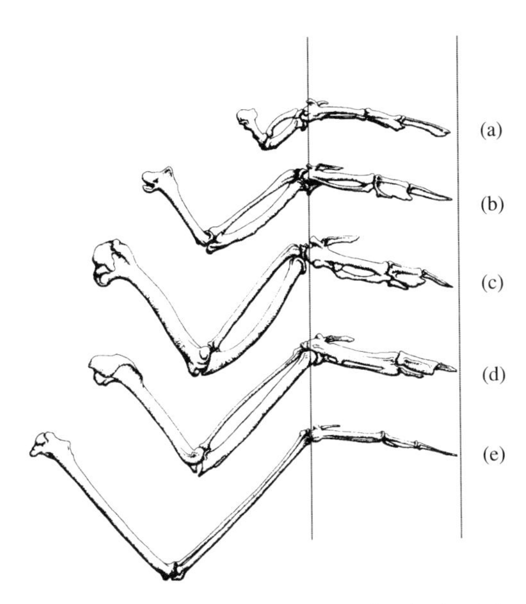

Fig. 2.2 Relative dimensions of the skeleton of the forelimb of five species: (a) Calliope hummingbird; (b) Rock dove; (c) Blue grouse; (d) European starling; (e) Laysan albatross. The skeletons of the hand are drawn at the same length (from Dial (1992)).

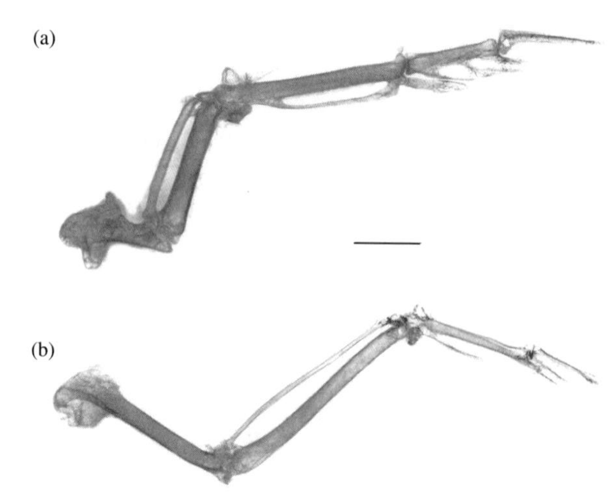

Fig. 2.3 Actual dimensions shown by X-ray pictures of the skeletons of the common swift (a) and a songbird, the European goldfinch (b). The scale bar is 1 cm.

1979). General descriptions of their origin and insertions can be found in handbooks on bird anatomy usually together with a description of the specific action (i.e. Proctor and Lynch 1993). Muscle activity during flight of only 18 of these muscles has been seriously studied using electromyogram (EMG) techniques. Results based on these studies regarding the timing of muscle activity during wing beat cycles are summarized in Chapter 7.

The shape of a wing is only marginally determined by the internal anatomy; it is the feathers that make a wing fly.

2.2.2 The external shape of bird wings

Very different kinds and sizes of feathers are implanted in the skin of the wings where rows of follicles follow well-defined tracts (Lucas and Stettenheim 1972). The large flight feathers in the wing are the remiges, wing coverts are termed tectrices. The primary remiges (in short primaries) are 9–11 strong feathers found in the hand part of the wing (grebes are exceptional with 12). These usually have asymmetric vanes with a narrow leading edge (outer) vane and a wider (inner) vane forming the rear part or trailing edge. The asymmetry is stronger towards the outer feathers. The secondary remiges (secondaries) form the larger part of the surface of the arm wing. Their number varies greatly between 6 (usually overlapping) in hummingbirds, 9–11 in songbirds, and 11–15 in pigeons, 25 in large vultures, and up to 40 in the albatrosses. Primaries are stiffer and more pointed than the secondaries. The remiges forming the bastard wing or alula are small versions of the primaries. Tertial remiges (tertials) cover the space between secondaries and the body. The shape of the arm part of the wing is formed by rows of lesser and greater coverts, covering the propatagium (forming the leading edge) and the follicles of the remiges.

The variation among birds makes it a practical approach to concentrate on the design of one particular wing first and use its features in comparison with other wings. The wing of the northern goshawk serves as an example in Fig. 2.4. Contour feathers cover the front part of the arm wing and the proximal part of the hand wing. These provide the rounded leading edge shown in cross sections (a), (b), and (c). A row of greater coverts covers the implants of the primaries; a row of secondary coverts does that with the implants of the secondaries. Towards the leading edge of the wing, on both

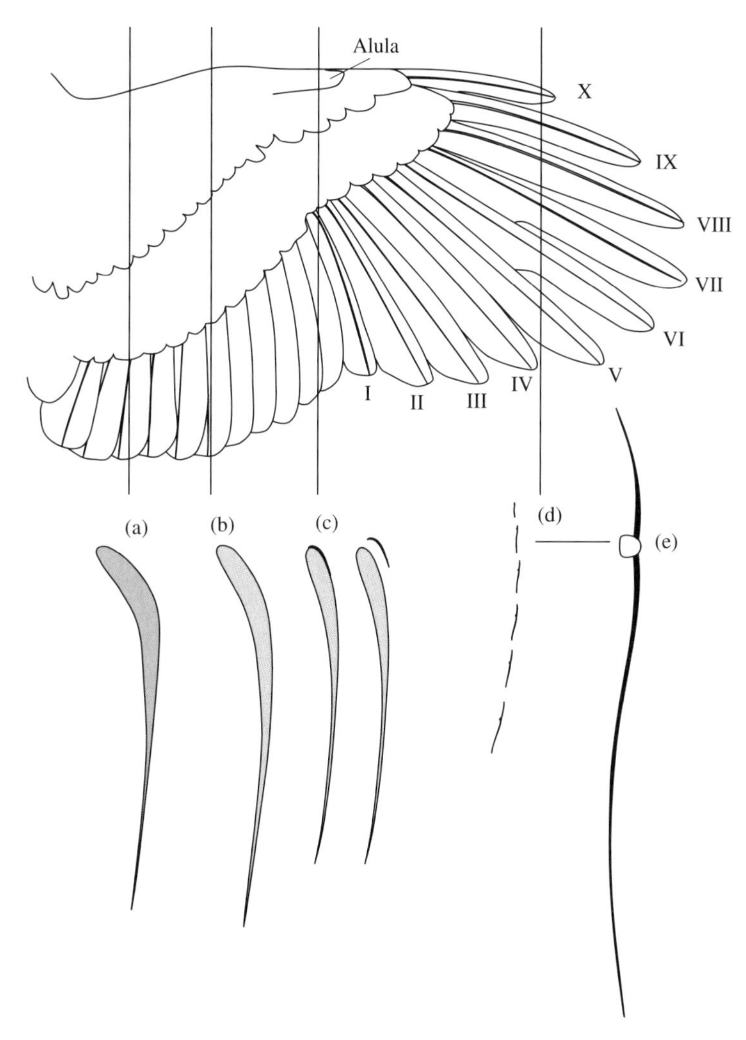

Fig. 2.4 Drawing of the dorsal side of the wing of a goshawk (modified from Herzog (1968)) with cross sectional profiles at four positions ((a)–(d)). The cross section through primary IX in (d) is enlarged to show its actual shape (e) (see text for more details).

the dorsal and ventral side, rows of increasingly smaller so-called marginal coverts overlap each other like the tiles on a roof.

Symmetrical tips of 11 secondary remiges form the sharp trailing edge of the arm wing. The profile of the cross sections (a), (b), and (c), through the arm has a rounded leading edge and is highly cambered. The leading edge resembles that of classical aerodynamic profiles used in aircraft design, but the extreme camber makes the cross sectional profiles substantially different from most man made wings. Profile (c) is situated at the transition from arm to hand wing. It also has a rounded leading edge and a sharp trailing edge. The picture shows a cross section through the remiges of the alula on top of the leading edge in two positions. In the section on the left the alula is kept close to the wing; the right-hand drawing shows the situation with the alula extended.

Cross sections through the hand wing differ fundamentally from classical aircraft profiles because they have a sharp leading edge formed by the narrow outer vanes of either primaries X, IX, or VIII. In Fig. 2.4(d), primary X forms the leading edge. There is some distance between the sections through the primaries here caused by a combination of spreading of the hand feathers and by the emargination of these feathers. Emargination is the term used to indicate that the distal part of the primary vanes decreases rather abruptly in width. This may occur at the narrow leading edge vane, at the wide trailing edge vane, or at both vanes simultaneously. Spreading of the hand wing and emargination forms the slots near the wing tip seen in many groups of birds.

Each of the feathers in cross section (d) is a more or less an independent wing section illustrated by the magnification of the cross section through primary IX in Fig. 2.4(e).

The hand wings of albatrosses and of the southern and northern giant petrel deviate fundamentally from the goshawk wing example and therefore receive special attention in a later paragraph.

There are surprisingly few accurate measurements of sizes of arm and hand wings in the literature. In most birds, the hand wing is longer than the arm part but there are many exceptions especially among large soaring birds. The longest relative hand wing lengths are found in swifts and hummingbirds where the extended length of the arm skeleton is extremely short compared to the hand wing skeleton.

2.2.3 Hummingbird and swift wings

Hummingbirds and swifts deviate so substantially from the general description that they require a separate paragraph. The internal wing design of hummingbird wings will be described first followed by brief comments on that of the swift. The hummingbird configuration is described accurately by Stolpe and Zimmer (1939) and a more recent analysis does not seem to exist. The relative dimensions of the bones of the wing skeleton resemble those of swifts but are very different from other birds (Fig. 2.2). During hovering flight the main axis of the bird is obliquely downward and the wings beat in an approximately horizontal plane. The arm is extremely short because the humerus and radius and ulna are short and kept in a fixed sharp-angled V-shaped position during flight (Fig. 2.5). This angle cannot be enlarged in a stretch because nerves and blood vessels run straight from the shoulder to the hand. The hand wing is relatively the longest found in birds. Hertel (1966) indicated that the hand wing of a hummingbird occupies 81% of the wing length against 41% in the case of a buzzard. There are only 6 partly overlapping secondaries in the arm; 10 long primaries form the main surface of the wings.

The sternum bears a substantial carina. The main flight muscles, the pectoralis and the supracoracoides, occupy about 27% of the body mass, the pectoralis being only 2 times as big as the supracoracoides (these figures are 18% and 12 times in passerines respectively, according to Greenewalt (1975)). An extremely long scapula supports the shoulder joint; it runs down the body to almost reach the pelvic girdle. The humerus is very short and has a bizarre shape; it is kept in an almost vertical position during hovering flight. The articulating surface with the shoulder joint is not at the terminal position of the humerus but there is a condyle at the inner side of the proximal end. The condyle in this position is a unique character of hummingbirds. The tendon of the supracoracoid muscle contains a sesamoid bone (Fig. 2.5). It is attached to the outer part of the humerus head, runs through the triosseal canal, and from there down to the muscle on the sternum. Contraction of the supracoracoides will cause adduction and rearward rotation around the

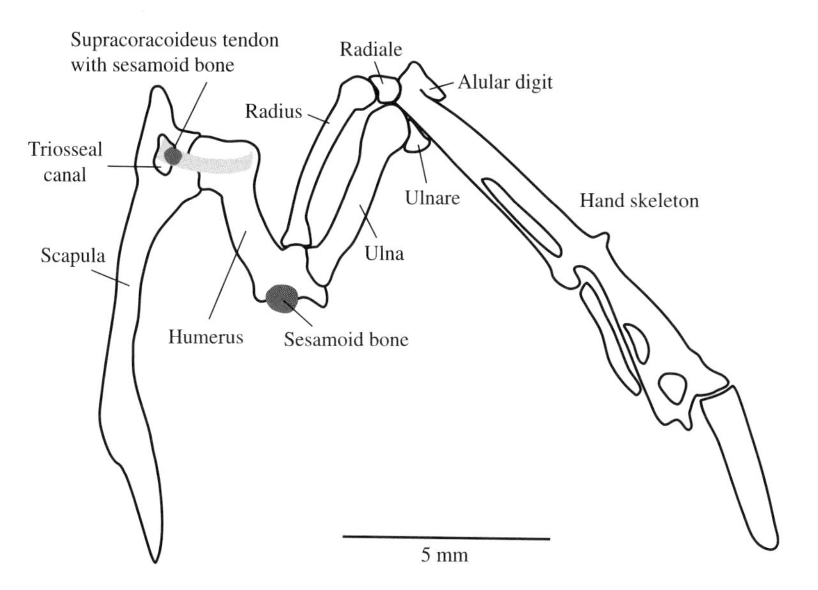

Fig. 2.5 The special skeletal structures of an amazilia hummingbird right wing in dorsal view. The drawing is based on photographs taken from an alizarine stained skeleton of a cleared specimen using the procedure of Taylor and Dyke (1985).

length axis (supination) of the humerus. This rotation causes in fact the back (up) stroke of the wing. Pronation of the humerus by the pectoral muscle inserting on the front part of the humerus head will result in the forward (down) stroke.

The elbow joint is peculiar too because it is obviously not designed to stretch. The muscles of the arm wing are extremely well developed and encapsulate the joint, keeping it in folded position. The extensor muscle (the scapulotriceps) has changed its function. The tendon contains a large sesamoid bone on the rear (upper) side of the elbow (Fig. 2.5) which determines its working direction. The sesamoid bone sits in a dent in the distal end of the humerus. Its presence causes the extensor muscle to rotate the ulna and radius backwards (upwards) instead of stretching the arm.

The capacity to rotate is even bigger in the complex wrist joint between the hand skeleton and the radius and ulna. The alula digit is reduced and immobile. (Hummingbirds have no alula.) The primaries are firmly attached to the bony elements, supported by cartilage and connective tissue. The pectoralis powers the hovering wing beat during the forward stroke and the supracoracoideus during the backstroke. These muscles rotate the vertical triangle formed by the V-shaped humerus and radius and ulna. The hand wing is attached to this triangle at the wrist and follows the movement. Combined rotations of radius and ulna and of the wrist joint enable the extreme rotation of the wing plane during the backstroke where the wing is used in upside down position.

The wing of the common swift deviates less from the basic bird wing design than the hummingbird wing does. The hand wing is still extremely long covering 75% of the total wing length. The elbow is less confined in its movements than that of the hummingbirds and the humeral joint has the familiar egg shape and allows normal vertical flapping motion. The swift has 11 primaries. Number XI at the leading edge is only 2 cm small, stiff, and almost without vanes. It supports the base of the longest primary number X. There are seven secondaries in the short arm part of the wing. The alula consists of two or three feathers with a total length of about one-eighth of the wing length. All these features witness the fact that the swift is an extremely agile flyer flapping its wings mainly vertically and not horizontally. The description of the swift in the *Birds of the Western Palaearctic* (Snow and Perrins 1998) emphasizes its extreme flight capacities: 'Flight dramatic, showing complete mastery of open air space and marked ability in gliding, wheeling, diving, accelerating or stalling, and climbing; wing-beats rapid and made usually with wings in distinct backward curve...'.

2.2.4 The wings of albatrosses and giant petrels

Large oceanic birds that spend nearly as much time on the wing as the swift rely heavily on the extremely long arm wings for the generation of lift. Fast gliding in high winds is their speciality. The most extreme dynamic gliders among birds, the albatrosses and giant petrels, are believed to be capable to lock the wings in stretched position and in doing so avoid spending muscular energy to fulfil that task. Both the source of this knowledge and the mechanism behind it are difficult to track down. Hector (1894) gives, after 're-examining the wing of a large albatross in the flesh', the following record of his findings:

> The extensor muscular tendon, instead of being attached as in other birds only to a fixed process at the distal extremity of the humerus, is also attached by a subsidiary offset to a projecting patelloid bone which is articulated with the process, and thence proceeds to the radial carpal bone, and thence onward along the radial aspect of the manus, where it expands into fibrillae that embrace the quills. When the wing is fully extended the thrust of this projecting process on the elbow joint causes a slight rotation of the ulna on the humerus, so that the joint becomes locked, which renders the wing a rigid rod as far as the wrist joint. At the same time the slight play permitted by the articulation of the patelloid bone on the process allows of the transmission of the muscular pull from the shoulder to the manus without unlocking the joint.

This description does not explain clearly what is supposed to be happening. Yudin disagrees with the idea that sesamoid bones are involved and offers an alternative explanation. He presented the theory in 1954 at the international ornithological congress in Basel under the name K. Joudine. It was published in the proceedings in French (Joudine 1955). It appeared in Russian in the *Zoologiceskij Zurnal* in 1957 under the name K. A. Yudin (Yudin 1957). The locking mechanism he describes is shown in Fig. 2.6a. Tube-nosed birds have a bump in the saddle on the proximal end of the ulna. The sliding radius finds a stable position on each side of the bump.

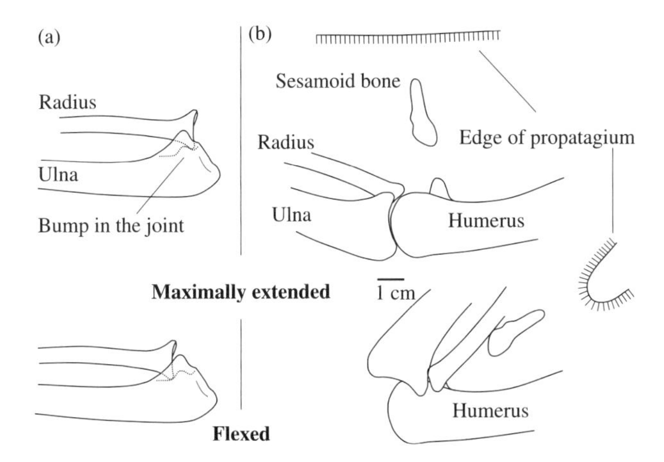

Fig. 2.6 (a) The locking mechanisms in the wings of albatrosses and giant petrels according to Joudine (1955). The position of the radius with respect to the ulna in extended and flexed position and the bump in the joint are indicated. (b) Outlines of the skeleton from X-rays of the elbow joint of a wandering albatross in extended and flexed position (right wing in ventral view). The radius seems to snap in the distal-most position during extreme flexion only, securing the wing while folded.

In fully flexed position the radius has moved to the distal-most point pushing via the radiale the carpometacarpus in the folded position. The results of the X-ray studies summarized in Figure 2.6b imply that Yudin's locking mechanism does not keep the extended wing stretched but the flexed wing folded. Pennycuick (1982) found a lock at the shoulder joint of albatrosses and the southern giant petrel. It consists of a fan-shaped tendon running from the carina to the deltoid crest on the humerus. This tendon is superficial in the wandering albatross, the black-browed albatross, and the light-mantled albatross and deeper inside in the southern giant petrel. By manipulating dead animals Pennycuick found that the shoulder joint came up against a lock when raised to the horizontal position after the stretched wing had been moved forward to the fully protracted position. The lock no longer operated when the humerus was retracted a few degrees from the fully forward position or when the tendon was cut.

The hand wing of albatrosses and giant petrels deviates from that of all other birds because the structure of the primary feathers is different (Boel 1929). This will be further discussed in Chapter 3.

2.3 *Dynamic wing properties*

Wing shapes may differ among species but the change in form during a wing beat cycle is more dramatic. Before take-off, the wings are neatly folded against the body. They unfold and stretch at the onset of flight, flex partly during each upstroke, and extend fully before the beginning of the downstroke. The principle movements of a wing as a whole allowed by the structure of the shoulder joint are up and down, for and aft, backward rotation (supination) and forward rotation (pronation). The head of the humerus in most birds has not a ball but an egg-shape, which reduces the freedom of movement. The range of possible movements is also limited by ligaments around the joint. Usually, forward and backward rotation of the humerus in a horizontal plane around a vertical axis through the joint is allowed. When the wing is extended the humerus can move up and down and rotate around its lengthwise axis. The angle of the upward movement can be more than 90°, whereas the downward movement is usually restricted to less than 35°. Pronation is commonly much more restricted than supination.

The dimensions may be different but the mechanism moving the wing is surprisingly uniform among flying birds. Here we are first concentrating on the mechanics of the principle movements of the upper arm, the forearm, and the hand. These movements are flexion and extension of the wing and circumduction of the hand. A feeling for the basic mechanisms will enable us to appreciate what is known about the wing beat dynamics in greater detail later.

2.3.1 The drawing parallel action of the radius and ulna

The distal head of the humerus forms the elbow joint with the proximal endings of the ulna and the radius. When the wing is stretched the shape of this joint severely limits dorsoventral rotation of the forearm with respect to the upper arm. Freedom of movement in the horizontal plane allows stretching and flexing of the forearm. During these movements the radius shifts parallel to the ulna inducing flexion and extension of the hand. The parallel shift has long been attributed to the shape of the distal head of the humerus. A knob on the head was thought to push the radius in outward direction. A close examination by Vazquez (1994) however, showed that the shape of the humerus condyls in the plane of interaction with the radius and ulna were circular in flying birds. Rotation around a circular knob will not result in relative shift of the bones involved.

During wing flexion the drawing parallel action of the radius and ulna is caused by collision of bulging muscles of the forearm and the upper arm when the elbow is flexed to angles smaller than 60° (Fig. 2.7(a)). The pressure of the abutting muscles dislocates the radius from the end condyle of the humerus

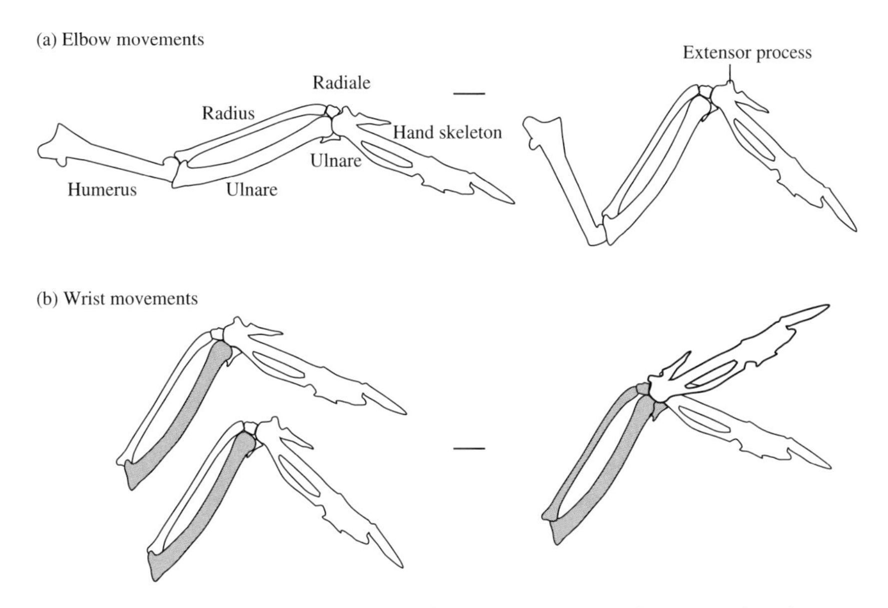

Fig. 2.7 Movements of the skeleton of a pigeon wing (right wing, dorsal view). (a) Automatic flexion of the hand via the radius. During flexion of the elbow the angle between humerus and ulna decreases. When the angle reaches about 60°, muscles of the forearm collide against muscles of the upper arm. The abutting muscles dislocate the radius and push it against and along the ulna in the direction of the wrist. The action results in flexion of the hand shown on the right. (b) Hand flexion occurs via two joints in the wrist. (Not moving parts are grey.) The first joint on the left involves rotation of the radius, the radiale, the ulnare, and the hand skeleton around the ulna head. The other wrist joint allows rotation of the hand skeleton around a joint with the radius, the ulna, the radiale, and the ulnare as is illustrated on the right. The scale bars are 1 cm (based on Vazquez (1994)).

and pushes it against the ulna. The shape of the facets at the position where radius and ulna meet moves the radius distally towards the wrist.

During wing extension elbow and wrist movements are also coupled. When the elbow angle widens the radius will slide along the ulna because collateral ligaments attach it to the humerus. The distal end of the radius pulls via the radiale on the frontal edge of the carpometacarpus, extending it. The automatic action due to the drawing parallel system is enhanced via tendons by muscle activity. The extension of the hand, for example, is enlarged by a pulling action of the propatagial tendon. The wrist moves away from the shoulder by the widening of the elbow and one slip of the propatagial tendon pulls at the extensor process on the carpometacarpus and another slip at the radiale and at the end of the radius (Fig. 2.7(a)). A tendon slip of the biceps muscle also pulls at the extension process of the carpometacarpus. More muscles and tendon complexes play a role in these complicated movements. The relative contributions are difficult to assess by dissecting and manipulating dead wings. Understanding muscle function in flight requires EMG techniques combined with high-speed (preferably X-ray) cinematography. An overview of what is known about the role of wing muscles in flight will be given in Chapter 7.

The effect of the drawing parallel system on the movements of the hand becomes clear when we study the multiple joints of the wrist (Fig. 2.7(b)). Five bony elements are connected in the wrist: the radius and ulna, two carpal bones (the radiale and the ulnare) and the hand skeleton. The shape of the connections provides and restricts the freedom of movement between the hand and the forearm. Vazquez (1994) distinguishes two distinct joints. The movement of the radius, both carpal bones and the hand around the ulna head defines the first joint. In the second one the hand flexes and extends with respect to the other bones (Fig. 2.7(b)).

During the downstroke, the plane of the hand is parallel with the plane of the wing. The wing is stretched and the hand cannot flex dorsally or ventrally about the wrist. The only movement possible is flexion in the plane of the wing. During the downstroke, the primaries of the hand wing attain the greatest vertical velocity and inflict large rotational forces on the wrist. The position of the supporting skeleton in the hand wing is close to the leading edge. The primaries form a large surface behind the skeletal support causing a strong pronation tendency during the downstroke resulting in forward rotational forces on the wrist joint. These forces are counteracted by an interlocking mechanism formed by ridges on the carpometacarpus, the ulnare, and the ulna in the joint. Supination of the hand wing around the skeleton in the leading edge is prevented by the radiale in articulation with the radius and the carpometacarpus where a ridge stops the movement.

In most birds the wrist joint can change from a stiff construction into a flexible one during the early stages of the upstroke. This is caused by the ulnare gliding along the winding articular ridge of the ulna to its other extreme position. Due to this action the hand can rotate over 90° with respect to the plane of the arm wing. Some birds show a backward and upward hand flick also during vertical take-off and landing. A similar movement is also made when the bird folds its wings into the rest position. This aspect of the wrist function was investigated by Vazquez (1992) using the mallard as example of a configuration, which occurs in most orders of flying birds.

2.3.2 The hand wing

The hand skeleton consists of the carpometacarpus, the alular digit and two to three other (major and minor) digits at the tip (Fig. 2.1). The alular digit is supported by 1 or 2 phalanges. The terminal one bears a claw in some orders. The phalangeal joints are saddle joints allowing 2 degrees of freedom. The joint between the alular digit and the carpometacarpus is more complex. It allows the alula to be abducted and adducted from and to the leading edge of the wing as well as be moved up and down. The joint also allows supination in the up and pronation in the down position. The major digit at the distal end of the hand skeleton has 2 or 3 phalanges and a claw in some groups. It attaches with a rather complex joint to the carpometacarpus. The finger can be slightly curved and stretched in the plane of the hand. Movements perpendicular to that plane are limited. The minor digit is a somewhat triangular platelet. Its joint with the carpometacarpus is cylindrical only allowing abduction and adduction. A ligament connects this bone with the first phalange of the major digit. This connection limits the curvature of the major digit and hence the risk of overstretching during the downstroke (Sy 1936).

2.4 *Scaling wings*

The following summary of scaling of dimensions important for flight in birds is based on surveys of Greenewalt (1975); Rayner (1988); and Norberg (1990). The results of scaling exercises are never precise but provide some feeling for the order of magnitude of the relevant dimensions.

Wingspan ranges between about 8 cm in 2 g hummingbirds to values of more than 3 m for 10 kg albatrosses. The relationship with body mass is allometric; the span increases approximately with mass to the power of 0.4. The isometric exponent of one-third (based on mass being proportional to length cubed) would predict a wingspan of 1.7 m for a 10 kg albatross if compared with a hummingbird. The exponents within various functional or taxonomic groups of birds vary around the value of 0.4. That among hummingbirds is exceptionally high with a value slightly above 0.5. The scatter in general is large. The span of 1 kg birds, for example, varies between 0.5 and 1.7 m.

The special geometric position of the hummingbirds becomes even more apparent when we consider wing areas. An isometric relationship would relate wing areas with a two-third power of mass. Most birds have an exponent in the order of three-quarter but that of hummingbirds is about 1. The wing area of a 3 g hummingbird is about 10 cm^2 and that of an individual belonging to a species which grows to up to 12 g, reaches 40 cm^2. Again, the variation found among birds in general is large. The areas of 1 kg birds show an 8-fold variation.

Wing loading (body weight divided by wing area) is low and practically constant at 20 $N m^{-2}$ for hummingbirds but varies greatly among other groups. Isometrically, values would have to increase with body mass to the power of one-third. In reality the exponents are lower varying between 0 for hummingbirds, 0.22 for passerines, and 0.29 for ducks and other shore birds. Large auks have the highest wing loadings. Murres of about 1 kg reach, for example, values of more than 230 $N m^{-2}$. This is more than twice as much as values found for other birds, ducks, for example, of the same size. The high value must be a reflection of the capacity to use the wings for underwater flight. Auks do not stretch the wings fully during underwater flight but nevertheless the structures have to move a medium of a factor 1000 denser than air. Flight in air of these birds demands high wing beat frequencies to compensate for the relatively small wing dimensions. The wing loading of birds of prey is usually low around 30 $N m^{-2}$, indicating their capacity to carry large prey.

Aspect ratio (AR = span squared divided by the area of the wings) is large for birds with large spans and narrow wings such as albatrosses (AR: 14) and swift (AR: 10) and low for, for example, pheasants (AR: 5) with short broad wings. It is a shape factor which is more or less independent of body mass. What it tells us about flight performance depends on which parts of the wing contribute to the span and area of the wing. In general, birds with high aspect ratios are fast, low drag gliders. Their manoeuvring capacity may differ considerably depending on how much of the span is taken by the hand wings. It is less than 50% in albatross and about 75% in the swift. This partly explains why albatrosses are not good at low-speed manoeuvring, especially not under difficult wind conditions. Birds with a low AR are capable to glide slowly and take-off steeply or are good at complex manoeuvring at short range. A broad hand wing usually forms the larger part of the wing area in this category.

Preliminary measurements of the proportion of the wing length occupied by the sharp leading edge of the hand wing show large differences with rather good correlation with the predominant flight behaviour. The hand wings of notoriously soaring birds (i.e. buzzards and storks) occupy between 40% and 45% of the total length. That proportion is slightly less than half the total wing length in extreme gliders (albatrosses). Song bird hand wing lengths cluster around 70% whereas more agile faster flyers, including the swift and the peregrine falcon, reach usually 75%. We saw that the hummingbird wings are most extreme with a hand wing length of more than 80% of the total wing length.

2.5 *Attempts to a functional interpretation of bird wings*

Aerodynamic interpretation of bird wings is extremely difficult not only because of their complexity including the differentiation in an arm and hand part and the presence of the bastard wing, but also because of the highly dynamic shape which changes drastically during the stroke cycle.

A simple approach to study wing function is to remove various parts of the wing and observe the flight performance of the victims in some way or another. Pettigrew (1873) removed half the secondaries and one-fourth of the primaries of the house sparrow to reach the conclusion that the maiming did not impair flight. His paper does not indicate how the removal took place and which feathers were removed. Lilienthal (1889) did experiments with pigeons. He did not remove the feathers but tightened some groups together as shown in Fig. 2.8. The drawing shows the most extreme case where the bird could still repeatedly fly high and fast. Boel (1929) refers to experiments by C. Richet whose pigeons were capable of apparently normal flight with all secondary, tertiary, and 3–4 proximal primaries removed. Much more recently Brown and Cogley (1996) reached the same conclusion as Pettigrew using the same species. They removed all secondary and tertiary feathers with their coverts and most proximal primaries, leaving only the six distal-most primaries on the wing. This treatment did not have a noticeable affect on the distance flown in a windless corridor. Even after additional removal of 8 and 16 mm of the tip of the remaining primaries the distance flown hardly decreased during repeated tests. Cutting 24 mm of the remaining primaries obviously was a dramatic change because the birds would only fly less than 10% of the distance they flew under the other conditions. The same feather treatment was applied to birds of which the propatagium was severed by an incision perpendicular to the leading edge. About 50% of the surface of the arm wing remained after this mutilation. The performance of the birds was not affected in the cases where the 6 primaries had their full length or where 8 mm was removed from the top. The average distance flown dropped dramatically when 16 and 24 mm were

Fig. 2.8 Lilienthal's (1889) Fig. 23 showing his most extreme experiment where the pigeon was still capable to fly fast and high.

removed from the tip of the primaries. I must admit that I do not like this type of experiment, but since the birds have been sacrificed we better use the results to improve our understanding of avian flight. Strangely enough, Brown and Cogley used an extremely simple two-dimensional steady state aerodynamic computer model to reach the unjustified conclusion that the cambered propatagium is the major lift generating component of the wing proximal to the wrist. They ignored their own important finding that the 6 remaining distal-most primaries were sufficient to let the birds fly repeatedly over the distance they would normally fly even after additional removal of 8 mm of the feather tips and decreasing the surface of the propatagium by 50%. These experiments tell us that the distal-most primaries play a dominant role in generating lift and thrust during flapping flight in birds, not more and not less.

2.6 *Tail structure and function*

The tail is supported by a few caudal vertebrae and the pygostyle, a fusion of the last vertebrae of the vertebral column (Baumel 1979). There is considerable variety in tail shapes and sizes. Hypotheses on the relationships between form and function of flight-related aspects are discussed in the last paragraph.

The anatomy of the tail of birds is complex and derived. Tail feathers, termed retrices, are implanted on a broad stubby tail supported by highly modified vertebrae. The tail of the pigeon is studied in great detail by Baumel (1988) and by Gatesy and Dial (1993) The functional morphology of tails of flying birds is rather uniform; the pigeon tail in Fig. 2.9 can therefore be used as a model for most species. The moveable part of the tail skeleton is composed of five, six, or even seven caudal vertebrae ending in a pygostyle. More cranially some caudal vertebrae are fused with the synsacrum. The concave anterior and convex posterior globular surfaces forming the articulations between the free vertebrae allow movements in all directions. The pygostyle consists of a vertebra type body extending caudally into a vertical plate. Its connection with the last free vertebra is a horizontal hinge joint with a transverse hemi cylindrical notch in the anterior part of the body of the pygostyle. On each side of the pygostyle the recticial bulbs form the seat of the 12 rectrices. The bulbs are fibro adipose structures, partly encapsulated by a striated muscle, the bulbi rectricium. Sockets on each side of the caudal vertebral column form joints in which the bulbs can move. Six pairs of muscles connect the vertebrae, the pygostyle, and the bulbs to the pelvis, synsacrum, femur, and vent. The bulbi recticium are responsible for spreading the tail fan by pulling the calami of the rectrices together. The other caudal muscles function to hold and move the adjustable tail fan. EMG recordings of these muscles during take-off, level flight, and landing are discussed in Chapter 7.

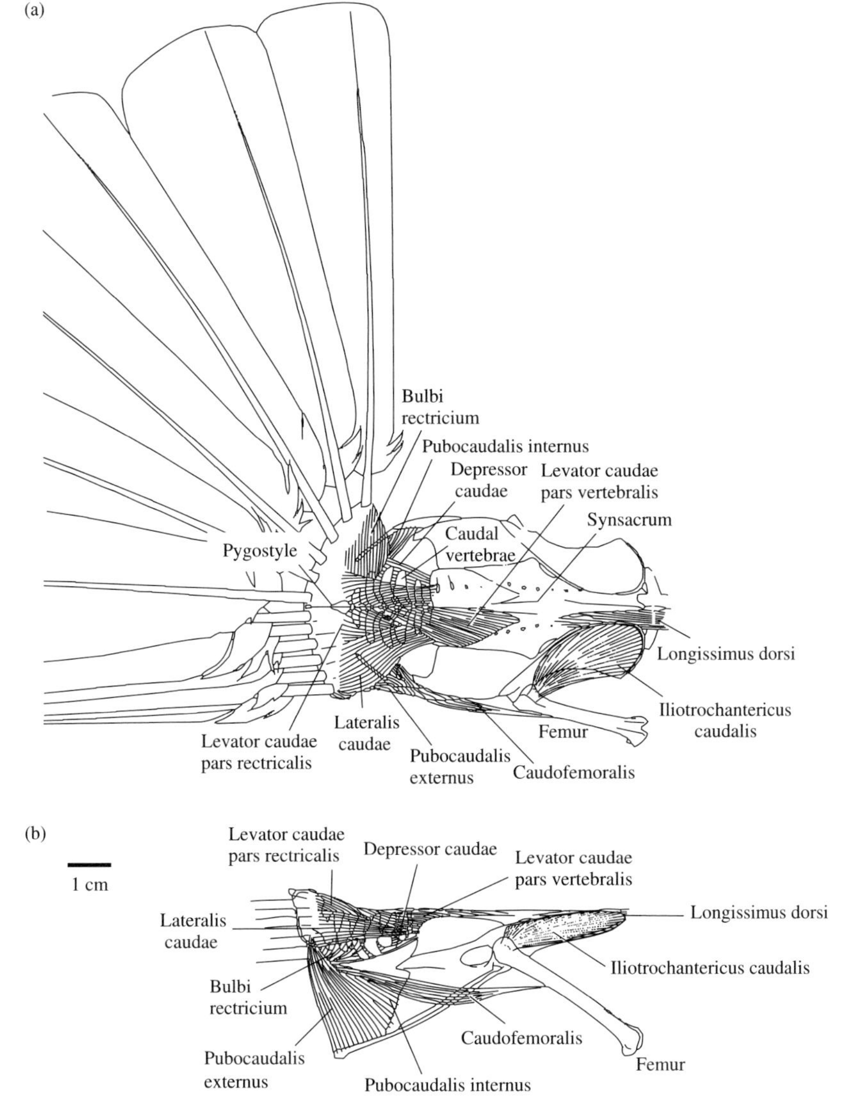

Fig. 2.9 Morphology of the pigeon tail as a model for bird tails in general: (a) dorsal view; (b) lateral view (Baumel 1988; Gatesy and Dial 1993, with kind permission of The Company of Biologists).

Up to 24 rectrices in the tails of birds differ in length and shape to form an almost infinite number of tail designs. Not all the diversity is flight-related. Ornamental tails play an important role in reproductive behaviour.

The shape and size of the tail vary considerably among birds. The variety might be even greater than that of wings. However, not all of the variation is related to flight. Many extreme tails belong to males only and are obviously intended to impress females in some way or another (i.e. to show that he is even strong enough to fly with a handicap). If we concentrate on the tails of females and on those of species without sexually dimorphic tails, the variety is still large but reducible to a few general patterns. However, little of it has been functionally explained. Tail shapes not only vary among species, but may also change quickly in time due to spreading and closing of the tail fan. The left and right halves are usually symmetrical but differences in the amount of spreading and tilting can create high degrees of asymmetry. The outer tail feathers can have a narrower outer and a wider inner vane but most other feathers are left and right symmetrical. Table 2.1 shows that most major groups of birds have 12 rectrices, some have twice as many, whereas 8 seem to be a minimum. Ducks, geese, swans, and pelicans may have up to 24, whereas grebes have no functional rectrices. In most flying birds, the vanes of the outermost rectrices are usually asymmetrical with the outer ones much narrower than the inner vanes. In many birds the feathers are equally long. The folded tail in that case has a narrow rectangular shape, when spread it forms the segment of a circle. In forked tails the rectrices become increasingly shorter towards the centre. Deeply forked tails have an inverse circular shape when spread. Shallow fork tails may show a straight trailing edge in the spread out position. Birds like the common magpie, mousebirds, pheasants, gannets, some doves, and cuckoos have a wedge-shaped tail with slightly longer central feathers and shorter outer ones. Such a tail is slender spade shaped when spread out. Extremely elongated central feathers occur in both sexes in very distant groups as tropic birds, skuas, bee-eaters, and some species of sandgrouse and in a few single species, for example, the South American long-tailed tyrant and the secretary bird.

Variation within one functional group can be large as well. Among aerial feeders as swallows, martins, and swifts the tail shapes range from short square or pointed stubby ones to extremely long-forked tails such as those of the barn swallow. Some swifts have hair like shafts sticking out (i.e. the brown-backed needletail). Stiff shaft elongations are found in various tree

Table 2.1 Numbers of rectrices in the tails of some major groups of flying birds (based on Lucas and Stettenheim 1972; Van Tyne and Berger 1976).

Vestigial	Grebes
8–10	Cuckoos
8–12	Herons
8–14	Rails, gallinules
10	Swifts, hummingbirds, nighthawks (but the common swift has 12: polyrectricity)
12	Woodpeckers, trogons, kingfishers, parrots, macaws, typical owls, pigeons, doves, cranes, sandpipers, plovers, gulls, terns, alcids, songbirds
12–14	Hawks, eagles, osprey, falcons, caracaras, cormorants, new world vultures
12–18	Quails, pheasants
12–24	Ducks, geese, swans
16–20	Loons
22–24	Pelicans

creeping birds. Tail feathers of the southern emu wren seem to consist of a shaft with loose barbs.

Flight-related functional explanations of tail configurations are usually fairly general, and rarely if ever backed by experimental proof. They require insight into the aerodynamics of tails. Chapter 4 shows how far that insight goes.

2.7 *The rest of the body in relation to flight*

We have not yet paid attention to the role of the body, the head, and the hind legs in flight. In flying birds we expect the body and head to form a well-streamlined body of rotation with a rounded leading surface and a pointed trailing end. The largest diameter should be situated at approximately one-third of the length and the ratio of diameter over length ought to be between one-quarter and one-fifth. Such a body, the fuselage in aircraft terms, is optimal in the sense that it offers the smallest drag for the largest volume. Caylay first described it in 1809 (see Gibbs-Smith 1962). The shape of the head and body of a starling, for example, is close to this ideal if we exclude the sharp beak at the point.

Birds with long necks either stretch these during flight, as storks and swans do, for example, or keep the necks folded as pelicans and herons do. The neck length determines the position of the heavy head with respect to the centre of gravity. Birds can be expected to use the stretch ability to adjust the position of the centre of gravity. Windhovering birds may use the possibility to stretch or contract the neck to keep the head in a fixed position with respect to the ground (see Chapter 6). In Chapter 1 we saw how Borrelli worried about the effect of sideways movements of a long neck and head during flight (Fig. 1.2, picture 6). Beaks form the leading structures in the flight direction of the flying birds. The existing variety of bizarre bill shapes among flying birds leaves the suggestion that there is no heavy aerodynamic penalty connected to the possession of extravagant frontal parts. Aerial feeders such as the swift have a small beak but open it wide to catch insects. Other birds carry substantial objects, large prey items or nesting material in their bills. Serious investigations into the effects of structures up front are however lacking and we have no idea how large a handicap these represent and how birds adapt their flight to cope.

Take-off and landing require an undercarriage. Birds use legs and feet to push off or even run some distance prior to take-off and to absorb excess forces during landing. Legs can also be used to dynamically control the position of the centre of gravity. Some birds tuck them away under the feathers; others stretch them rearward underneath the tail. Birds of prey carry their victims underneath close to the body or in the claws at the end of stretched legs. The osprey is well known for carrying big fish in the head first streamlined position. I saw how a tawny eagle did that with giant mole rats in the Bale Mountains of Ethiopia. Legs and feet are of course important devices during take-off and landing in many species. They are used as airbrakes by many birds, especially those with webbed feet, sea gulls and cormorants, for example, can be seen using this trick. Note that if the drawing on the cover is a landing cormorant, the artist forgot to indicate the webs between the toes. It is an understandable omission because while watching a cormorant's landing action the extreme spreading of the toes is more impressive than the presence of the webs in between. Wilson's storm petrel sticks its webbed feet in the water to use them as a sea anchor during soaring close to the water surface. Gliding against the wind, the bird is blown backward with respect to the water. This produces a hydrodynamic drag force on the feet, which can balance the drag force on the body and makes the backward speed of the bird smaller than the wind speed creating a horizontal wind component relative to the bird, which generates lift on the wings. The bird operates as a kite, where the tension in the string counterbalances the aerodynamic drag on the kite (Withers 1979).

2.8 *Summary and conclusions*

The highly derived internal anatomy of avian wings shows a common pattern which is interesting from a comparative anatomical point of view. However, specific features deserve more attention in future research on bird flight.

The outside of the wings of all flying birds consist of two distinct parts: the arm wing and the hand wing. Cross sections through the arm wings have classical aerodynamic profiles with a round leading edge, a cambered shape, and a sharp trailing edge. The hand wing consists mainly of the primary feathers. The leading edge of that part is sharp because it is formed by the narrow vane of the outermost primary. Cross sections through the hand wing are usually flat or slightly curved, the leading edge and the trailing edge are both sharp. Feather emargination and spreading of the primaries create slots near the wing tip of many species of larger birds.

The hand wing in most birds takes up more than half the total wing length. Extremely specialized flyers (i.e. hummingbirds and swifts) have the longest hand wings and almost exclusively use these to fly. Albatrosses and giant petrels on the other side of the scale have long arm wings and are able to lock the wings in extreme stretched position during gliding.

Both the possible movements of the wing as a whole with respect to the body and the freedom of movement inside the wing is restricted but not to the same extend in all species. The shoulder joint allows the largest freedom of movement in most groups. The folding and stretching is reasonably well understood although this knowledge is based on a few species only. The wrist joint is complicated by the possibility to change its dynamic characteristics depending on the configuration of the bony elements.

Scaling of the dimensions of the flight apparatus of birds provides insight into differences among functional groups. Hummingbirds and auks do not obey the rules that seem to emerge for other groups. We have to be aware that in some cases similar dimensions are based on different morphologies and may require different functional explanations.

Experimental testing of the influence on flight performance by removing various parts of the wings provide a remarkably consistent conclusion: the distal-most five–six primaries are crucial for the ability to fly. Other parts of the wing hardly affect that ability.

Bird tails are unique, extremely derived structures among vertebrates mainly consisting of a few caudal vertebrae, a pygostyle, rectricial bulbs encapsulated in muscles and up to 24 rectrices. The tails are left right symmetrical and can be spread and folded and tilted sideways. The shape varies greatly and depends on the amount of spreading and on the distribution of the feathers of various lengths. There are functional explanations for some tail shapes but evidence that these reflect the truth is usually circumstantial.

The shape of bird bodies, including the head, is usually fairly accurately streamlined offering the largest volume for the lowest drag. Some birds with bizarre beaks probably evolved under selection pressures where aerodynamic design did not play an important role. Legs and feet are important during take-off and landing, they can operate as air brakes, carry load, regulate the position of the centre of gravity and serve (in Wilson's storm petrel) even as sea anchors.

Our insight in the functional morphology requires a closer look at the unique structures which made flight possible in this group of animals in the first place: the feathers.

3 *Feathers for flight*

3.1 *Introduction*

Feathers are the hallmark of birds. Birds are unique among the flying groups of animals because the capacity to fly is exclusively based on the highly complex modified scales. Feathers come in various shapes and sizes and serve a multitude of functions. We are particularly interested in the structure and function of flight-related feathers. The anatomy is amazingly intricate and functional explanations cannot yet be given for many details. The morphology of flight-related feathers is therefore interesting especially in combination with what we know about their mechanical and aerodynamic properties. We saw in Chapter 2 how feathers form the lift and thrust generating surfaces. Detailed understanding of the structure and mechanical properties of the feathers as the main constructional unit is required to start to appreciate how these interact with the air. Feathers are dead structures used in complex functions. The connections between the feathers and the living part of the birds are made of skin, tendons, muscles, and nerves. To understand flight we need to know how these work together as a functional unit in interaction with the rest of the flight apparatus and the central nervous system.

This chapter begins with a description of the complex macroscopic and microscopic structures of contour feathers. Feathers must be hard and strong as well as light and elastic. These seem incompatible design constraints. Results of measurements of the mechanical properties of the shaft and vanes are presented and discussed. Microscopic structures are used for classification purposes; we want to know the functions of the sometimes bizarre shapes. The phenomenon of primary feather emargination is also included in this chapter.

Some feathers in tails deviate from the normal pattern and we need to know if and how these deviations are related to flight.

Finally, the feather implantation and the connection with muscles, tendons, and the nervous system are discussed. The role of feather follicles and associated muscles and nerves in the detection of movements is of great interest. We expect feedback systems involving tactile sensors and motor nerves to be essential for flight control.

3.2 *General description of contour feathers*

An outgrown feather is a dead, extremely complex epidermal structure mainly consisting of the protein keratin. The usual form has a shaft along the entire length and two vanes, one on either side, along the distal part of the shaft. The shaft or quill of a contour feather is named calamus close to the bird and rachis over the distal part where the vanes are attached to each side. The following general description of contour feathers is mainly based on Lucas and Stettenheim's (1972) standard work. See Appendix 3 for the definitions of the anatomical terms. Figure 3.1 illustrates the description but it is also very illuminating to have a few large feathers handy while reading it.

3.2.1 The shaft

The calamus, emerging from a follicle in the skin, is a tubular hollow structure, covered by a dry sheath originating from the one that covered the developing feather. There is a small pit in the lower pointed end, the inferior umbilicus, which is closed from the inside in the outgrown feather. It is the place where the artery entered during the stages of development. The superior umbilicus is another remnant of the blood supply system in the early stages of a feather. It is situated at the point where the calamus changes into the rachis. The calamus is usually transparent and dried remains of the early vascular system around the axial artery can often be recognized inside. The next part of the shaft, the rachis, is solid, surrounded by a stiff wall, the cortex, and filled with spongy tissue (pith). The large cells of the pith are filled with air, which makes the rachis appear white and non-transparent. Cross sections reveal the shape of the rachis. On the dorsal side, it is smooth and convex; in many cases it appears to have longitudinal stripes. These are parallel cortical ridges projecting into the pith on the inside. The sides of the rachis are flat or slightly convex, usually thinner than the dorsal and ventral side (Hertel 1966; Oehme 1963). The proximal part of the rachis has a depression in the middle of the ventral side, forming a groove.

The calamus of flight feathers is relatively long. In swan primaries, it may occupy up to 30% of the length. It can be slightly thicker than wide giving it an oval cross section. The rachis of large birds can have a lumen over several centimetres from the base to the tip. Substantial internal cortical ridges provide stiffness.

3.2.2 The vanes

The vanes, consisting of rows of parallel barbs, start just above the superior umbilical where they are commonly fluffy in appearance. The most proximal barbs are not well attached to each other and are indicated as plumulaceous or downy. More distally, the vanes are well structured and composed of interlocking pennaceous barbs. Feathers with mainly pennaceous vanes are usually not straight but curved downwards and sideways. In a lateral direction, away from the rachis, vanes may curve either downward, upward or first downward and then upward. In many primaries of large birds the outer or leading edge vane curves downward whereas the inner or trailing edge vane bends downward near the rachis and sharply upward near the edge. The narrow leading edge vane of asymmetrical flight feathers is always stiffer than the wider vane forming the trailing edge. This difference is related to differences in barb spacing. Barbs occur in nearly equal numbers on both sides of the rachis and the spacing varies typically between 0.1 and 1 mm.

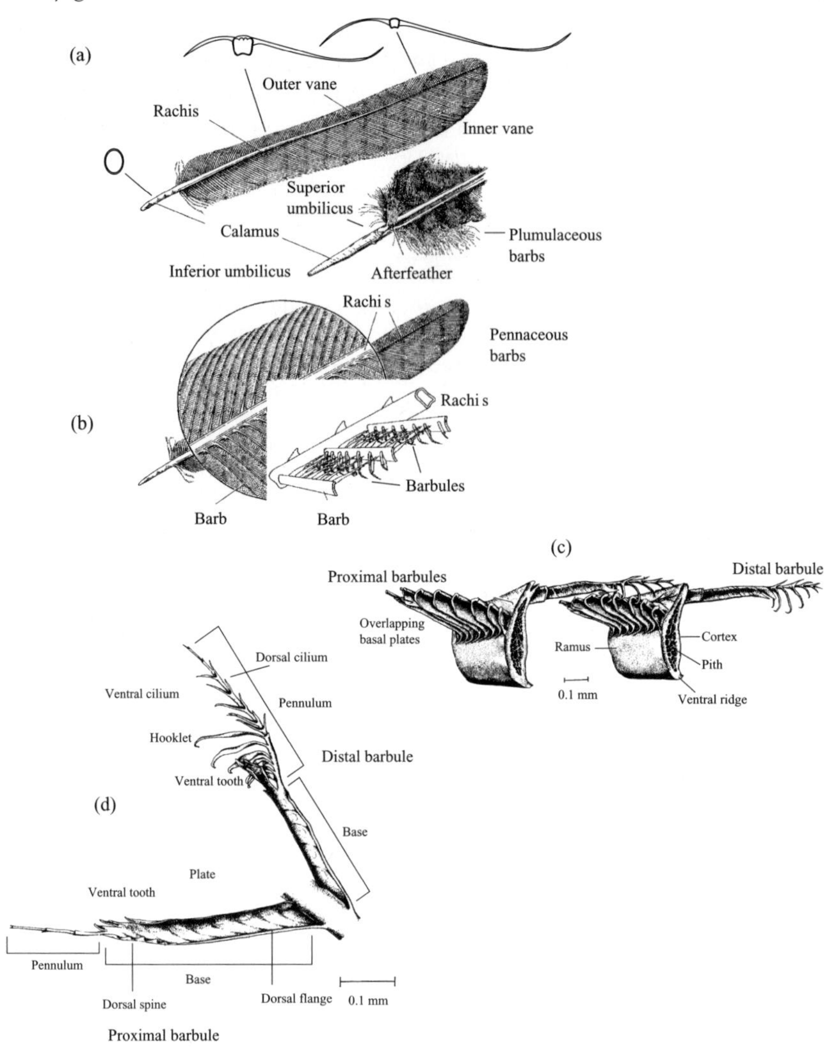

Fig. 3.1 The structure of a flight feather. Details are explained in the text (based on Ennos *et al.* 1995, with kind permission of The Company of Biologists; Lucas and Stettenheim 1972; Van Tyne and Berger 1976).
a. Overview of the topography of a primary bird feather with cross sections at three indicated positions and an enlargement of the proximal part.
b. Magnified picture of the pennaceous barbs and connecting barbules.
c. Sections of two successive barbs illustrating the connection between proximal and distal barbules.
d. Detailed representation of the structures forming the proximal and distal barbules.

They tend to be closer together towards the tip of the feather. The length of the barbs varies greatly, within one vane, between inner and outer vanes and among feathers. The barbs are connected to the rachis under acute angles pointing in the direction of the tip. Barbs are usually cambered on cross section, especially those of the outer vanes of the primary feathers. The insertion of the base of the barbs to the rachis is inclined towards the tip of the feather, that is, the dorsal rim is closer to the tip than the ventral. This oblique insertion vanishes distally to become dorso-ventrally straight near the tip. The branching angles of the outer vane barbs of pigeon wing primaries decrease from 40° near the base to 20° near the tip. For the inner vanes, these figures are larger at 47° near the base and 35° near the tip. The average branching angle of the outer vanes of the primary feathers decreases towards the distal-most feathers near the leading edge of the wing. A general rule seems to be that the larger the vane asymmetry the greater the difference between the branching angles of the outer and inner vane barbs (Ennos *et al.* 1995).

On cross sections (Fig. 3.1(a)), vanes of primaries are usually thinner than the shaft. The upper surface is level with the dorsal surface of the shaft. On the lower face the shaft sticks out and forms a rim. The primaries of albatrosses and giant petrels differ from those of other birds because their vanes are thick and form a smooth surface on the top and the lower face of the feather (Boel 1929). Stacks of these primaries in the hand wing can form a reasonable conventional aerodynamic cross-sectional shape. Due to this phenomenon albatrosses and giant petrels lack sharp leading edged hand wings.

3.2.3 Microstructures

The fine structures of the feathers are not easily visible to the naked eye. Mascha (1904) describes the delicate structures of wing feathers is great detail. Sick (1937) provides even more detail in an article of 166 pages on the form and assumed function of the microscopic structures of feathers.

Each barb consists of a ramus fitted with barbules on both the proximal and distal upper sides (Fig. 3.1(b) and (c)). The ramus tapers from base to tip. The shape of the cross section varies enormously not only along each ramus but also among contour feathers on one bird and among birds. Cross sections reveal that the ramus has a cortex surrounding a medulla filled with pith cells. The medulla is not continuous with that of the rachis. In owls and goatsuckers, it consists of a single vertical layer of pith cells. The cortex has a dorsal, a ventral ridge, and a ledge on each side dorsally from where the barbules emerge.

The barbules of the plumulaceous barbs are simple stalks of single cells forming a row of nodes and internodes (Fig. 3.1(d)). Pennaceous barbules are far more complex; each one consists of a base and a distal pennulum. The ones on the distal side of the barbs differ markedly from the proximal barbules. The distal barbules usually have a simple plate like base serrated near the end into teeth and spines. They possess a complex pennulum fitted with a variety of outgrowths called barbicels. Straight barbicels have unfortunately been termed cilia (they are dead structures and unlike real cilia cannot move); longer barbicels with hooks at the end are the hooklets. The base of proximal barbules is a curved plate with a dorsal flange ending in dorsal spines and ventral teeth. The pennulum of proximal barbules is commonly simple. Hooklets of the distal barbules interlock with the dorsal flanges of the proximal barbules.

Rami of barbs of primaries are often fitted with expanded ventral ridges (tegmen) with fringes on the lower side (Fig. 3.1(c)). The dimensions of the ridges of ventral rami vary considerably. In most birds ridges are small or absent. Moderately wide ventral ridges are found in pelicans, herons, storks, vultures, hawks, sandpipers, plovers, and sand grouse where they overlap slightly and provide a velvety appearance. There are very wide tegmen in certain remiges and rectrices of loons, albatrosses, ducks, geese, swans, eagles, some owls, and some galliform birds. Among the members of this last group, there is a strange dichotomy. Ridges are wide in Western capercaillie, Eurasian black grouse, willow grouse, some pheasants, grey partridge, and turkeys, but absent in quails, guinea fowl, red-legged partridge, common pheasant, and chicken. Large obliquely expanded tegmen cover the space between the barbs probably functioning as flap valves during the downstroke by preventing air from passing upwards, forming air filled chambers. In some birds, the structure of the tegmen causes a glazed sheen on the underside of the feather. It is difficult to understand why some groups have ventral ridges expanded to form tegmen and others not.

Pennaceous barbules of flight feathers are fairly constant in size and spacing. For example, the distal barbule from the middle of the inner vane of the primary of a Eurasian griffon vulture is only 3.5 times as long as the length of a similar barbule of the sword billed hummingbird. The two species vary by a factor of 1000 in body mass and 10 in body length. Distances between barbules vary between 20 and 30 μm on the distal and between 30 and 40 μm on the proximal side. Angles between barbules and barbs range from 29° to 58° on the distal and from 10° to 41° on the proximal side. However, there are marked differences between inner and outer vane barbules. The base of the inner vane is shorter and relatively wider, the pennulum is longer, and there are fewer hooklets and more cilia. Lobate proximal dorsal cilia are only found on the inner vanes. There are no functional explanations available for most of the complex microstructures of feathers let alone for the variability among these structures.

3.3 *Mechanical properties of feathers*

Feathers must be light, strong, and wind tight. Light because the plumage is part of the mass of the bird that has to be lifted against gravity during flight. Wing feathers must be light because they are accelerated cyclically and rotate at some distance around the shoulder joint. The amplitude of the flapping wing, and hence the magnitude of inertial forces, increases with the distance along the wing from the shoulder joint and with the mass at that distance (see Box 1.1). The demands on the primary feathers to be light are therefore extremely high. Simultaneously, they can be expected to experience the highest cyclic loadings. Strength is not a simple property in relation to a structure as complex as a feather. The shaft should be stiff to a certain extent, but also elastic to avoid buckling and fracture. The barbs supporting the vanes are loaded differently from the shaft. Barbules and especially the pennula with hooks and cilia are vulnerable structures with probably the least resistance to abrasive wear. Daily preening sessions are required to arrange the microstructures properly but will also cause a certain amount of wear. The timing of moult is most likely related to wear of the smallest feather structures. These microstructures also play an important role in making the feather wind tight. This is most important in the primaries since they form a single layer with some overlap of the vanes of neighbouring feathers when the wing is stretched. The transmissivity of the distal part of the hand wing depends entirely on how air tight the vanes of the primaries are.

The complexity of the anatomy makes it difficult to fully assess the mechanical properties of feathers. A survey of the relevant literature will provide a feeling of the state of the art of our knowledge on the subject.

3.3.1 Hard and elastic

Feathers are made of keratin with a molecular mass (the sum of the mass of the atoms a substance is made of) of about 10 kDa (1 kDa is 1000 daltons; the dalton is a unit of mass nearly equal to that of a hydrogen atom which is about 1.66×10^{-27} kg). Keratin is the name of a large family of complex proteins commonly found in integument structures of vertebrates. Feather keratin is lighter than that of bills and claws of birds which have a molecular mass in the order of 15 kDa. Molecular mass differences reflect differences in hardness, which is somehow proportional to wear resistance. The resistance to indentation as a measure of hardness of European starling primary feather shafts is about half that of the beak (Bonser 1996). Melanin, a black polymeric pigment, is regarded to improve the hardness of feathers. It has a molecular mass of about 180 Da which is ten times that of water. Melanin granules are embedded in keratin in black feathers. This would imply that black feathers are heavier and stronger than white feathers. There is a lot of circumstantial evidence to support this hypothesis. Extremely aerial birds such as the common swift and the frigate birds have black wings. There are several examples showing increased wear of albino feathers in comparison with normally pigmented counterparts. Tests using primary feather shafts of the willow grouse have shown that the hardness of the dorsal part of the shaft containing melanin can be up to 40% higher than that of the white ventral part (Bonser 1995).

How difficult it is to break a feather is determined by its toughness, defined as the amount of energy per unit cross-sectional area required to crack it. Toenail clippers were used by Bonser *et al.* (2004) to measure feather toughness. The force applied to bring the sharp blades of the toenail clippers towards each other appeared to increase linearly until the blades made contact. Small squares of keratin taken from the dorsal part of the proximal end of ostrich feathers were tested. Both the force and the displacement were recorded while the clipper was cutting through the specimen. The amount of work done was calculated from the area under the force curve during the displacement of the clipper through the 0.5 mm thick feather, corrected for the linear force increase to move the clipper blades without a specimen (force in *N* times displacement in metres gives the work done in Joules). To obtain the toughness, the amount of work required to cut a specimen was divided by its cross-sectional area. The average toughness while cutting a specimen in longitudinal direction of the shaft varied between 3 and 8 $kJ\,m^{-2}$, and between 11 and 18 $kJ\,m^{-2}$ while cutting in transverse direction. Ostrich feathers, and most probably those of all other birds, are obviously built to avoid breakage across the shaft. No other measurements of feather toughness have been done so far.

Elasticity is another important property of structures used under extreme dynamic loads. Bonser and Purslow (1995) did tensile tests on small strips of keratin cut of the rachis of primary feathers to determine the stiffness along the primaries of a number of species. As expected, stiffness decreased towards the tip of the feathers mainly depending on the surface area of the local cross section and on the Young's modulus of the material. (The Young's modulus is the theoretical force per unit cross-sectional area required to stretch a sample to twice its original length. Most materials will not allow that to happen and will break or permanently deform when stretched too far, long before reaching the double length. In those cases, the modulus is calculated from the measured gradient of the stress (load per unit area)—strain (elongation per unit length) curves where the deformation is purely elastic and the curves linear.) The mean modulus found was 2.5 GPa (1 GPa (giga pascal) = $10^9\ N\,m^{-2}$), varying from a value of 1.8 in the grey heron to 2.8 in the primaries of the tawny owl. No relation with body mass could be detected. The change in the Young's modulus from base to tip has been tested in primaries of a swan, a goose, and an ostrich (Cameron *et al.* 2003). The modulus of the swan rachis increased from 2 GPa near the base to about 4 GPa at the tip; in the case of the goose these figures were somewhat higher varying from about 3 to 5 GPa. The average value of the modulus of the ostrich primary was about 1.5 GPa and there was no increase towards the tip of the rachis. In the flying birds the increasing Young's modulus from the base to the tip indicates that the absolute stiffness does not decrease from base to tip as dramatically as the decrease in cross-sectional areas would imply. In flight, feathers are bent and store elastic energy due to the stiffness of the rachis. The question remains as to how it is possible that the Young's modulus of the tested feather keratin increases towards the tip. It could be caused by changes in the structure of the rachis. Bonser and Purslow (1995) discuss the possible role of outer and inner sheets of rachis keratin in which the fibres are orientated differently. In the outer layer the keratin fibres are orientated circumferentially, whereas they are running parallel to the longitudinal direction of the shaft in the inner layer. The proportion of longitudinally aligned keratin increases towards the tip because the outer layer becomes thinner in that direction. X-ray diffraction patterns reveal that feather keratin fibres consist mainly of helically arranged keratin molecules. Cameron *et al.* (2003) show that the keratin fibrils become more longitudinally aligned towards the tip of the primary feathers in swans and geese. This change in molecular alignment correlates with the change in Young's modulus and probably causes it.

In the pigeon, the outermost feather of the wing is equally stiff in dorso-ventral as in lateral directions. Its flexural stiffness in the lateral direction is higher than that of the other primaries that are less stiff laterally than dorso-ventrally. The resistance to dorso-ventral bending of the shaft of the primary feathers from the pigeon increases with the weight of the bird (Purslow and Vincent 1978). Interestingly, measurements by Worcester (1996) show that among species, larger birds have more flexible primaries than smaller birds.

The mechanical properties of vanes have been the subject of only a few studies. The resistance to forces in a direction perpendicular to the surfaces of the vanes was studied by Ennos *et al.* (1995). In general, the vanes of outer primaries are stronger than those of the inner primaries and secondaries. In most wing feathers vane resistance to forces from below is about 1.5 times as large as resistance to forces from above. This is not the case for the vanes of the outermost primaries where the resistance is equal. Although we have to keep in mind that this has only been substantiated in pigeon wings.

Butler and Johnson (2004) tested 302 barbs of the wide inner vane of a primary of the osprey mechanically by pulling at a few centimetre long piece of each barb until breakage. The osprey primaries have black and white bands running perpendicular to the rachis. On average the barbs did break when extended to 6% of the unloaded length. The breaking force increased from about 0.5 N proximally to a plateau of about 1.7 N halfway towards the distal end of the feather. The distal-most barbs tested (at about 0.95 the length of the feather) were breaking at about 0.8 N. The cross-sectional areas of the barbs more or less followed the breaking force changes along the feather by increasing distally up to maximum values at about 0.7 of the length of the feather and decreasing from there towards the distal end. The breaking stress, defined as the breaking force per unit cross-sectional area of cortex material was more or less the same along the feather at a value of 0.28 GPa. Surprisingly, no differences were found between the breaking stresses of black and white barbs.

3.3.2 Strain measurements *in situ*

Results of the measurements treated so far were obtained using feathers removed from the birds. Corning and Biewener (1998) measured strains on the shafts of pigeon's primary feathers in flight. Small strain gages were glued to the dorsal surface of 5 primaries and one secondary feather at about 2 cm distal of the calamus.

A strain gauge is a small metal strip. The electrical resistance of the strip varies in proportion to the amount of fractional change in length (the strain). The length change and the change in resistance can be very small. Measurements are made by using the strain gauge as a resistor in a Wheatstone bridge. The voltage measured across the bridge varies with the resistance in the strain gauge. The measurement is calibrated by applying known forces.

Upward bending of the feathers compressed the strain gauge and downward bending stretched it. Only relative measurements of dorsal and ventral bending strain of the feathers could be recorded because the effect of the strain gauge itself and of the glue could not be accounted for. Figure 3.2 shows typical results of the strain on the 9th primary pigeon feather during a short flight over 9 m along a straight path. The flight speed is slow at about $5\ \mathrm{m\,s^{-1}}$.

The pigeon used 18 wing beat cycles. Figure 3.2(a) shows that the strain from upward directed lift force is slightly more than two times as high as the downward strain. One indicated wing beat sequence is enlarged in Fig. 3.2(b). It starts when the wings are flicked backwards and upwards. The feathers are bent down and the strain on the shaft is negative. In the next phase, the wings are extended in preparation of the downstroke. The positive strain, due to upward bending of the feathers, increases throughout the downstroke and into the following forward swing, reaching a maximum value just before the start of the upstroke. The slow speed, the large angle between the body and the horizontal and the extreme kinematics of the wing beat cycle indicate that the strain recordings are probably also rather extreme. Comparative measurements between pigeon primaries showed that in general distal feathers have higher peak strains. Primary 8 experiences the highest values. These are 2.5 times higher than the strain on a secondary feather. Surprisingly, the peak values of primary 9 were lower than those of its more proximal neighbour.

3.3.3 Transmissivity

Pressure differences between the air above and below a wing are essential for the generation of lift. Wings need to be airtight to maintain such pressure differences. Flight feathers (i.e. primaries, secondaries, and tail feathers) are considered impervious to air because of the enlarged curved basal plates on the proximal and distal barbules. Contour feathers do not have these and are permeable. Transmissivity of air for flight feathers and coverts of the kestrel have been compared by Müller and Patone (1998). Transmissivity is expressed as the volume of air passing per unit time for each unit pressure

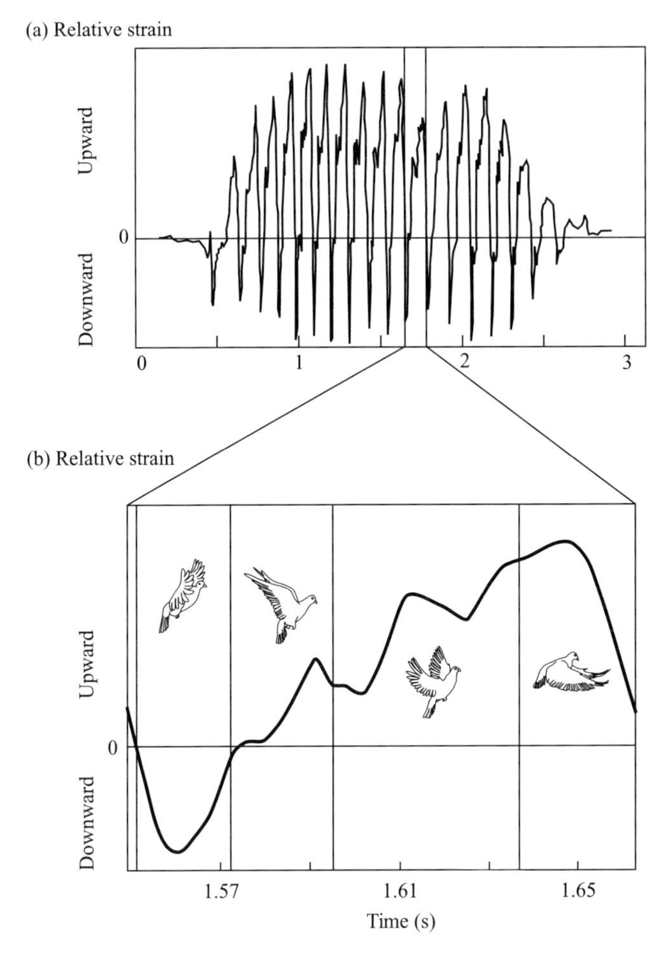

Fig. 3.2 Relative dorsal and ventral bending strain on the 9th primary feather of the pigeon recorded during a short flight. The graph is enlarged in the lower panel to show the strain in relation with a wing beat cycle (from Corning and Biewener 1998, with kind permission of The Company of Biologists; see text for explanation).

difference. Pressure differences across the vanes used during the tests varied from 390 to 1800 Pa, where the first figure can be considered realistic, and the second one excessive. The flow of air through the vanes increased linearly with increasing pressure difference. On average, the transmissivity from dorsal to ventral was 10% higher than in the opposite direction. The most striking result was that the narrow outer vanes of the primaries, secondaries, and coverts were on average 10 times more transmissive than the inner vanes. The difference is largest in the secondaries where 2.3×10^{-3} $m^3\,s^{-1}\,N^{-1}$ leaked through the outer vane against only 0.12×10^{-3} $m^3\,s^{-1}\,N^{-1}$ through the wide inner vane. Müller and Patone believed that this phenomenon helps to press individual feathers towards one another during the downstroke. The less transmissive inner vane is pushed against the outer vane of the overlying feather when the pressure comes from below.

3.4 *Functional interpretation of flight-related microstructures*

3.4.1 Zones of overlap between primaries

Modified distal barbules occur at zones of overlap between primaries. Figure 3.3 illustrates the overlap configuration in the pigeon. Within each zone of overlap, sub-zones have been distinguished. Outside the overlap area the distal barbules are not modified, they have a short pennulum bearing a few ventral hooklets and a single dorsal cilium. A large zone on the upper part of the inner vane bears distal barbules with elongated pennula, lobate dorsal cilia, ventral cilia, and an increased number of hooklets. The first one

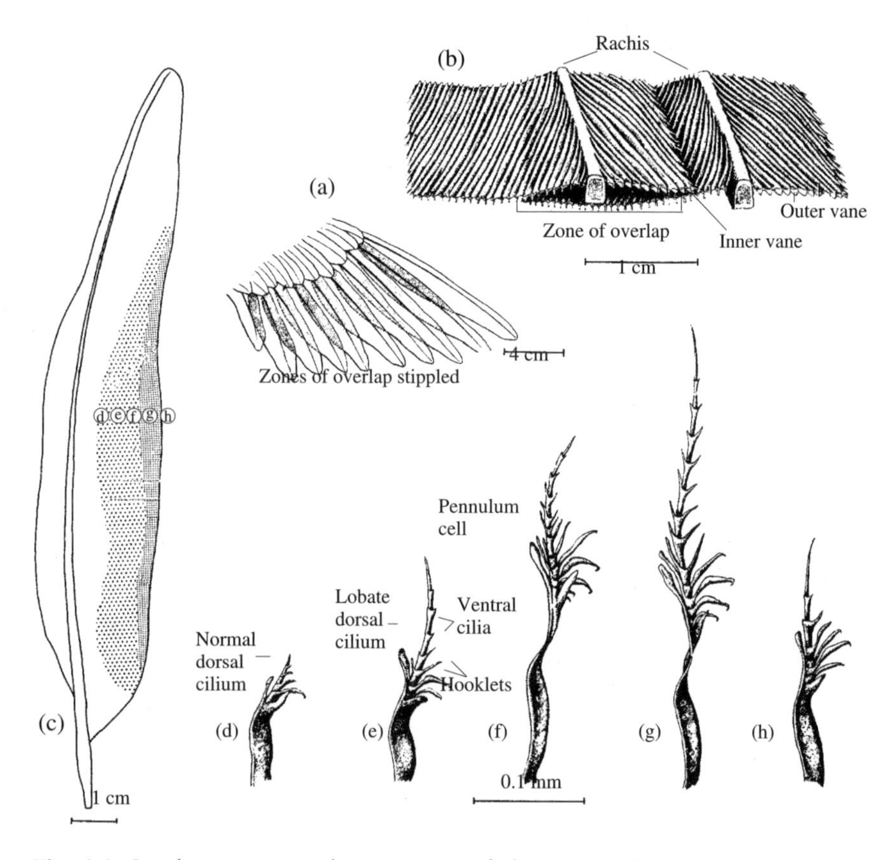

Fig. 3.3 Overlap zones on the primaries of the pigeon (based on drawings by R. B. Ewing in Lucas and Stettenheim 1972, figure 174): (a) Dorsal view of the hand wing. The zones of overlap of the primaries are stippled. (b) Sections of two overlapping primaries seen from the distal end. (c) Primary 8, the zone with modified barbules is indicated. (d) Unmodified barbule. (e)–(h) Modified lubrication barbules.

or two dorsal cilia are usually lobate in shape and enlarged. Towards the edge of the vane the distal barbules are shorter again, still larger than the unmodified barbules but the pennula contain a few nodes only. There is a gradual changeover in the shape of the distal barbules when passing from one zone to the next. The pennula and the dorsal cilia are in contact with the rami or, if present, with the ventral ridges of the rami of the overlying feather. The pennula might even extend between the rami and contact the basal lamellae of the proximal barbules. Descriptions in the literature are obscure at this point. This is probably partly caused by the ideas about the function of the structures in the zones of overlap. The authors first describing the anatomy of the zones of overlap were convinced that the special structure of the distal barbules was meant to create friction. Sick (1937), for example, describes how the areas of overlap in the fully extended wing of the blackbird coincide with the areas occupied by the modified distal barbules and concludes that this makes it *kaum zweifelhaft* (hardly doubtful) that these structures create friction between the feathers. Graham (1931) claims that he can feel the braking effect of the friction zones when he tries to open a wing with the feathers pressed together. Lucas and Stettenheim (1972) do not argue about the function but simply translate the German term *Reibungsradien* to 'friction barbules'. In fact, there is no real evidence that the minute structures on the upper part of the distal barbules cause friction and it would be surprising if they actually did. In flight most birds open and close their wings during every wing beat cycle. The energetic costs to overcome frictional forces during wing extension and flexion would be extremely large. Extensor and flexor muscles situated in the wing would have to deliver the forces to perform the action. Intrinsic wing muscles should be as small as possible to minimize the mass of the wing and hence the moments of inertia of the flapping wing.

The most detailed account of the friction story is given by Oehme (1963) using the blackbird and the starling as examples. He focuses in particular on two primaries of the blackbird. Primary 7 of the left wing is overlapped by primary 6 as indicated in Fig. 3.4(a) and (b). When the wing is spread and ready for the downstroke the edge of the outer vane of 6 touches the upper part of the inner vane of 7 in the zone fitted with the longest pennula on the distal barbules. With the feathers pressed together, the barbs of the outer vane of 6 resist further spreading and sideward motion because they are imbedded among the long pennula on the surface of the inner vane of 7. These run roughly in the same direction as the barbs of the narrow vane of 6 as illustrated by Oehme's figure (Fig. 3.4(b)). The wing is fully stretched and ready for the downstroke in the configuration shown in Fig. 3.4(a). Stretching the wing further as Oehme tried is unnatural and one can feel indeed friction while doing so. The structural evidence given about the overlap zones can also be used to tell an alternative story in which friction between feathers does not play a role. Imagine that during the end of the upstroke just before the beginning of the downstroke when the wing is stretched the feathers are not pressed together while they are brought in the

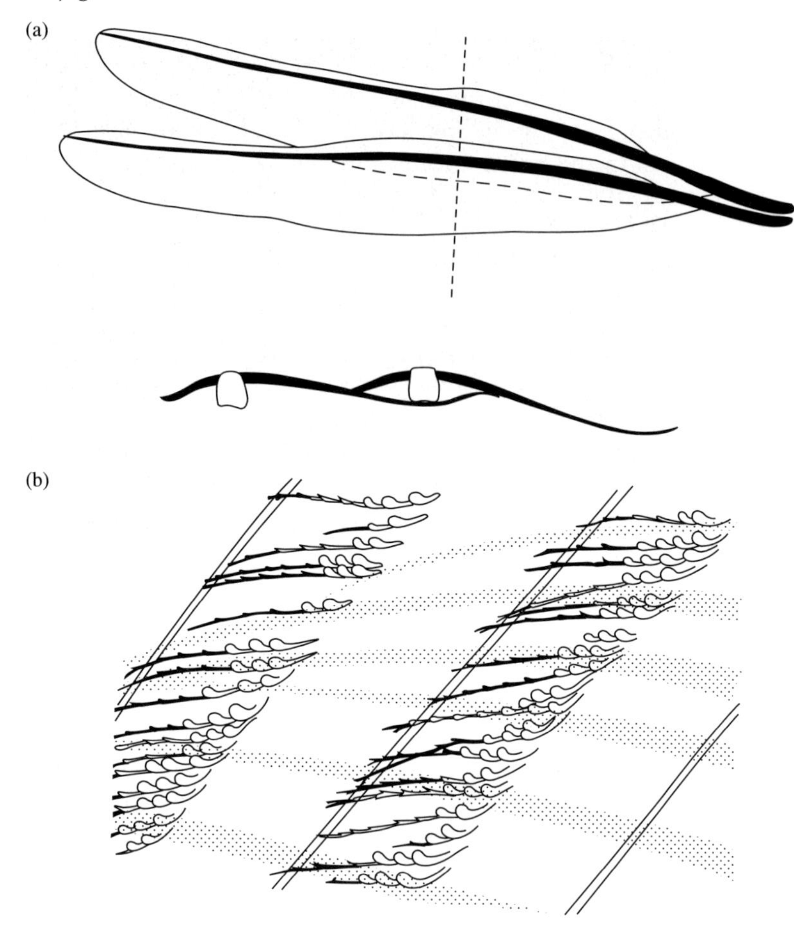

Fig. 3.4 Interaction between two overlapping feathers of the pigeon (from Oehme 1963, with kind permission). (a) Primaries 6 and 7 in the position taken when the wing is fully stretched. The cross section shows the overlap between the vanes. (b) Enlarged picture of the elongated pennula in the overlap zone of primary 7. The barbs of the overlapping outer vane of primary 6 are indicated as stippled areas.

stretched position. A wing of a freshly dead blackbird can be stretched easily if the feathers are not pressed together. During the downstroke, pressure differences between the ventral and dorsal side of the wing will press the feathers together in the position as indicated by Fig. 3.4(a). The pressure locks the feathers in place. High pressure from below attempts to spread the feathers during the downstroke but this is counteracted by the locking mechanism. The orientation of the pennula with respect to the rami of the overlying feather shows that the main direction of the locking mechanism is sideways. Although experimental evidence is still needed, it seems more appropriate to use the term locking barbules instead of friction barbules.

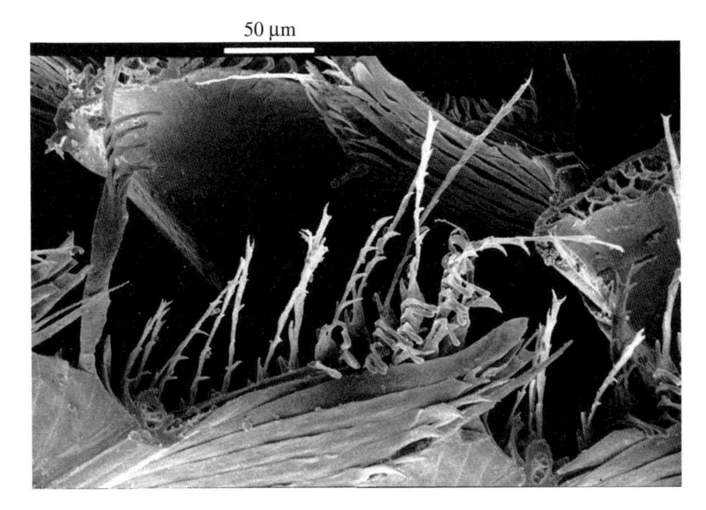

Fig. 3.5 SEM photograph of a cross section through primary feathers IX and VIII of the common swift.

The structure of pennula on the distal barbules in the overlapping zones varies among species. Hawks and falcons have long dorsal cilia appearing as a forest of spines. They form a thin layer of mainly air filled space between the feathers and this probably decreases the rubbing surface and hence friction. Here the lobate dorsal cilia on the long pennula in the zones of overlap may serve as dry lubricants, facilitating the folding and spreading of the wings.

The outermost primaries of the swift also interlock under slight pressure from below. A SEM picture (Fig. 3.5) of the primary IX and VIII (X is the largest outermost primary) gives an indication of how the mechanism works. Hooks on the pennula of the distal barbules of the underlying feather stick out dorsally and will attach to the ventral ridges of the barbules of the upper feather when the two are pressed together vertically. A slight shift of the two feathers with respect to each other will release the locking hooks when the pressure is reduced.

It may not be an easy task but the challenge of future research is to find out what these functions actually are. The locking mechanism and dry lubrication are just two possibilities. Since feathers are dead structures worn parts cannot be replaced, feathers have to be replaced as units. The wear of microstructures might be an important reason why birds must moult cyclically. There is no proof yet that these alternative points of view are correct but they are more realistic than the standard functional explanation for which there is no proof either.

3.4.2 Sound reducing structures

Some owls and caprimulgiform birds have a woolly surface at the zones of overlap formed by long pennula of the distal barbules. The current idea

Fig. 3.6 SEM picture of the elongated pennula on the dorsal side of primary IX of the barn owl. The scale bar is 0.1 mm.

about the configuration shown in Fig. 3.6 is that it quiets the sounds of extension and flexion of the beating wings of these night hunting birds. This supports the lubrication theory. Flight sound reduction in owls is also achieved by a toothed leading edge of the outermost primary feathers. The ends of the barbs of the narrow vane bend forward and have very short barbules creating a saw-toothed leading edge (Mascha 1904 and Fig. 3.7). This probably causes micro-turbulent flow decreasing the sound effects of larger scale air movements, but again there is no real proof for these generally accepted ideas.

3.4.3 Emarginated primaries

Chapter 2 showed how one or both vanes of primaries are often narrowed towards the tip, starting at some point along the length. The change in width can be gradual or abrupt and substantial or barely noticeable. The tips of such emarginated or notched feathers do not overlap when the wing is stretched forming slots separating the distal parts of the feathers of the hand. Large soaring birds set extreme examples but many extremely diverse groups have emarginated feathers and others do not. The distribution of emargination among specialist groups does not give away the function. Emarginated feathers are not common among seabirds and wing propelled divers. Extreme flyers such as swifts and hummingbirds do not have them. Only the narrow leading edge vane of most passerines is emarginated. Trailing edge emargination is extreme in some Galliformes (i.e. Western Capercaillie), birds of Prey (Falconiformes), corvids (Corvoides), and storks (Ciconiidae). In many ducks (Anatidae), only the first feather shows emargination of the wider

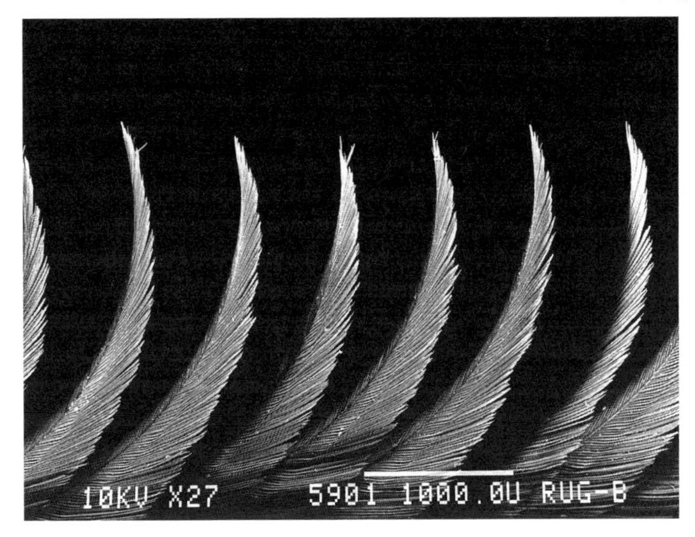

Fig. 3.7 SEM picture of the serrated leading edge along primary IX of the barn owl. The scale bar is 1 mm.

rear vane. All birds soaring in thermals have extremely slotted wings due to primary spreading and heavy emargination. Pennycuick (1973) suggests a relation with the necessity to make tight turning cycles. Ground dwelling birds in dense vegetation like pheasants might benefit from heavily slotted wings during vertical take-off but it is not clear how.

Many birds, even those without emarginated feathers may show some slotting during the wing beat cycle especially during extreme manoeuvres or take-off. Extreme spreading of the hand wing feathers can also result in wing tip slotting especially in birds with pointed primaries. Gaps in the wings occurring commonly during the moulting periods can give the impression of wing slots but are of course a completely different phenomenon. The free narrow tips of the feathers are often curved backwards. In soaring flight of heavy birds, these are obviously carrying weight because they are bent upwards. Each feather probably will generate lift and may be acting as a delta wing. Blick *et al.* (1975) measured the vorticity behind a plain and a slotted wing tip and found that the maximum vorticity of the plain wing tip was an order of magnitude higher than that of the slotted tip. The slotted wing used was a wooden rectangular wing model with 5 Canada goose primary feathers glued to the tip to provide the slotting. We have no idea how realistic this comparison was. Flow visualization could provide real insight into the function of this common phenomenon.

A number of functions have been assigned to slotted primaries ranging from a delay in stall under high angles of attack, via reduction of drag, increase of lift, and storage of elastic energy to an increase of longitudinal stability (see Norberg (1990) for an overview and discussion). Graham (1931) considered the space between the raised alula and the hand wing a wrist slot. He saw the slots as anti-stalling devices operating at high angles of attack of the air onto the wings. The function of the wing tip and wrist slots is still not clear. The spread out wing tip feathers of birds usually soaring in narrow thermals could have a sensory function in detecting the outer boundary of a thermal. Drag reduction, increase of lift, and increased longitudinal stability has been suggested. Unfortunately, there is no direct evidence for any of the proposed functions.

3.5 *Tail feathers*

Rectrices can show special adaptations but we are often not sure about the function that required the adaptation. Stiff shaft elongations sticking out beyond the vanes are found in various tree creeping birds for obvious reasons. But some swifts have spine like shafts sticking out as well (i.e. the brown backed needletail) and these are not used for support during perching. There are rumours about an aerodynamic function: the spines probably aid to guide the air flow from the rear of the bird and decrease drag by doing so. Few birds have wire tails consisting of a shaft with loose barbs (i.e. the Southern emu wren). A plausible explanation is not available.

Tubaro (2003) discovered an interesting difference in the shape of the rachis of outer tail feathers comparing related species with a deeply forked and a more square tail. The thickness of the rachis along the feather as a fraction of the maximum thickness at the base is consistently smaller in birds with deeply forked tails. Maximum differences found are in the order of 20–25%. This relationship was tested in 11 species pairs from 7 orders. Extreme values are found among hummingbirds, nightjars, kingfishers, flycatchers, swallows, and martins. The shafts of the outer tail feathers of magnificent frigatebirds are over the entire length relatively thinner than the outer tail feather shafts of the neotropical cormorant. A functional explanation is difficult to give without further measurements. The higher relative rachis thickness of outer rectrices of square tails suggests that these have to cope with larger forces compared to the deeply forked configuration. A study of the flow patterns generated by a closed delta wing compared with a V-shaped wing might show the significance of this finding. Tubaro believes that the difference reflects the idea that the rachis of the outer rectrices in species with deeply forked tails has not been selected to resist lift forces but to serve as a male ornament. There is no proof for this hypothesis in his examples because the comparisons are between species pairs and not between males and females of the same species.

3.6 *Feather muscles and nerves*

A complex array of feather muscles is attached to every flight muscle follicle. The muscles are smooth and placed in series with short stretches of tendon. The feather muscles form complex networks, which probably indicate that groups of feathers rather than individuals are affected by the activities of the feather muscles (Homberger and de Silva 2000). The main actions are erection and depression of feather fields. Smooth muscles are notoriously slow but persistent and in combination with tendon could serve to provide tension and resistance against erection and depression. Lucas and Stettenheim (1972) showed that muscles interconnect the follicles of the primary wing feathers of chicken and turkey.

Contractions of these rotate the follicles and the feathers (Fig. 3.8). It would be no surprise if all birds have these muscles inserting on their primaries. The direction of rotation suggested by the orientation of the muscles is forward, bringing the narrow vane down. These smooth muscles probably operate by resisting rotation in the opposite direction. Suggesting that during flight aerodynamic forces tend to rotate the feathers backward, pushing the narrow vane up. The view that primary feathers act as independent lift generators by cutting the air under fairly large angles of attack to generate a stable leading edge vortex over the length of the feather is consistent with this view. The leading narrow vane would be pushed upward rotating the feather and bringing the trailing vane down. The smooth rotating muscles on the follicles of the primaries would counteract this movement and hence control the angle of attack of the feathers. The secondary feathers of the arm part

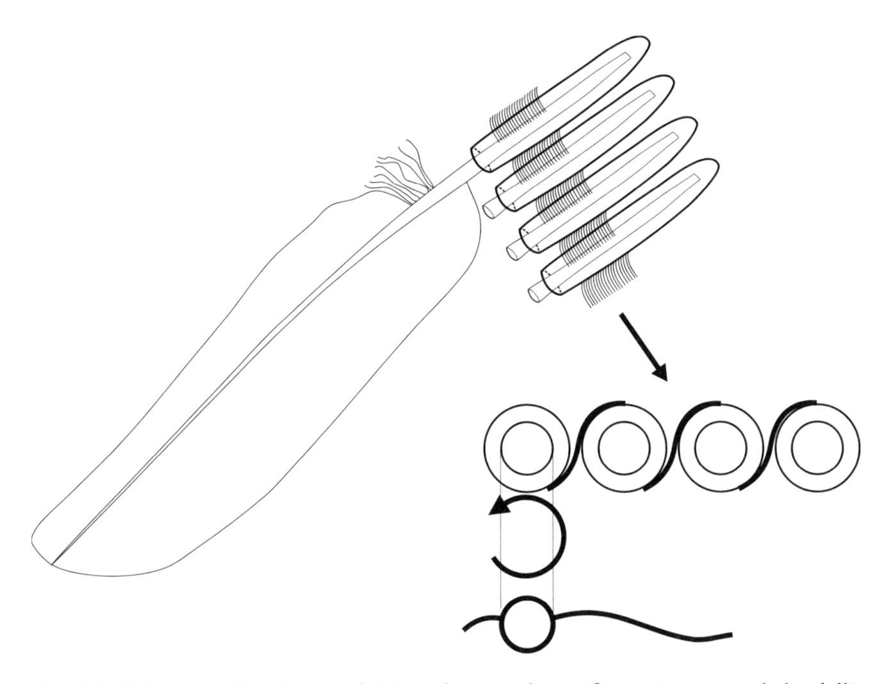

Fig. 3.8 Schematic drawing explaining the muscle configuration around the follicles of primaries. The insertion of the muscles causes counter-clockwise rotation during contraction counteracting the clockwise rotation induced by the air hitting the outer vane under a small angle of attack.

of the wing do not have this arrangement and they do not need it because their function in flight is very different. In that part of the wing, the coverts on the propatagium form the leading edge. We saw in Chapter 2 how the symmetric tips of the secondary remiges form the sharp trailing edge of the classical wing profile of the arm region of the wing.

3.6.1 Detection of movement

There is a dense network of nerves around feather follicles and in the inserting smooth muscles. Proprioceptive mechanoreception is an obvious function for such an anatomical configuration. Specialized feathers on the head and breast have been shown to act as indicators of wind speed and direction. Little is known about similar senses in the wings. The innervation of the wings consists of two main systems: the dorsal branches of the nervus radialis and the ventral nervous network coming from the nervus medianoannularis which are the 12th and 13th spinal nerve respectively (Baumel 1979). Necker (1994) studied the spinal cord in search of areas and nerve tracts involved in flight control. He found that nerve fibres coming from mechanoreceptive feather fields are projected in the nucleus proprius of the spinal chord. These projections connect to other neurons up and down the spinal chord and to the cerebellum. Direct or indirect connections between these incoming nerves and the dendrites of motor neurons have not yet been found. Brown and Fedde (1993) recorded spikes of neural activity in the radial nerve near the humerus of a chicken while manipulating the alula, feathers, and follicles manually and by blowing air through a tube onto the wing. Activity could be registered when the alula was extended, and when the mechanoreceptors associated with wing coverts were stimulated. The small filoplumes growing at the edges of the follicles of secondary feathers have a sensory function and cause discharges in the radial nerve when the associated feathers are moving. No activity in the radial nerve could be registered when primary feathers were manipulated. This does not indicate that there is no mechanoreception in the follicles or intrinsic muscles of secondary or primary feathers. Investigations involving the whole nervous system of the wings are needed.

Necker (2000) summarizes the types of structures in birds that could be excited by mechanical stimuli. Free nerve endings are usually thermoreceptors but could also serve to detect motion although electrophysiological evidence is lacking in birds. Herbst corpuscles, small complex bodies around nerve endings, are more likely candidates to detect tactile stimuli. These are widely distributed in the skin, are associated with feather follicles and with the muscles of the feathers. Flying birds are supplied with a larger number than non flying birds. There is no information available about the distribution on the wings. Herbst corpuscles are sensitive to vibrations at high frequencies. They hardly react to frequencies below 100 Hz but have a low excitation threshold for vibrations between 300 and 1000 Hz. At these frequencies the amplitude required for excitation is less than 0.1 μm.

Ruffini corpuscles are axon endings in close contact with collagen fibres, probably serving as stretch receptors and numerous in-joint capsules. They react to a stimulus with a regular firing response in the nerve. Such a response has been detected in the afferent nerves of chicken wings with an increased activity with increasing elevation of the covert feathers.

Our understanding of flight control in birds is extremely limited. A lot of work lies ahead before one can even start to understand the sensory and motor pathways involved in the complex movements required by flight.

3.7 *Summary and conclusions*

A general picture of the structure of flight feathers is well established although based on a limited number of species studied. Scanning Electron Microscopic (SEM) techniques should be used more widely to study specific features in more detail and quantitatively. The primaries of extreme flyers, for example, albatrosses, hummingbirds, swifts, frigate birds, and auks can still be expected to hide many unknown features.

Our knowledge of the mechanical properties of feathers is extremely limited especially regarding relevant demands in flight. Measurements of the breaking forces of single barbs is interesting from a comparative point of view but adds little to the insight in the physical strength of the feather as a whole. Much more emphasis should be put on the strength of the weakest parts of feathers. How big are the forces a hooklet at the end of a pennulum can take and how strong are these structures in relation to the forces applied during flight? It is also important to obtain more information on the wear of microstructures in relation to moult.

Microstructures require much more attention because they serve a lot of functions. We only just start to understand some of those including the differential dorso-ventral transmissivity for air, the interactions between overlaying feathers such as locking mechanisms and sound reduction.

Emarginated primary feathers are present in many species of birds with very different flight behaviour. There is a lot of speculation about the functions of emarginated feathers but in the end we need to conclude that there is little evidence supporting the repeatedly published hypotheses.

The existing knowledge about the shape and strength of tail feathers and the related functions is in fact still anecdotal. Lots of the statements in taxonomic publications are based on generally accepted ideas for which no experimental proof has been obtained.

Feathers are connected to the follicles in the skin. A very complex system of muscles and nerves interconnects the implant patterns. The nervous system of the wings originates from two spinal nerves. In the spinal column projections of these nerves run up and down and eventually connect to the cerebellum. The wiring of the flight-related motor and sensory nervous systems and the physiological software remains virtually unexplored.

4 *Aerodynamics*

4.1 *Introduction*

A flying bird generates lift forces to counteract gravity and thrust forces to overcome drag. The magnitude of these forces can be crudely approximated using elementary physical principles. Steady flight in still air at a uniform speed and at one altitude is the simplest case. It requires balanced forces where lift equals weight and thrust equals drag as well as balanced moments of these forces about the centre of gravity. Under these relatively simple conditions the magnitude of the mechanical power involved in the generation of lift and thrust in relation to speed can be estimated. The power to generate lift is inversely proportional to flight speed and the power needed for thrust increases with the speed cubed. The total mechanical power is the sum of the lift and thrust powers and hence follows a U-shaped curve if plotted against speed. A U-shaped power curve implies that there are two optimal speeds, one where the power is minimal and a higher one where the amount of work per unit distance reaches the lowest value. The question is, does this U-shaped power curve really exist in birds?

Flow visualization is needed to really appreciate that which is happening between bird and air. How does a bird extract lift and thrust forces from the interaction with such a thin medium? We start to concentrate on the wake behind a flying bird because the wake shows what the bird did to the air. Subsequently, we want to know exactly how that wake is formed by asking the question, what happens at the wings in flight. The problems involved in direct measurements have not yet been solved. We therefore used particle image velocimetry, a quantitative method to study flow patterns, to study bird wing models in a water tunnel at realistic *Re* numbers. Gliding bird wings are the simplest case. Results of preliminary tests using a transparent model of an arm wing section of a fulmar and models of swift wings illustrate two important aerodynamic principles involved in bird flight.

Relatively little is known about what happens between birds and air during flapping and manoeuvring. Does flapping preclude the presence of attached flow even on the arm wings of all birds? What is the function of the alula? Novel aerodynamic mechanisms of insect flight probably provide some insight into what could happen in birds, but direct evidence showing that birds use similar mechanisms is not available yet.

Bird tail form and aerodynamic function received little attention so far compared to wings. A brief summary of a number of aspects will be given in the last paragraph before the summary and conclusion of this chapter.

4.2 *Rough estimates of forces and power*

In the simplest case (Fig. 4.1) a bird weighing W (N) flies at a constant altitude through still air at a constant speed v ($\mathrm{m\,s^{-1}}$). Newton's laws predict

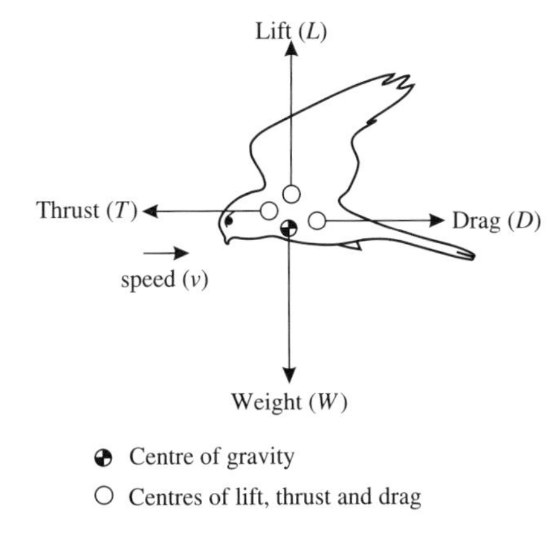

Fig. 4.1 Four resultant forces in a vertical medial plane through a kestrel flying at one height and at uniform speed. There is no evidence that the positions of the points of application of the forces are as indicated. Under steady flight conditions average clockwise and counterclockwise moments of rotation are supposed to cancel out.

that, on average over time, the total vertical component of force L must equal W, and the sum of the drag forces D must be balanced by the generated thrust T. A complication is that the four forces are the resultants of several parallel forces and most likely have different points of application as already suggested by Leonardo da Vinci. Moments of rotation will be restricted to the median vertical plane if we assume lateral symmetry. In the real bird in flapping flight the centres of lift and gravity move about and are probably never in the same position. We assume for simplicity that these centres are in a fixed position and coincide. This means that the sum of head-up and head-down pitching moments around the centre of gravity is close to zero and can be ignored.

A bird uses its wings to generate lift by accelerating masses of air downwards. It generates thrust by accelerating air backwards. Rough estimates of the mechanical energy involved in the thrust and lift forces at uniform speeds, can be obtained by looking at the vertical and horizontal forces separately. (Box 4.1 explains this approach more formally using equations.)

We assume that the mass of air affected by a bird per unit distance flown is approximately the mass of a circular cylinder of air, with a diameter equal to the wingspan b, along the flight path (Fig. 4.2). The surface area of the cross section of that cylinder is the square of the radius times π : $\pi(\frac{1}{2}b)^2$. This is also the volume of the cylinder per unit distance flown. This volume per unit distance multiplied by the density of air (ρ) provides an estimate of the mass of air involved per unit distance flown ($\mathrm{kg\,m^{-1}}$). Since the bird flies forward at v $\mathrm{m\,s^{-1}}$, the mass of air affected per unit time, or mass flux ($\mathrm{kg\,s^{-1}}$), will therefore equal the mass per unit distance times the velocity.

4.2.1 Lift counteracting weight, induced power and induced drag

For the generation of lift the affected air mass is accelerated down by the beating wings. The initial vertical velocity is zero because the oncoming airflow reaching the bird at v m s^{-1} is horizontal. Due to the acceleration, the vertical velocity increases from 0 m s^{-1} to a maximum value of say w m s^{-1}. The air is left behind moving obliquely downwards as is indicated in Fig. 4.2. The downward momentum left behind per unit time is the mass flux times the vertical velocity of the air (kg s^{-1} m s^{-1} = N). Newton's third law tells us that this downward force given to the air by the flapping wings equals the reaction force L on the bird in upward direction. The vertical kinetic energy given by the bird to the air is one-half the product of the affected mass of air and the downward velocity (w) squared, (using the kinetic energy equation explained in Chapter 1). To obtain the downward kinetic energy per unit time the mass term must be replaced by the mass flux (Felix Hess personal communication; Sunada and Ellington 2000). Kinetic energy per unit time is expressed in J s^{-1} = W, the unit of power. The term 'induced power', P_i, is commonly used for the power required to generate lift. It equals the vertical kinetic energy per unit time given to the air.

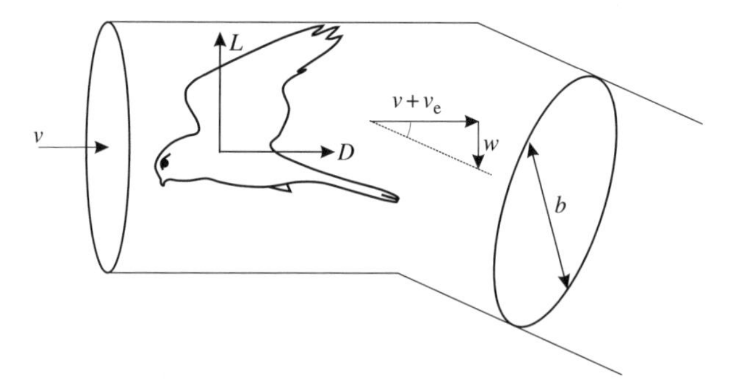

Fig. 4.2 Impression of a hypothetical circular cylinder of air deflected obliquely downward and accelerated backward due to the action of a flying bird. See text for further explanation.

Box 4.1 ***Mass flux model to obtain estimates of lift and drag during flapping flight***

Starting point Fig. 4.2: a bird flies with uniform velocity v at one altitude. Centre of lift and centre of mass are assumed to coincide. Under these conditions lift force L is equal and opposite to weight W and the total thrust, T, equals the total drag force, D.

The mass of air, m, accelerated per unit distance flown is approximately that contained in a circular cylinder of air, with a diameter equal to the wingspan b, along the flight path:

$$m = \pi\left(\tfrac{1}{2}b\right)^2 \rho \quad (\text{kg m}^{-1}) \qquad (4.1.1)$$

where ρ is the density of air. The bird flies forward at v m s^{-1}. The mass of air affected per unit time (the mass flux, $\dot{m}$) will therefore be in the order of:

$$\dot{m} = \pi\left(\tfrac{1}{2}b\right)^2 \rho v \quad (\mathrm{kg\,s^{-1}}) \tag{4.1.2}$$

The air mass reaches a vertical velocity of w m s^{-1}. The downward momentum left behind per unit time equals the reaction force L:

$$L = \dot{m}w \quad (\mathrm{kg\,s^{-1}m\,s^{-1}} = \mathrm{N}) \tag{4.1.3}$$

The vertical kinetic energy of the air is $\frac{1}{2}mw^2$. Replacing m by $\dot{m}$ gives the downward kinetic energy per unit time (Sunada and Ellington 2000) which is equal the induced power P_i:

$$P_\mathrm{i} = \tfrac{1}{2}\dot{m}w^2 \quad (\mathrm{kg\,s^{-1}(m\,s^{-1})^2 = J\,s^{-1} = W}) \tag{4.1.4}$$

If L equals W, equation (4.1.4) can be rearranged using (4.1.2) and (4.1.3) to:

$$P_\mathrm{i} = \frac{1}{2}\frac{W^2}{\pi((1/2)b)^2 \rho v} \quad (\mathrm{W}) \tag{4.1.5}$$

Induced drag (D_i) is caused by the generation of lift. The induced power required to overcome that drag at forward speed v is:

$$P_\mathrm{i} = D_\mathrm{i} v \quad (\mathrm{W}) \tag{4.1.6}$$

Equations (4.1.4) and (4.1.6) combined give an expression for the induced drag:

$$D_\mathrm{i} = \tfrac{1}{2}\pi(\tfrac{1}{2}b)^2 \rho w^2 (\mathrm{N}) \tag{4.1.7}$$

Thrust

During flight at uniform speed air is accelerated backwards to a higher speed $v + v_\mathrm{e}$ behind the bird. The thrust gained equals:

$$T = \dot{m}v_\mathrm{e} \quad (\mathrm{N}) \tag{4.1.8}$$

The thrust power is the increase in kinetic energy:

$$P_\mathrm{t} = \tfrac{1}{2}\dot{m}(v + v_\mathrm{e})^2 - \tfrac{1}{2}\dot{m}v_\mathrm{e}^2 = \dot{m}vv_\mathrm{e} + \tfrac{1}{2}\dot{m}v_\mathrm{e}^2 \quad (\mathrm{J\,s^{-1} = W}) \tag{4.1.9}$$

During horizontal flight at uniform velocity $D_\mathrm{d} = T$. The drag force can be described as:

$$D_\mathrm{d} = \tfrac{1}{2}\rho v^2 A C_\mathrm{d} \quad (\mathrm{N}) \tag{4.1.10}$$

A represents a relevant area of the bird. The drag coefficient C_d depends on the choice of the area and on several unknown factors. The energy required to balance the drag forces is:

$$P_d = D_d v = \frac{1}{2}\rho v^3 A C_d \quad (\mathrm{W}) \tag{4.1.11}$$

The total mechanical flight power (P_{tot}), is the sum of P_i and P_d.

During horizontal flight in still air lift force L equals on average the weight of the bird W. This implies that W is also equal to the product of mass flux times the downward velocity w or, in other words, that w equals the weight of the bird, W, divided by the mass flux. If we now calculate the kinetic energy using W divided by the mass flux instead of w, we will find that the induced power is proportional to the weight of the bird squared and inversely proportional to the mass flux and hence inversely proportional to flight speed v. This result indicates that at low speed the induced power remains high and birds have to generate a lot of energy to remain airborne.

Changing the direction of the air by accelerating it downward to generate lift causes not only lift but also resistance which has been given the term induced drag (D_i). The power required to deflect the air downwards and to overcome that drag at forward speed v is the product of drag and velocity. This is the same induced power, P_i, determined before as the kinetic energy given to the air per unit time. Hence, the induced drag, D_i, equals P_i divided by v. That makes the induced drag, D_i, equal to one-half the product of the affected mass of air per unit distance flown and the downward velocity squared (w^2). The vertical velocity w and the wing span b are obviously the dominating factors.

4.2.2 Thrust

During flight at uniform speed air meets the bird at the flying speed v. In order to keep flying at the same speed, the bird must generate an amount of thrust equal to the total drag on body and wings in horizontal direction. Therefore the wings accelerate the air under their control backwards to give it a higher speed $v + v_e$ behind the bird. The thrust force gained equals the mass of air affected per second (the mass flux) times the speed increase v_e reached by the accelerated air in the opposite flight direction. The kinetic energy per unit time given to the air is the difference between the total kinetic energy per unit time of the mass of air moving at $v + v_e$ relative to the bird minus the kinetic energy per unit time that would have been present without the acceleration. This difference is the thrust power P_t.

During flight at uniform speed the horizontal drag forces on the body, the wings, and the tail add up to the resultant drag force D_d which must be equal and opposite to the thrust force. The air is pushed aside by the passing bird causing a build up of dynamic pressure. Bernoulli's law predicts (Chapter 1) that this pressure is proportional to $\frac{1}{2}\rho v^2$ ($\mathrm{N\,m^{-2}}$). The drag force equals that pressure times some surface area and a drag coefficient (see Marey's definition in Chapter 1). For the area, either the frontal or the total surface area of the bird can be taken into account. The drag coefficient depends on the choice of the area, on the *Re* number (see Chapter 1) and on several unknown factors such as the shape of the bird and the roughness of the surface area. The shape of a bird in flapping flight changes continuously and so will the drag coefficient.

The power required to balance the drag forces, P_d, is force times velocity. The drag force is proportional to the velocity squared implying that P_d is proportional to the speed cubed and at higher speeds is hence the major component of the total mechanical flight power (P_{tot}), which is the sum of P_i and P_d.

This coarse analysis predicts that the power required for flight as a function of speed follows a U-shaped curve (Fig. 4.3). Such a curve implies that there is one speed at which the flight power is minimal (the minimal power speed, v_{mp}) and a higher speed where the amount of work per unit distance covered is minimal (the maximum range speed, v_{mr}). The maximum range speed can be found by drawing the tangent to the curve from the origin of the graph. It is the speed where a minimum value is found for the ratio of power ($W = J\,s^{-1}$) over speed ($m\,s^{-1}$). This ratio is the amount of work per unit distance ($J\,m^{-1}$).

How good is this approximation of the power required for bird flight? Experimental evidence to show that the U-shaped curve actually exists requires the use of a variable speed wind tunnel in combination with some way to measure energy expenditure at a range of speeds. There are now several studies completed that meet these requirements (see Chapter 8). In fact only the very first variable speed wind tunnel study ever, by Tucker (1972) measuring a budgerigar, showed a clear U-shaped relation between speed and energy consumption. The shape of curves from later studies differed by being

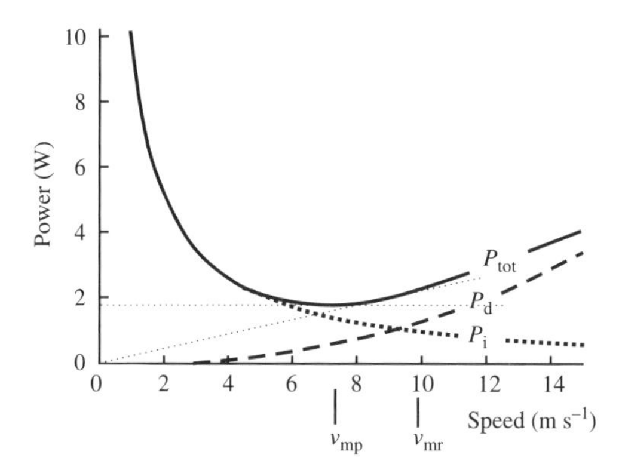

Fig. 4.3 Hypothetical mechanical power curves as a function of speed predicted by theory. The total power follows a U-shaped trajectory. Minimum power speed and maximum range speed are indicated.

flatter or J-shaped, in some cases due to the lack of data points at extremely low and high flight speeds. Evidence that birds actually have distinct minimum power and maximum range speeds is weak and experimental evidence controversial (see i.e. Alexander 1997; Dial *et al.* 1997; Tobalske *et al.* 2003; Welham 1994). One important point, which could well be the reason for the controversies regarding this issue, is the fact that birds, unlike aircraft, can change the way they fly easily, either gradually or abruptly for every speed.

We need to know what actually happens during the dynamic interactions between a flying bird and the air before we are able to understand the aerodynamics of bird flight. A crude approximation assuming the deflection of a cylinder of air is not good enough. What are the mechanisms used by birds to accelerate air down and backwards in order to fly? Precise visualization of the flow is required to answer this question.

4.3 *Visualization of the wake*

The wake behind a bird flying in still air shows the result of the interaction between the bird and the air, reflecting the reaction forces generated by the action of the bird. The earliest report on wake structures behind flying birds came from Magnan *et al.* (1938), who used tobacco smoke to show that pigeons in slow flight produce a smoke ring during every downstroke. Kokshaysky (1979) managed to visualize the wake during short flights of the chaffinch and the brambling qualitatively. The birds were forced to fly through clouds of wood and paper dust between perches in small enclosures. Multi-flash pictures (Fig. 4.4) were taken when the birds were forced to fly

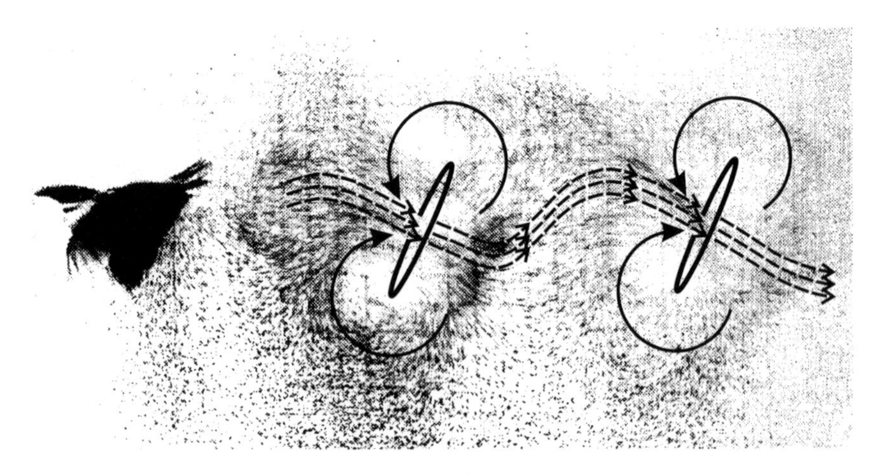

Fig. 4.4 Interpretation of the wake structure behind a small bird in flapping flight. A chaffinch is flying through a cloud of dust. Vortex rings are formed during the downstrokes. The upstrokes drag the dust upwards but do not generate vortices. A jet of air undulates through the centres of the vortices of two downstrokes (with Nature's permission based on a picture in Kokshaysky (1979)).

in the dark through the dust cloud to the opposite perch. The pictures show that every downstroke produces a closed vortex ring and that the upstroke hardly contributes to the wake. The starting vortices of both wings (see Chapter 1) interconnect above the body and form the upper part of the ring. Tip vortices from both wing tip generate the side parts of the vortex ring and the lower part is formed by the stopping vortices of both wings at the end of the downstroke. The starting, trailing, and stopping vortices close into one ring-shaped structure. The plane of the ring is not vertical but at an oblique angle. A jet of air can be seen winding through the centres of the subsequent vortex rings on average at an obliquely downward angle. During the upstroke the wings are folded close to the body and do not seem to generate vortices.

Quantitative studies to trace the wake of flying birds were undertaken by Geoff Spedding. In a first series of experiments birds were trained to fly down the length of a cage evoked by a light switch under reduced light level conditions. A cloud of neutrally buoyant soap bubbles (approximate diameter 2–3 mm) filled with helium was used to visualize the flow. Two horizontally orientated cameras took series of pictures in a stereoscopic set-up. An example is given in Fig. 4.5. High-speed cine films at 200 frames per second of the same flight behaviour were taken soon after the bubble

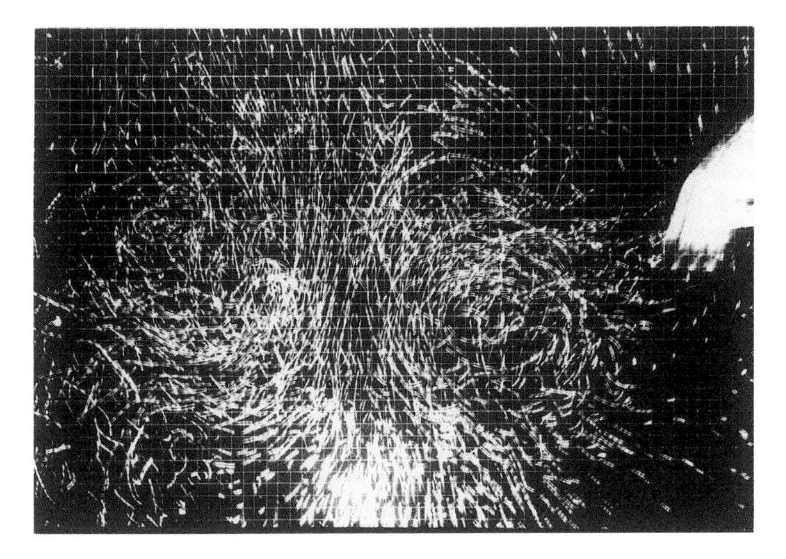

Fig. 4.5 One picture of a stereo pair showing a vortex ring structure in the wake behind a pigeon in slow (2–4 m s^{-1}) flight (Spedding *et al.* 1984, with kind permission of The Company of Biologists). The flow is visualized by taking images of helium bubbles in a cloud. Four flashguns fired in sequence and four subsequent images, 5 ms apart, of each bubble are shown. Each bubble reflects light off the front and back surface and appears as a double streak. The duration of one flash was approximately 3.5 ms, long enough to depict the moving bubble as two little strips. The direction of movement can be detected because the brightness of the bubble images in a row decreases in time.

experiments to match vortex structures with wing beat kinematics. Vortex theory (Rayner 1979*a,b* and *c*) was used to estimate momentum and energy in the wake. Spedding *et al.* (1984) analysed the wake of a pigeon in slow flight using this technique. Pigeons in slow flight generate vortex rings during the downstroke similar to the rings seen in the wakes of the finches. The orientation of the plane of the rings is almost horizontal at about 11°. The calculated momentum in the wake was only about half the amount required to support the weight of the pigeon. The next bird studied using the same technique was the Eurasian jackdaw in slow flight (Spedding 1986). Results were very similar to those obtained with the pigeon: ring-shaped vortex rings from the downstroke at a small angle with the horizontal. Once again there is a large discrepancy between the momentum measured in the wake and that required for weight support. Only 35% of the required momentum seemed to be present in the rings. Geoff Spedding assumes that the vortex ring model is probably too simplistic. The shape deviates from that of a real ring and there is the possibility that not all the vorticity rolls up into the ring and this is not taken into account. The flight speeds of the pigeon and the jackdaw were very slow at 2.4 and 2.5 $\mathrm{m\,s^{-1}}$ respectively. The common kestrel studied by Spedding in 1987 flew at 7 $\mathrm{m\,s^{-1}}$, close to its normal cruising speed. The wake turned out to be very different. There were no separate rings but a continuous pair of trailing vortices with constant circulation (Fig. 4.6). The core of the circulation is 6.6 cm wide during the upstroke and the vortices run close together obliquely upwards. During the downstroke the trailing vortices follow the path of the wing tip outwards and inwards. The core of the vortices is 3.2 cm narrow during this phase. The calculated downward wake momentum balanced the weight almost exactly.

Colin Pennycuick, more than 40 years the undisputed godfather of bird flight studies, used his expertise to design and build a large recirculating wind tunnel in Thomas Alerstam's department of Animal Ecology at the University

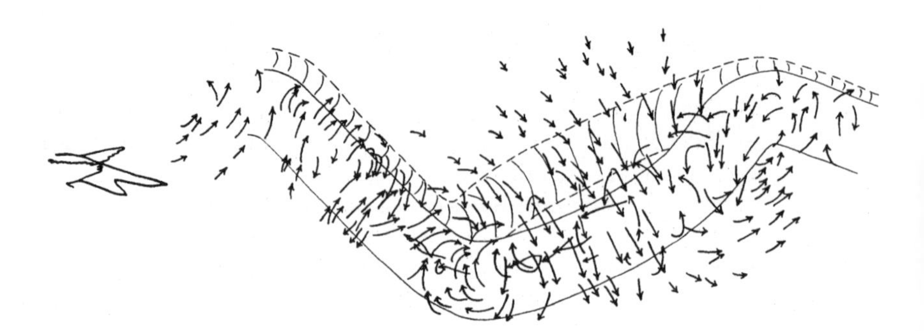

Fig. 4.6 Spedding's (1987) interpretation of the airflow behind a kestrel in flapping flight. Vortex structures caused by one downstroke and the subsequent upstroke are shown. (Reproduced with kind permission of the author and The Company of Biologists.)

of Lund (Sweden). The tunnel was designed to accommodate a bird in free flight (Pennycuick *et al.* 1997). It has an octagonal test section 122 cm wide and 108 cm high and the maximum velocity is 38 $m\,s^{-1}$. Geoff Spedding was invited to develop, in close cooperation with the local specialist Anders Hedenström and PhD student Mikael Rosén, a method to visualize the flow disturbances induced by a steadily flying bird at various speeds. The bird used in the experiments was a thrush nightingale, a nocturnal long distance migrant. The training lasted several months before the actual experiments. The bird had to fly at a position near the centre of the test section in low light conditions with an upstream light as the only reference point. Finally, the thrush nightingale flew steadily at speeds between 4 and 11 $m\,s^{-1}$. Kinematic analysis (Rosén *et al.* 2003) showed, surprisingly, that the wing beat frequency was about 14 Hz at all speeds. How did the bird manage to fly faster? Wing tip amplitudes varied only slightly over the range of speeds without a detectable trend. The downstroke fraction of the wing beat cycle decreased steadily with increasing speed but only from about 0.5 to 0.45. Since the cycle duration was constant, this means the downstroke velocity increased and the upstroke velocity decreased. There was also a slight increase in maximum span, but none of these variables are substantial enough to explain how the bird varied its speed. The investigators assumed that aspects of the kinematics that could not be measured, for example, changes in rotational movements of parts of the wings or in angles of attack, must be responsible for the velocity changes. To visualize the flow, fog particles were introduced into the tunnel. Behind the bird a vertical pulsating laser sheet parallel to the flow illuminated the fog particles (Spedding *et al.* 2003). The thin light sheet was positioned in successive experiments in three span-wise positions to capture the flow behind the wing tip, half way down the wing and behind the body, respectively. The time delay between the laser pulses varied between 100 and 500 μs depending on the flow speed. Successive images of the fog particles in a sheet were imaged digitally and analysed using digital particle image velocimetry (DPIV) (Stamhuis and Videler 1995; Stamhuis *et al.* 2002). Two successive images of the particles in the flow were required. The time delay between the images had to be short because the same particles had to appear in both images. In the second image the particles were displaced relative to their position in the first image if they were moved by the flow. By comparing the images, the displacement of the particles could be detected and showed the direction and the instantaneous velocity of groups of particles in a small area. The instantaneous velocity is the displacement distance divided by the time between the images. The field of view of the camera is therefore divided into a large number of equally small areas. A vector is calculated showing the average displacement over the time between two successive images and the average direction of the displacement in each area. The resulting vector field is a quantitative two-dimensional representation of the wake in which vortex structures can be detected and analysed.

The wake pattern behind the thrush nightingale in the Lund wind tunnel changed gradually with increasing speed. At 4 $m\,s^{-1}$ the downstroke resulted in a distinct vortex ring in the wake similar to that behind the pigeon and the jackdaw, again at a small angle from the horizontal. The upstroke also produced a vortex ring which increased in size with speed. At the highest speeds the undulating wing tip vortices become the more important features in the wake. These move up and down and spread and close continuously in steady flight. The discrete ring structures vanished and the wake resembled that of the kestrel in Speddings' earlier study shown in Fig. 4.6. The wake structure obviously depended on the flight speed but probably also on the flapping style. Rayner (1995) expects that birds with shorter wings use vortex ring gaits at all speeds and that longer winged birds change gait with speed. The latter category does not fold the wings so much during the upstroke in flights at intermediate and higher speeds, generating aerodynamic forces continuously. This would result in a continuous vortex trail as the one shown by the kestrel and the thrush nightingale at high speeds (a concertina vortex trail). At low speeds longer winged birds change their kinematics by drawing the wings up in feathered position close to the body followed by a wide spread-out beat during the downstroke. This type of slow flight would generate a wake consisting of distinct vortex rings.

Although the wake represents a true picture of the forces exerted by the bird, it does not show exactly how the wing movements generate the lifting and thrusting forces. We have to study the flow right at the wings to discover how birds really fly.

4.4 *The flow near a steadily gliding wing*

The only way to fully appreciate the interaction between a bird wing and air is to visualize the flow at the wing in free flight quantitatively. The experiments in the Lund wind tunnel show that the ideal experiment is within reach but it has not yet been done.

We applied DPIV to quantify flow phenomena but in water instead of air (Stamhuis and Videler 1995; Stamhuis *et al.* 2002). DPIV in water is technically less demanding mainly because the flow velocity can be slower at the same *Re* number and seeding the flow with neutrally buoyant particles is easier. Our water tunnel has a 50 cm long test section, 25 × 25 cm across. The re-circulating flow is made laminar by leading it through several straightening structures and can be varied in speed from 0 to 1 $\mathrm{m\,s^{-1}}$. The water is seeded using neutrally buoyant PVC particles, approximately 50 μm in diameter. Bird wing models or sections of bird wings tested under the same *Re* number in water as they would normally experience in air, will provide an accurate indication of the velocities and directions of the flow. These tests are as close to the real thing as we can get at this moment in time.

Using this approach we managed to show that the wings of gliding birds can keep the animal airborne by at least two flow patterns, both generating lift and drag: conventional attached flow and leading edge vortex (LEV) flow. To study the conventional attached flow we look in detail at the interaction between a section of a wing with a round leading edge and a sharp trailing edge. We saw in Chapter 2 that the arm wings of most birds show that kind of cross-sectional profile. Large oceanic birds have long arm wings and are adapted to use the conventional principle predominantly.

LEV flow can probably be generated by most hand wings because cross sections have a sharp leading edge (see Chapter 2), hand wings can easily be kept in swept back position and the angle of attack on the hand wings can be readily varied by most birds.

A model of a swift wing will be used to illustrate the use of the LEV by gliding birds. Another swift wing model, with a variable sweep-back angle, shows how a conventional flow pattern can change gradually into a LEV with increasing sweep-back angles and how both patterns can simultaneously be present on a wing with a moderately swept back hand wing.

4.4.1 The conventional flow around an arm wing

Fulmars are specialized gliders with straight narrow wings. The arm wings are relatively long. The cross-sectional shapes along the arm wings have a rounded leading edge and a sharp trailing edge. Bird arm wings are usually much more cambered than aircraft wings. Camber is most extreme close to the body and gradually decreases towards the wrist. The high camber is probably an adaptation to generate lift using conventional attached flow at low speeds.

We choose a wing sectional profile of a northern fulmar taken from a position near the end of the arm wing just proximal of the carpal joint. The chord length of the wing at that position was 12.5 cm. A 20 cm long transparent Perspex wing was made with a chord length of 9.3 cm. It had a uniform cross-sectional profile along its length, mimicking the chosen fulmar profile. The model was made transparent to be able to shine a laser light sheet straight through and to visualize the flow completely around the wing in a plane parallel to the flow near the middle of the model in the centre of the tunnel. The model is uniform over its entire span and we therefore do not expect changes in the flow in the span-wise direction across the channel. The *Re* number based on the chord length and a water flow velocity of 0.5 m s^{-1} was about 4.65×10^4. The equivalent air speed of the real wing would have been 5.6 m s^{-1} which is about 20 km h^{-1}. This is an extremely low flight speed. However, the results show that even at that speed the wing can create a lift generating flow pattern. The angle of attack, measured between a straight line drawn to touch the underside of the wing near the leading and trailing edge, and the horizontal is about 6°. Figure 4.7 shows the wing cross section in the vector field. The presence of the wing generates the pattern we see and produces the local changes in the velocity and the changes in the direction of the flow. The highest velocities are found above the wing just behind the round leading edge. From there backwards the flow nicely follows the curvature of the cross section downward. Below the wing the velocities

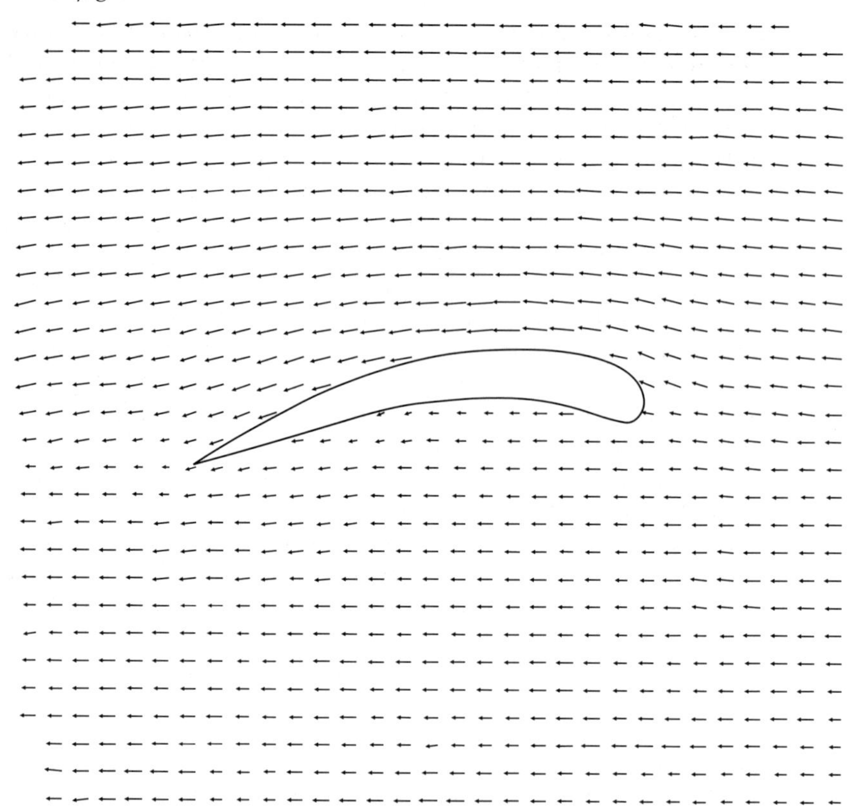

Fig. 4.7 Results of a pilot experiment showing the interaction between a Perspex model of a wing section taken from the arm wing of a northern fulmar and water flowing at 0.5 m s^{-1}. Neutrally buoyant particles were illuminated by a thin laser light sheet parallel to the flow in the centre of a 25 × 25 cm re-circulating water tunnel. Two successive digital images, 0.004 s apart, provided the direction and distance of displacement of the particles on which the equally distributed velocity vector diagram is based. Interpretation of the flow pattern is given in the text.

are reduced, most strongly in the area of camber. It can be clearly seen that the flow anticipates the presence of the wing by going up in front of it. This flow is termed the up-wash. Newton's laws are used to explain what the pattern tells us about the forces involved. In the free undisturbed flow in the water tunnel above and below the region affected by the wing the flow is straight from right to left and the velocity is uniform. Every change in the direction of the flow and every change in velocity requires a force exerted by the wing on the flow. There will be equal reaction forces exerted by the fluid on the wing in opposite direction. The up-wash in front of the wing will result in a downward force on the wing. Next we see how the upward directed flow returns to horizontal over the top of the wing. The change in direction results in an upward reactive force on the wing and that happens again over the rear part of the wing where the water goes downward. The flow follows the curved convex shape of the upper part of the wing because of viscous forces in the layer close to the surface. At the surface of the wing the flow velocity is zero and increases away from the surface to reach the free flow velocity. Under laminar conditions the thickness of that layer (which effectively is Prandtl's boundary layer explained in Chapter 1) is the chord length divided by the square root of the *Re* number (Lighthill 1990). Our flow vector field is too coarse to show the about 0.4 mm thin layer. The shear between the slow flowing water close to the surface and the faster water further away from the surface gives the flow the tendency to bend towards and stick to the surface of the wing (the Coanda effect, Anderson and Eberhardt 2001). We not only see directional changes but also velocity differences in the flow caused by the presence of the wing. The flow stagnates underneath the wing and accelerates over the curved top side where it reaches the highest values. The flow is pushed up by the rounded leading edge and if it would continue to move in that obliquely upward direction a void would develop behind the highest part. We saw why the fluid follows the surface and instead of creating a void it rushes down the rear part. This causes the fastest velocities above the highest part of the wing. So the flow is forced down and there will be an upward directed reaction force on the wing. The total lift effect is reflected by the velocity differences between the faster flow above the wing and the reduced flow speed below it. The velocity differences result in a pressure difference which sucks the wing in upward direction consistent with Bernoulli's law. This is not a separate force but the same reaction force because the different velocities are caused by the presence and the shape of the wing in interaction with the horizontally approaching fluid. Changes in flow direction, stagnation, and velocity changes are all part of the effect that the presence of the wing obstacle with its special shape has on the fluid.

The velocity pattern in Fig. 4.7 is relative to the static wing and the ground. To obtain an impression of the flow relative to the wing moving at an average velocity of 0.5 $m\,s^{-1}$, we must subtract that speed from every vector. If we do that, velocities below the wing will become negative and point forward in the direction of the leading edge of the wing. Above the wing the relative speed is positive because the vectors are larger than the average speed. The total pattern shows the circulation around the wing of the bound vortex (see Chapter 1). The counterclockwise circulation is consistent with the up-wash in front of the wing. Birds flying close to the surface of the earth, either over flat land or water, can make use of a phenomenon called ground effect if the distance to the surface is about half the span or less (Anderson and Eberhardt 2001). In our pilot experiment with the transparent fulmar wing section the bottom of the tunnel was about 12.5 cm underneath the 20 cm long wing section which is probably too far to benefit from this effect. But the question is what causes it? In ground effect the circulation around a wing is reduced due to the close presence of the static surface. This is reflected in a reduction in the up-wash flow. We saw that up-wash causes a depressing reaction force on the wing which adds to the weight that needs to be compensated by lift during level flight. A reduction of up-wash decreases the lift force required. Both lift and drag force can be reduced by decreasing the angle of attack of the wing which makes flight in ground effect easier.

The quantitative flow pattern around the cross section in Fig. 4.7 allows the calculation of lift and drag per unit span of that particular section. However, to know the forces on an entire bird wing we need to know the lift and drag characteristics of all cross-sectional profiles along the wing. The wing in our experiment differed slightly from a section of a wing of a bird because it was locked up between the walls of the tunnel and there is no free wing tip where pressure differences result in a wing tip vortex.

The force on a flying wing is proportional to the dynamic pressure $\frac{1}{2}\rho v^2$ ($\mathrm{N\,m^{-2}}$) (Chapter 1). This is a point force. Multiplied by the chord length it becomes proportional to the force per unit wing length. Note that it is still proportional and not equal. We need coefficients, one for drag and one for lift, to be able to calculate the forces exactly. That means that we are back at square 1 because the coefficients must be measured for every wing shape at each position along the wing and for every angle of attack. That is relatively easy for an aircraft wing with a uniform cross-sectional shape over much of the wing length, but not for bird wings where the shape changes drastically along the wing.

Increase of the angle of attack of a conventional wing can increase both lift and drag until a rather sudden change in the flow occurs where the fluid does not follow the curved upper part of the wing but separates from the wing surface at some distance behind the highest part of the cross section. At that instant the lift vanishes almost completely and the drag on the wing increases abruptly. The phenomenon is feared in conventional aircraft where it is known as 'stall'. As mentioned in Chapter 1, flow separation is not always detrimental. Delta wings can control it and use it to generate lift and drag.

4.4.2 Leading edge vortex flow

Gliding birds often can be observed to keep the hand wings in swept-back position. That is frequently the case just prior to landing. It gives the hand part of the wing the swept-back shape of a delta wing. We saw in Chapter 1 how delta wings generate LEVs. The sharp margin of the stiff narrow vane of the outermost primary feather is the leading edge of the hand wing (Chapter 2). We tested models of the wings of the common swift in our water tunnel to find out if LEVs are likely to develop on swept-back wings and tried to obtain an indication of the magnitude of the required sweep-back angle.

The common swift is one of the most extremely aerial bird species. It spends almost its entire life airborne, roosting, mating, and foraging in the air; it only lands to breed. Swifts migrate over thousands of kilometres twice a year between winter-feeding areas and breeding grounds (Bäckman and Alerstam 2002; Lack 1956). Swifts are capable of speeds of over 17 $\mathrm{m\,s^{-1}}$

Fig. 4.8 An adult common swift in gliding flight showing the torpedo-shaped body and the scythe-shaped wings with relatively short arm wings and long slender hand wings (picture J. F. Cornuet). Cross-sectional profiles of the arm and hand wing at the positions indicated are drawn. The inset is a SEM picture of the sharp leading edge of the hand wing (the white scale bar is 100 μm).The pointed barbs of the narrow anterior vane of primary X form a serrated sharp cutting edge (from Videler *et al.* 2004).

(61 $\mathrm{km\,h^{-1}}$) but usually fly slower (Bruderer and Boldt 2001). Average gliding speeds under conditions without wind vary between 8 and 14 $\mathrm{m\,s^{-1}}$ (Oehme 1968).

An adult swift has a streamlined body; a short forked tail (closed during fast flight) and long curved scythe-like wings (Fig. 4.8). The arm wing is very short, the skeleton of the arm being much shorter than that in songbirds of the same size. Only seven, partly overlapping, secondary feathers form the main surface area of the arm. Cross sections of the arm wing have a rounded leading edge and a sharp trailing edge (Fig. 4.8). The hand part of the wing, starting at the wrist, is particularly long occupying about 85% of the total wing length. Eleven rather stiff primary feathers form the hand wing. The most distally implanted one (P XI) is only 3 cm short and lacks proper vanes. The vanes of primaries P X to P I are well developed and the total feather length gradually decreases from 14.5 cm (P X) to 5 cm (P I). Microstructures (pennula) connect primaries VII to X tightly together when the wing is fully extended and under pressure during the downstroke and during glides. This turns the hand wing into a slightly cambered plane with a sharp leading edge and a sharp trailing edge (Fig. 4.8). The hand wing bends slightly downward towards the wing tip. The narrow outer vane of primary X (Fig. 4.8, inset) is the sharp serrated leading edge of the hand part of the wing. The average wing chord in the flight direction is approximately 5 cm.

Lift generating mechanisms of a physical model of a swift wing in fast gliding posture (60° sweep of the hand wing leading edge) were studied in our water tunnel. The flow patterns in water and in air are the same as long as the flow is studied at the same Reynolds (*Re*) number. An average speed of 11 $m s^{-1}$ (the mean measured value for free gliding swifts under conditions without wind) and a wing chord length of 5 cm were used to estimate the *Re* number at 3.75×10^4 in air of 20°C, at sea level. The model wing was scaled up 1.5 times to an average wing chord of 7.5 cm. To obtain the same *Re* number in water, the speed required was about 0.5 $m s^{-1}$.

The wing plan form of the physical model was cut out of brass plate (0.9 mm thick). The cambered arm wing and the wrist joint were thickened and shaped with epoxy resin; the hand wing was cambered and curved to approximate the shape of an actual wing. As in the real wing, the arm wing had a rounded and the hand wing a sharp leading edge.

The wing model was mounted stationary against the wall in the middle of the test section at an angle of attack of 5° with regard to the oncoming flow relative to the arm wing chord. This small angle of attack is already sufficient to produce stable LEVs at the hand wing. To study the flow patterns quantitatively, the wing was illuminated using a 3 cm thick laser light sheet perpendicular to the flow, in four planes successively. Digital video images of the displacement of the particles in the light sheet were taken at 25 frames per second from the rear through a peek window downstream of the wing. Again, the direction and the velocity of the particles in successive frames were analysed using DPIV (Videler *et al.* 2004).

The results shown in Fig. 4.9 reveal the presence of a prominent LEV on top of the hand wing. In this figure, the vortex centre is indicated as a dot, and the vortex core shape is depicted as a loop at the level of maximum vorticity. The LEV core diameter increases from the wrist (Fig. 4.9(a)) towards the wing tip (Fig. 4.9(c)), indicating that the LEV has a conical shape. The vortex centre is located above the wing and follows the wing just inward of the leading edge towards the tip. Two centimetre behind the wing (Fig. 4.9(d)), the centre is still in a position inward from the wing tip and the core diameter is still increasing. The maximum downwash flow velocity increases along the wing (with the LEV strength) and is twice as high at the wing tip (Fig. 4.9(c)–(d)) compared to the position just behind the arm wing (Fig. 4.9(a)). This shows that the lift increases along the wing backward. The maximum downward velocity component at the wing tip is about 10% of the free flow velocity. The speed of the rotation in the vortex increases with increased free flow velocities of the water tunnel, but the diameter of

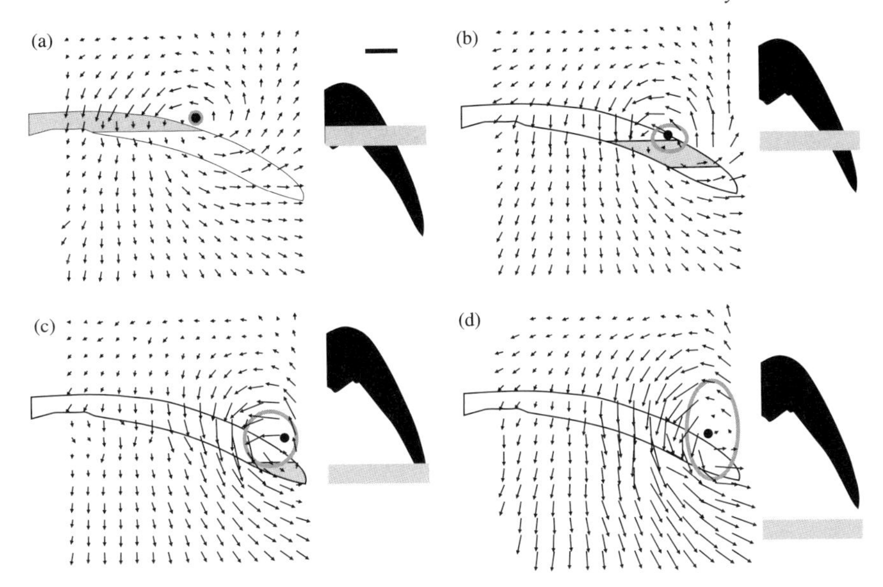

Fig. 4.9 Flow patterns created by a swift wing model in a water tunnel. Flow velocity vector maps based on particle displacements in four planes perpendicular to the flow filmed from the rear at three positions on the hand wing ((a), (b), and (c)) and at one position just behind the wing (d). The outline of the wing is indicated in each case. A 3 cm thick laser sheet was used for illumination of the particles. Grey areas on the wings in (a), (b), and (c) indicate the position of the laser sheet. The grey bar in each of the small pictures of the wing model in top view on the right gives an indication of the position and width of the laser sheet in top view. The scale bar (top right in (a)) is 5 cm and relates to all wing model drawings in top view. In all four panels the vectors are drawn at the same scale. The LEV centre is indicated with a black dot; the grey loops around that dot represent the core diameter of the vortex at the level of maximum vorticity in each velocity diagram. The LEV core diameter increases from wrist (a) towards the tip ((b)–(c)–(d)), indicating the conical shape of the LEV (from Videler *et al.* 2004).

the conical vortex does not change. An impression of the total flow on the wings of a gliding swift is given in Fig. 4.10. In the wake behind the bird the LEVs will generate two trailing vortices. These are not distinguishable from trailing vortices generated by the wing tip when there is conventional attached flow on the wings because both vortices have the same rotational sense. Studies of the flow in the wake behind a bird are therefore not suitable to discover the lift generating mechanism used by the wings.

LEVs represent a robust lift producing aerodynamic flow system producing lift over a wide range of angles of attack. Leading edge flow separation almost instantaneously creates the LEV and a resulting aerodynamic force normal to the wing chord. At high angles of attack the drag component of this force is large. Many birds use swept-back hand wings during gliding. The sweep-back angle may vary and is shown to be related to gliding speed

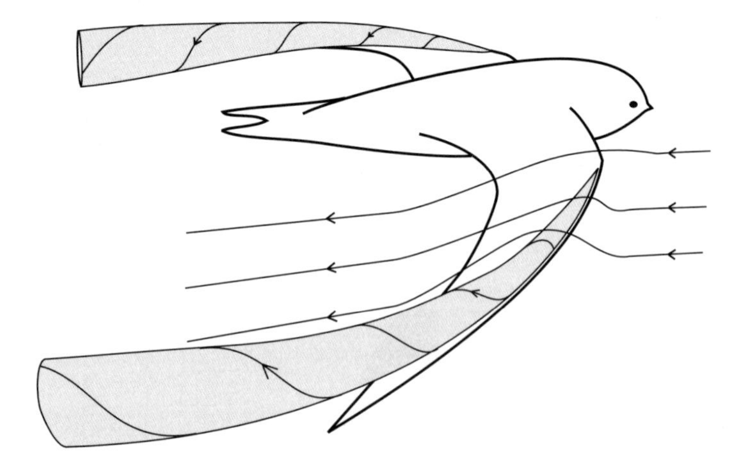

Fig. 4.10 Artist impression of the conical LEVs on the wings of a swift in gliding flight. The oncoming flow is deflected downwards by the attached LEV-system, showing the lift generating downwash. LEV separation starts at the wrists. From there the LEVs are attached over most of the wing length but start to go upward and inward approaching the wing tip and behind it (from Videler *et al.* 2004).

(Pennycuick 1968). Landing in most birds requires high lift and high drag at low speeds; LEVs on swept-back hand wings kept at high angles of attack provide these forces and make, for example, landing on a branch possible. Birds use the high lift to keep the right altitude, and use the high drag to brake during the approach glide.

The arm wing and hand wing in birds play different roles in flight. Arm wings use the conventional aerodynamic principle and hand wings induce LEV flow to generate lift.

4.4.3 The influence of the sweep-back angle on the flow at the wing

The air flow reaches a swept wing of an aircraft or bird in the flight direction at the flight speed. The sweep angle determines the magnitude of the normal and span-wise components of the air speed relative to the wing. The normal component perpendicular to the leading edge is equal to the flight speed times the cosine of the sweep angle. The span-wise component, parallel to the leading edge of the wing, is equal to the flight speed times the sine of the sweep angle. The span-wise component does not contribute to lift generation by attached flow but in case of a LEV it would move the rotating air in the direction of the wing tip. It would keep the LEV diameter small as long as it is on top of the wing and would enhance the shedding process at the wing tip.

To illustrate the effect of sweep on flow patterns on a bird wing, a swift wing model was investigated at low steady speeds in our re-circulating water tunnel. The model was made of a brass plate using the dimensions taken of

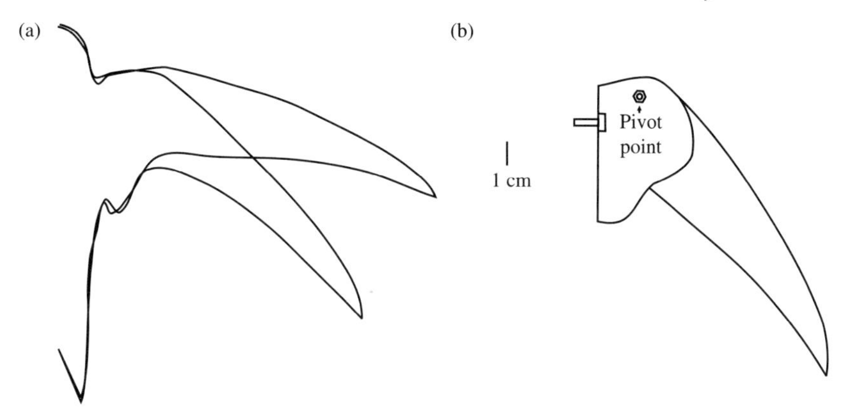

Fig. 4.11 (a) Sweep-back angles of the hand wings used by gliding swifts based on photographs taken in the field. (b) A brass real sized swift wing model with an adjustable hand wing. The model is used to test the effect of the sweep-back angle on the presence of a LEV on the wing.

a dead swift. The short arm part was slightly cambered following the centre of the natural cross-sectional profile of the real wing. The sweep of the hand wing was adjustable and could copy the sweep angles taken from pictures of gliding swifts (Fig. 4.11(a)). The model (Fig. 4.11(b)) was mounted against the wall in the centre of the measuring section of our water tunnel. The flow patterns in a plane formed by a 2 cm thick laser sheet perpendicular to the flow a few centimetres behind the model was visualized using DPIV. The displacement of neutrally buoyant particles in time was filmed from the rear using a frame rate of 25 Hz. The sweep-back angles tested were approximately 10°, 20°, 40°, 50°, and 70°. The geometric angle of attack of the arm wing was 12°. The velocity of the flow was kept constant at 0.2 $\mathrm{m\,s^{-1}}$ (the equivalent airspeed would be 2.8 $\mathrm{m\,s^{-1}}$). The *Re* number based on a chord length of 3.5 cm was in the order of 7×10^3.

Figure 4.12(a) shows that the flow just behind the straight wing shows a pattern consistent with attached conventional flow along the wing. There is down-wash and vorticity is shed along the entire wing length at a constant rate. A wing tip vortex is clearly present at the wing tip. The sweep-back angle of 20° (Fig. 4.12(b)) does not alter the general pattern of the flow; we see again down-wash, vorticity shed along the wing and a clear vortex at the tip. The velocity and vorticity maps are different at 40° (Fig. 4.12(c)). An ellipsoid vortex is present near the distal end of the wing. Its position is proximal from the wing tip. This is a transitional situation where the wing tip vortex merges with a newly formed LEV. At 50° (Fig. 4.12(d)) the LEV is clearly present proximal of the wing tip. In Fig. 4.12(e) with a sweep-back angle of 70°, the wing generates a fully developed LEV which contains virtually all vorticity. Its position is above the wing and away from the wing tip in the direction of the body.

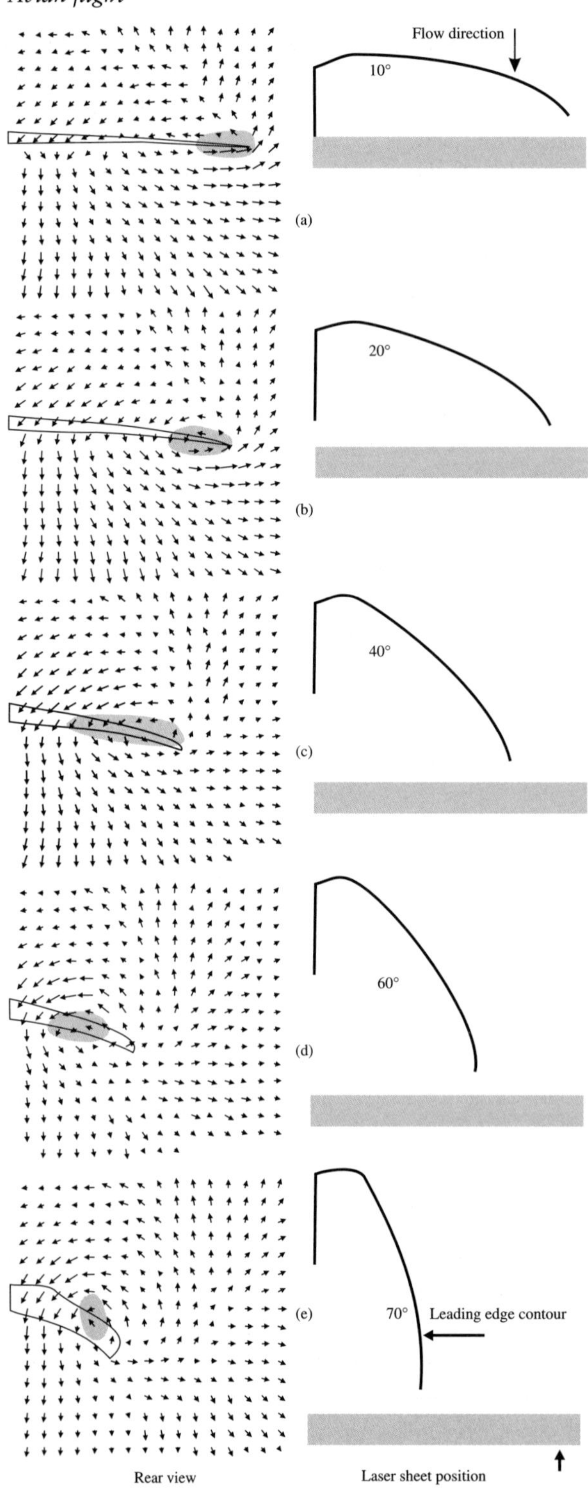

Fig. 4.12 Flow velocities in vector maps and the vortices on a brass swift wing model with different sweep-back angles, tested in a water tunnel. The pictures (a) to (e) on the left show the velocity vector field in a laser sheet perpendicular to the flow as seen from the rear (the vertical components of the water velocity are shown). The wing contour and the area with maximum left turning vorticity (grey area) are indicated. Diagrams on the right indicate the contour of the leading edge, the sweep-back angle, and the position and width of the laser sheet (grey bars) for each angle in (a) to (e).

This pilot experiment shows that a conventional flow pattern is found around the straight wing: the flow remains attached and a vortex emerges at the wing tip. Increased wing-sweep angles induce a leading edge vortex in the swift wing model even at low speeds and at low *Re* numbers. Bird flight watching shows that most birds keep the sharp leading edged hand wings in swept-back positions during a variety of flight manoeuvres. In particular during landing, high lift and drag forces are probably generated even at low speeds with a LEV on the hand wing.

4.4.4 Possible functions of the alula

The alula is situated at a position along the wing between the arm wing and the hand wing in most birds (Chapter 2). The conventional interpretation of its function is that of an extended leading edge slat in aircraft (Handley Page slat). This device delays flow separation at the leading edge and hence stall under high angles of attack. In aeroplanes it is mounted on top of the leading edge over a large part of the wing length. It is a plate with a curved cross section following the leading edge profile of a conventional wing. It extends over its entire length parallel to the wing. Nachtigall and Kempf (1971) used smoke to visualize the flow, in an attempt to demonstrate that the alula indeed delayed the break away of the flow over the upper part of the wings under angles of attack up to 50°. However, the evidence provided in the article is not conclusive because smoke trails on photographs are difficult to interpret in three-dimensions. The leading edge slat function is possible but the limited length of the alula reduces this effect to a very small part of the wing. Unlike the Handley Page slat, the alula is attached on one side and extends obliquely upward from the leading edge of the wing. Delaying stall is most probably not the main function of the alula.

The position of the alula at the end of the arm part of the wing suggests that extension could help to induce a leading edge vortex over the hand part of the wing (Jim Usherwood and Eize Stamhuis, personal communication). Alternatively, its function could be similar to that of the devices used in aircraft design to produce a small vortex at positions more than half way down swept wings. These devices are small fence-like surfaces mounted at the leading edge or a saw-tooth in the leading edge (Chapter 5, Fig. 5.5 and Barnard and Philpott 1997). The small vortex separates the attached flow system on the arm part of the wing and the leading edge vortex over the hand part and is shown to stabilize the position of the start of the LEV. The reader should be aware that there is no real proof yet for the ideas regarding the function of the alula or for the hypothesis that the hand part of the bird wing generates lift as a delta wing. However, the proof for the conventional story, even in relation to the arm part of the wings of birds, is also very thin.

We still do not know in detail how birds glide let alone how they produce lift and thrust in flapping flight. Dynamic changes in wing shape can drastically change the interaction between bird and air and we are very ignorant about the aerodynamic consequences of these changes.

4.5 *Aerodynamics of flapping flight*

4.5.1 Conventional flow

Stable conventional lift generating circulation around a static wing with a rounded leading edge (Fig. 4.7) takes time to develop after air flow above a minimum speed has started to interact with the wing under a small angle of attack (Chapter 1, Wagner effect). Can we expect built-up of circulation on a wing, which is not static with respect to the flow but flapping? Aircraft builders know that vertical movements may either seriously diminish or enhance the lift generating capacity of the wings. They use the 'Reduced Frequency' parameter which is the ratio of the amount of vertical movement over the forward velocity of a wing. It must be less than 0.5 to allow steady state attached lifting flow to develop. The arm wings of most birds will probably operate within that limit but it is less likely that this is also the case for the hand wing.

A quasi-steady or blade element approach has been used to calculate lift and thrust generated by flapping bird wings assuming conventional attached flow. To apply the method a wing is sectioned into narrow strips. For each strip the relative flow velocities and the angles of attack during a wing beat cycle are determined. The aerodynamic force per unit strip length on each of the wing elements is resolved into a lift component normal to the relative flow and a drag component in the direction of the flow. The instantaneous lift and drag forces per unit wing length can now be calculated if we know the lift and drag coefficients of each strip. These depend on the aerodynamic properties of the wing section and must be measured empirically for each section under all the occurring angles of attack during a wing beat cycle or can be approximated by comparisons with properties of documented aircraft wing profiles (there are huge databases on the world-wide-web). To determine the direction of the airflow with respect to each of the wing sections during one wing beat cycle is a tremendous task requiring three-dimensional analysis of the wing movements of a bird flying in still air. Oehme (1963) used three sections per wing for European starling and Eurasian blackbird and approximated the angles of incidence from kinematic analysis of free flying birds. Hummel and Möllenstädt (1977) did such a quasi-steady analysis at one instant just before the middle of the downstroke of a house sparrow which was filmed and analysed by Bilo (1971). The Reduced Frequency of 0.23 of the house sparrow was much lower than the upper limit of 0.5. Bilo (1980) however questions the quasi-steady approach despite this low figure. He argued that rotational and twisting oscillations with frequencies twelve times higher than the flapping frequency of about 22 Hz were superimposed on the basic movement of the downward striking wing. Distal parts of the wing vibrated at 260 Hz during the downstroke. The hand wings of the house sparrow oscillated at much higher velocities than the arm wing. Furthermore Bilo found that the angle of attack of the distal part of the wing decreased at a rate of 3600° s^{-1} shortly before the middle of the downstroke and increased immediately thereafter with 5700° s^{-1}. These arguments strongly suggest that steady state aerodynamics cannot explain the flight of smaller birds and that we should be looking for other means of lift generation. The blade element method could give reasonable results for large birds flapping at moderate to slow frequencies while keeping the wings fairly stretched even during the upstroke. It is surprising that (to my knowledge) nobody has done such a study yet.

4.5.2 LEVs on flapping bird wings?

In birds the hand wings are stretched during the downstroke and swept-back during the upstroke which leaves room for speculation about the development of leading edge vorticity and hence the generation of lift by the flapping hand wing. Charlie Ellington's group in Cambridge discovered a leading edge vortex that remained in position above the wing during the downstroke of a large-scale model hawk moth (Berg, van den and Ellington 1997, Ellington *et al.* 1996). Size and shape of the robot wings were similar to the hand wings of many species of birds. During the downstroke steady span-wise flow from the wing hinge to about three-quarters of the distance to the wing tip was observed. This flow might be generated by centrifugal forces. Due to the span-wise flow the LEV is continuously renewed while it spirals towards the point where it leaves the wing. The constant renewal keeps the LEV small and attached to the wing close to the leading edge. Ellington pointed out that in delta wing aircraft such as the Concorde the stable axial vortex flow along the leading edge is due to the back-sweep of the wings. The same explanation could be applicable to the LEV we found on the swift wing models. Stable LEVs have been detected on the flapping wings of free flying butterflies by Srygley and Thomas (2002), on the flapping wings of a fruit fly robot (Dickinson *et al.* 1999; Sane 2003) and on dragon fly wings (Thomas 2003).

LEVs may be present during flapping flight on the hand wings in birds, analogous to what has been found in insect flight and we have experimental evidence in support.

Other possibilities to generate lift have been described to generate wing bound circulation and lift in hovering insects, for example, clap, fling, and peel mechanisms (Ellington 1984; Maxworthy 1979; Usherwood and Ellington 2002*a*,*b*; Weis-Fogh 1973). It would not come as a great surprise if novel unsteady mechanisms were detected in birds. It is of course highly speculative but Bilo's observation of the high frequency oscillations during the downstroke of the house sparrow might turn out to be one.

The overall conclusion must be that we do not know exactly how birds fly.

4.6 *Tail aerodynamics*

Adrian Thomas devoted his PhD thesis, defended in Lund in 1995, to the study of the aerodynamic function of bird tails (Thomas 1995). His thesis was mainly used to construct the following overview of current ideas about form and function of tails in flight. The position of the tail plane in birds is behind the centre of lift and the centre of gravity. Forces on the tail will generate rotational moments around these points. The important forces involved are lift, drag and steering forces to the left and right. Tails have several degrees of freedom to move. They can be tilted upwards and downwards, spread and folded, and rotated to the left and right. Bird tails closely resemble delta wings in many cases and aerodynamic theories used for that type of wings most likely apply also to tails especially when spread out to some extend. The sharp leading edges formed by the stiff narrow outer vanes of the marginal feathers, will operate in the same way as suggested for the hand wings of most birds. Even at small angles of attack the air will separate and form a conical leading edge spiral vortex over the edges of the tail, one on the right and one on the left side. These vortices are stretched out areas of low pressure and this will create upward force aided by increased pressure below the tail. The longer the leading edge of the tail is, the higher the lift that can be generated by it. Drag consists of several parts: induced drag caused by the generation of leading edge vortex lift, profile drag which is proportional to the surface area opposing the oncoming air and drag due to surface friction in the boundary layer. Long stiff tails and probably also elongated central feathers will contribute to aerodynamic stability using controlled drag forces. Deeply forked tails provide high lift and low drag. Some highly agile birds (i.e., barn swallows and frigate birds) combine high aspect ratio wings with a low wing loading with large deeply forked tails. Many sea birds on the other hand have large wings and small stubby tails. These only aid to increase manoeuvrability at high speeds and will hardly contribute to the generation of lift. Sea birds probably can afford reduced manoeuvrability because they live in an environment free of obstacles. Woodland birds require sheer turning power to avoid collisions. They need a tail, which provides dynamic stability without the risk of damage. Therefore, tails of woodland birds are usually less extremely forked than those of open country birds. Eurasian sparrow hawks and the long-tailed hawk have long straight tails and share the habit of fast hunting in woodland; the latter species does that in African forest canopies. Barn swallows seem to use the variable span of their deeply forked tails to control the tightness of the turns. The largest turning radii are flown with the tail closed, tight turns with widely spread outer feathers. We must keep in mind that proof for form-function relationships of birds tails is usually circumstantial. Experimental research is urgently needed here.

4.7 *Summary and conclusions*

Our knowledge about the aerodynamics of bird flight is far from established. We are only able to estimate the parameters involved and their relative importance very crudely. Birds generate lift and thrust by accelerating masses of air downward and backward respectively. The kinetic energies per unit time given to the air are equal to the powers involved. The lift and thrust powers can be roughly estimated from the mass fluxes and the final velocities given to the air accelerated by the bird. The power required to generate sufficient lift to keep the weight up is inversely proportional to speed. Thrust power required to overcome drag forces increases proportional to the speed cubed. From this we expect a U-shaped relationship between flight power and speed. If that is the case, there will be a minimum power speed at which the total amount of energy required reaches a minimum and a somewhat higher maximum range speed where the amount of work per unit distance is minimal. It is not at all clear if the U-shaped curve actually exists in birds.

The interaction between the bird and the air is crucial in our understanding of the physics of the process. The wake behind a flying bird summarizes what happened during that interaction. Quantitative and qualitative studies, including particle image velocimetry, revealed that the wake may consist of vortex trails and rings. Depending on the flapping style, there are gradual changes with increasing speed. The wakes change from series of single vortex rings from the downstrokes via rings generated during both up and downstrokes, to continuous concertina vortex trails.

Particle image velocimetry of the flow around wing models in a water tunnel is used at the same *Re* numbers as in air to reveal the precise nature of the interaction between wings and fluid. Gliding is the simplest case. The flow around a conventional wing with a rounded leading edge and a sharp trailing edge shows how the presence of the wing changes the direction and velocity of the fluid locally. The reaction forces from the fluid to the obstructing wing have a large upward lift component and a small horizontal (drag) component if the cross section of the wing is well designed.

Bird hand wings have sharp leading edges and are often kept in swept-back position during gliding. A swift wing model is used as an example to show that hand wings can induce a LEV above the wing even under small angles of attack and at low speeds. The positive pressure below and the negative pressure in the LEV above the wing produce lift and a certain amount of drag. Another swift wing model shows how conventional attached flow is gradually replaced by LEV flow with increasing sweep angles.

Possible functions of the bastard wing or alula are discussed to reach the conclusion that we are still very ignorant about the function of this device which is abundantly present among flying birds.

The aerodynamics of flapping flight is even more enigmatic. Is it possible that attached flow develops on a flapping wing or does the Wagner effect preclude that? In large birds the arm wing probably has attached flow during flapping. Bird hand wings might use LEV to generate lift while beating the wings just as some insects do. Other lift generating mechanisms known from insects have not yet been investigated in birds.

There are many different tail shapes but virtually all could operate as a delta wing when spread. LEVs are probably the main aerodynamic feature of tails.

Our understanding of the flapping flight aerodynamics in birds is still in its infancy.

5 *Evolution of bird flight*

5.1 *Introduction*

Palaeontology is a difficult field of research. The ultimate goal is to discover the course of evolutionary processes using fossil evidence. This evidence is, by its nature, always circumstantial because experiments are impossible. This makes it difficult if not impossible to prove the hypotheses. That is probably the reason why there is usually a lot of controversy. Doctrines and opposing theories are defended more vigorously than in other disciplines. There is a strong link with evolutionary biology, another area that is rather inclined to dogmas. For example, the idea that the evolution of new species could only occur in areas separated by geographic barriers from that of the ancestors (allopatric or geographic speciation) was strongly advocated by Mayr (1963). It dominated the thinking of the second half of the twentieth century and still has many defenders, despite the increasing evidence (especially from studies in the aquatic environment) that sympatric evolution has been the common process. Geographic barriers are difficult to imagine for birds migrating around the globe. Sympatric speciation can be expected to be the ordinary pathway of evolution in birds with probably a few exceptions.

Debates about the evolution of birds and their typical form of aerial locomotion became fierce some 150 years ago when a few fossils of one single ancestor of modern birds, *Archaeopteryx lithographica* were discovered, and never stopped since. Discussions concentrated on the ancestral relationship of *Archaeopteryx*, on how it evolved and on the question of whether or not it was able to fly. *Archaeopteryx* still plays a key role in our understanding of the evolution of birds despite the recent discovery of many bird-like fossils. That is probably because these are either non-flying feathered dinosaurs or already possess the necessary key features for flapping flight shown by modern birds (Witmer 2002). The debate around the evolutionary mechanics of bird flight still concentrates on *Archaeopteryx* and therefore the main part of this chapter will be devoted to ideas that have been developed over the twentieth century and to a personal alternative approach.

During all those years, two opposing bastions existed regarding the origin and evolution of bird flight. One group of experts defends the hypothesis that bird flight started with tree-climbing animals that developed wings to glide down to the lower part of the trunk of neighbouring trees (the arboreal theory). Other specialists believe that flight started with a bipedal running animal using its arms to catch insects. Turning the arms into wings made it possible to make larger and larger jumps (the cursorial theory). The reasoning very often contains a high degree of teleological thinking where characters of species are treated as steps on the road to a proper flight apparatus owned by recent flying birds. Also, hypothetical scenarios not always try to obey the law of parsimony restricting the assumptions to the minimum needed to explain the facts. Since every species past and present is successfully adapted to its ecological niche, it is important to study form and function in the ecological context to understand the details of the adaptation. *Archaeopteryx*, for example, should not be studied as a stage in the evolution of bird flight but as an animal adapted to a certain environment in which it moved about, foraged, and reproduced.

My controversial hypothesis assumes that *Archaeopteryx* was a marine animal foraging on large mudflats along the Tethys Ocean. Its long legs and wings enabled it to run and slide over water and soft mud. The biomechanical details of this behaviour will be explained demonstrating the conditions *Archaeopteryx* had to fulfil to be able to use the same technique as the Central American Basilisk lizards. Subsequently, a scenario will be described around this special form of locomotion suggesting how it fitted in the presumed biology of *Archaeopteryx*.

During the past decades remains of a large number of other Mesozoic bird-like animals have been unearthed. Some of these are less and others more advanced than *Archaeopteryx*. Contrary to the expectations, most new fossils did not make interpretation of the evolutionary pathways easier. Recent articles and books on the subject of evolution of birds give a choice of cladograms based on various phylogenetic hypotheses. The problem is that virtually every newly discovered fossil represents a completely new branch of these trees. It is beyond the scope of this book to join the palaeo-taxonomical discussion. It therefore focuses on form and function in relation to the beginning of flight and refers the reader who is interested in this aspect to the cited literature including recent books on the subject: Chatterjee (1997); Chiappe and Witmer (2002); Dingus and Rowe (1998); Feduccia (1999); Gauthier and Gall (2001); Hou (2001); Paul (2002); and Shipman (1998). Mesozoic bird fossils have now been discovered in various parts of the world but China has the richest deposits. These fossils are at least several tens of millions of years younger than *Archaeopteryx* and witness the prolific radiation of Mesozoic birds between 150 and 65 million years ago. Similar structural adaptations to improve flight performance are found in several unrelated groups. Anatomical details make it possible to classify the Mesozoic bird fossils into a few large groups. Unfortunately these groups did not survive the mass extinction event 65 million years ago. A few small taxa with good flyers were hardly represented in the fossil records from before the disaster. These survived, taking the art of flight into the Tertiary to become the ancestors of modern bird lineages.

5.2 Archaeopteryx

Ten fossils of skeletal remains and one loose feather of *Archaeopteryx* were found so far in a 150-million-year-old (late Jurassic) marine sediments near Solnhofen in Germany. The Solnhofen environment was a tropical coral reef setting of lagoons speckled with low islands formed by coral and algal-sponge reefs. The limestone deposits are marine back-reef sediments of skeletal remains of mainly minute calcareous phytoplankton protists, for example, Coccolithophorids. The reefs were bordering the circum-global Tethys Sea. The temperature must have been in the order of 30° (±5°) Celsius. Blue-green algae and bacteria dominated the bottoms of the lagoons, with sparse evidence of epi- and endo-macrobenthos. The larger fossils, usually of nektonic nature, include evidence indicating mass mortalities of fish. The climate must have been dominated by monsoons with dry and rainy seasons. Easterly trade winds caused up-welling and plankton blooms in the summer. The land was flat and low. Evidence for the existence of cliffs is weak and disputed. Vegetation was sparse and dominated by seed-ferns, conifers, stem succulents, and mangrove type halophytes. There were no trees. Bushes with a maximum height of 3 m formed the highest vegetation. Fossils of land animals include small dinosaurs, pterosaurs, and large insects. Fossilization in the marine sediment occurred often fast with little signs of a long decomposition phase. The environment probably resembled the present situation in the Gulf of Cariaco (Venezuela) (sources: Buisonjé 1985; Viohl 1985).

The Solnhofen sediments produce high quality sandstone, which has been quarried since Roman times. The fine grain of the stone slabs is highly valued for lithography. This printing technique was at the peak of its use in the middle of the nineteenth century. Fossils in this type of marine sediment are preserved in very fine detail. The first skeleton of *Archaeopteryx* was discovered in the lithographic sandstone in 1855, but the fossil was interpreted as the remnants of a small pterosaur, *Pterodactylus crassipes*. The specimen was obtained by the Teylers museum in Haarlem (the Netherlands) where the American palaeontologist John Ostrom recognized it as a specimen of *Archaeopteryx* in 1970. The discovery of a feather in 1860 helped with the interpretation of the other fossils as bird-like creatures. The most beautiful one to date is the third or Berlin specimen (named after the city where it is kept), which was found in 1876 (Fig. 5.1). Several specimens where the imprints of the feathers were not obvious were at first categorized as theropods, small bipedal carnivorous dinosaurs. It was not a big mistake to misinterpret the fossils of *Archaeopteryx* for a theropod, because series of features indicate a close relationship with this group of lizard-hipped dinosaurs. Applying commonly used methods of comparative anatomy on the skeleton would classify *Archaeopteryx* as a (feathered) dinosaur. Thomas H. Huxley (1825–1895) emphasized the striking resemblance with the small bipedal dinosaur *Compsognathus* which was found in the same limestone deposits. Reptilian features of *Archaeopteryx* are most obvious in the skeleton of the head and in that of the rear part. The skull has dinosaur characteristics but it contained a relatively large brain. There is no bird-like beak but bony jaws with simple conical teeth. The vertebral column, the ribs and belly ribs (gastralia), and the tail skeleton are theropod features. The pelvic girdle has the characteristic theropod long pubis.

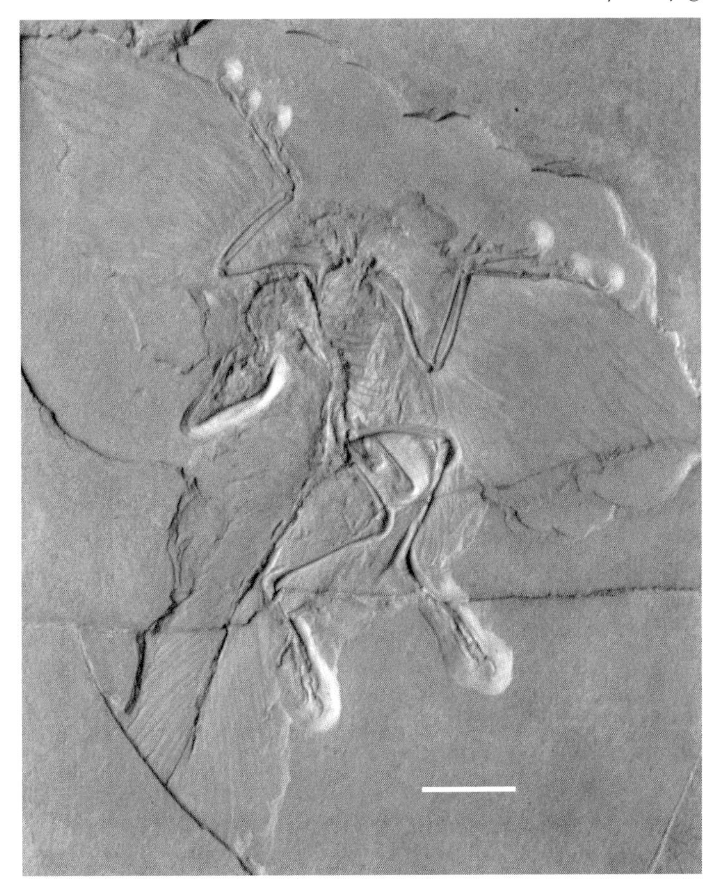

Fig. 5.1 The Berlin specimen of *Archaeopteryx*. The scale bar is 5 cm.

It was probably directed more backwards instead of vertically as is common in bipedal dinosaurs. The legs are long and strongly built. The feet have three long forward pointing toes and a short fourth one, the hallux, attached fairly high up the metatarsus. There are sharp nails on each of the last digits of the toes. The pectoral girdle consists of a wishbone (furcula, the fused clavicules), clearly present in the London specimen (found in 1861), shoulder blades (scapulae), and coracoids. A sternum was absent or cartilaginous in all but the fossil discovered in 1992 which is now kept in Munich. In the beginning of the twentieth century there was a great deal of emphasis on the presence of the wishbone. Critics of the relationship between dinosaurs and birds pointed at the absence of such a wishbone in theropods. However, furculae recently found in the fossil remains of bipedal dinosaurs such as *Velociraptor* and *Oviraptor*, revealed that the wishbone existed long before *Archaeopteryx*. The same is true for the presence of hollow bones. All theropods, including the largest such as *Tyranosaurus rex*, have them, indicating that this feature is not an exclusive property of animals attempting to fly. The glenoid fossa for the attachment of the humerus at the shoulder joint of *Archaeopteryx* seems to be directed laterally as in modern birds; scapula and coracoid meet at the joint but there is no triosseal canal. It is not at all clear whether or not the tendon of a supracoracoid muscle (the upstroke muscle in modern birds) passed the shoulder to insert on the upper part of the humerus. There is a prominent elongated deltopectoral crest on the anterior side of the humerus in all fossils hinting at strong pectoral muscles despite the apparent lack of an equally strong origin for this muscle. Radius and ulna are long and slender and do closely resemble the modern avian configuration, probably including the mechanical linkage system to flex and extend the entire wing (Chapter 2). *Archaeopteryx* could fold the hand part of the wings at least partly backwards along the underarm because the wrist contains two carpal bones that make this articulation possible. Vazquez (1992) points at slight differences between the wrist of *Archaeopteryx* and that of modern birds and argues that the *Archaeopteryx* condition does not allow full retraction of the hand part of the wing. The hand part occupies 40% of the length of the wing and is supported by three clawed fingers. The metacarpals seem to be un-fused. The finger claws in all fossil remains are remarkably sharp and pointed without signs of wear whereas the toe claws are worn. The best preserved claws are those of the Teylers museum (Haarlem) specimen. Figure 5.2(a) shows a worn toe claw and Fig. 5.2(b) a typical finger claw. There is a remarkable difference between the last joints of the finger and the toe bones. The phalanges of the finger nails have two sockets where the distal condyle of the penultimate phalanx could fit, the equivalent joints of the foot claws show only a single socket (Fig. 5.2(c)).

The primary feathers are one of the most striking features of the 150-million-year-old remains of *Archaeopteryx* because they stunningly resemble the primaries of modern birds consisting of a shaft with asymmetrical vanes. The vanes are made up of rows of barbs, each with numerous barbules probably attached to each other with hooks and grooves as in extant feathers (Griffiths 1996). The primary function of pennaceous porous feathers could have been water repellency. A porous surface repels water more strongly than a solid surface of the same material (Dyck 1985). Later the remiges probably became virtually wind tight and could form a functional wing. How did these structures evolve and what was the functional context in ancestors that were not capable of even the simplest form of flight?

5.3 *Landing after an arboreal or cursorial start?*

Since the end of the nineteenth century two conflicting scenarios, each with *Archaeopteryx* as the leading hero, dominated thinking about the origin of flight in birds. The arboreal theory predicts that flight started by gliding with

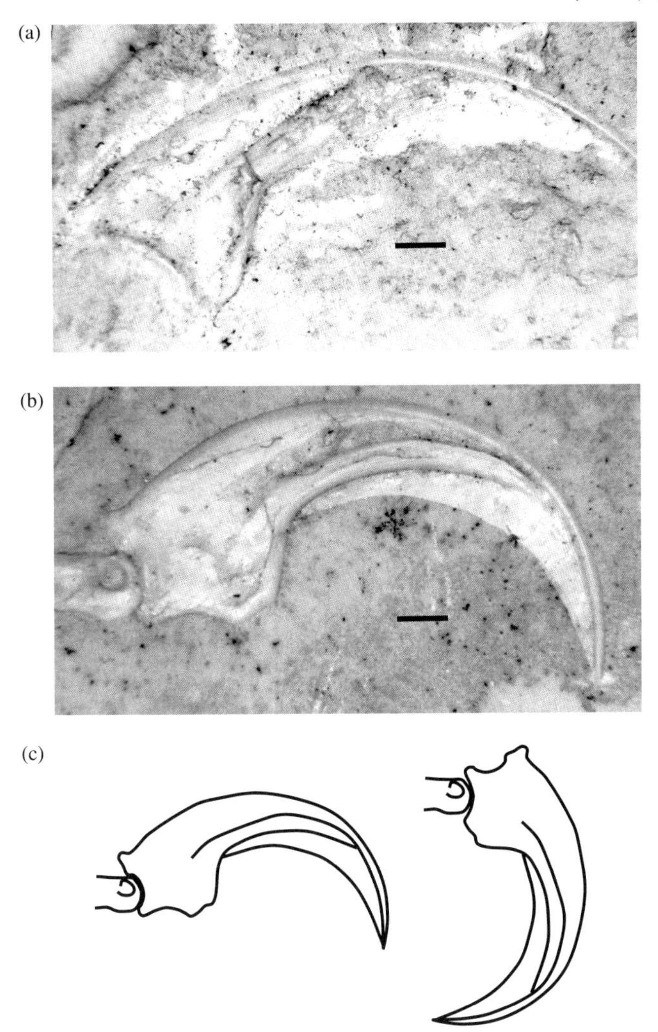

Fig. 5.2 Claws of the Teylers Museum specimen of *Archaeopteryx* (scale bars are 1 mm): (a) Claw on the last foot phalange. (b) Ultimate and penultimate finger phalanges and unworn claw. (c) Drawings of the outlines of the horny sheath and last phalange in (b). Left, the original position of the penultimate phalange with respect to the joint. Right, the distal condyle of the penultimate phalange positioned in the alternative socket of the bistable joint.

the use of wings from the height of trees down to the forest floor (a recent outline in favour of this hypothesis can be found in Feduccia 1999). The cursorial one depicts how fast running bipedal dinosaurs developed flapping wings for stabilization, gradually offering the possibility to take-off from the ground (Padian and Chiappe 1998, reviewed both hypotheses advocating the last one).

The arboreal approach required at least two skills: climbing trees and gliding. The main arguments for and against are:

Pro-arboreal:

- The claws on the hand and feet of *Archaeopteryx* have the typical design of the nails of tree climbing and perching birds (Feduccia 1993).
- Take-off, one of the most difficult parts of flight, becomes easy when performed from the height of a tall tree (Rayner 1985*a*).

Con-arboreal:

- The same shape of claws occurs in ground dwelling non-climbing birds and theropod dinosaurs (Peters and Görgner 1992).
- The manual claws lack evidence of wear at the sharp points (Peters and Görgner 1992; Wellnhofer 1985).
- The hallux of *Archaeopteryx* is too short to make the foot suitable for perching (Wellnhofer 1993).
- Aerodynamically the wings are not optimally designed to glide (Rayner 1985*b*).
- A soft and safe landing requires backward rotation (supination) of the wings using them as airbrakes. In modern birds this movement is powered by the supracoracoid muscle originating on the sternum and inserting on the top of the humerus via the triosseal canal (Poore *et al.* 1997). There is no trace of evidence that *Archaeopteryx* could use this type of equipment. To quote Balda *et al.* (1985): 'An animal the size of *Archaeopteryx* would need considerable morphological change to enter the gliding niche at sublethal speeds.'
- There were no trees to climb in the Solnhofen area nor in the wide neighbourhood since no fossil tree pollen were found in the limestone deposits. The maximum sized plants were multi-branched bushes, about 3 m high (Buisonjé 1985; Viohl 1985).

The cursorial hypothesis gained a lot of attention especially after Ostrom's (1979) ideas about the insect catching capacity of feathered hands of ancestors of *Archaeopteryx*. Arguments in favour are:

- Functional morphological features shown by the fossils indicate that *Archaeopteryx* was an agile bipedal runner.
- Independent locomotor activities of the fore- and hind-limbs already evolved in the theropod ancestors.
- Lift generating fore-limbs could have increased stability of fast running bipedal dinosaurs even if these were not yet suitable for flight (Balda *et al.* 1985).

In opposition, Rayner (1985*b*) argues that the ground speed required for an animal of the size of *Archaeopteryx* would have to be in the order of 6–7 $m\,s^{-1}$ which is about three times faster than the highest estimated running speeds of extant lizards and birds. Another argument against this hypothesis could be the following: The fitness increasing reason to run fast in order to fly could be to escape from predators or to catch fast flying insects. In both cases take-off would be disastrous because speed would reduce immediately due to drag not balanced by thrust. Quoting Rayner (1985*a*): 'A major limitation of the cursorial model is that no clear ecological advantage has been assigned to the running-leaping-gliding strategy: it is unlikely to be encouraged by escape from predators since direct running is faster, while no contemporary cursor travels at such high speeds while commuting for foraging sites.' Several scholars in the field have proposed variations of the two hypotheses. Thulborn and Hamley (1985) compared *Archaeopteryx* with herons suggesting that the wings could have been used during foraging in shallow water to shade the surface to improve submerged prey detection. Spread out wings also could have carried the animal from wave crest to wave crest extending the animal's feeding range.

The arborial and cursorial hypotheses do not use the knowledge about the environment in which the animals must have lived and died. The vast collection of contemporary fossils offers a unique opportunity to obtain insight into *Archaeopteryx* as part of an ecosystem. Scenarios should be based on facts and should follow the law of parsimony. All the *Archaeopteryx* remains are embedded in marine sediment and the most parsimonious explanation is that they are found on the spot where they died. There were no trees in the Solnhofen environment and suggesting that *Archaeopteryx* was an arboreal animal originating from a distant forested area is far-fetched.

Some anatomical features frustrate both the arboreal and the cursorial models. The interpretation of the function of the peculiar feathered tail, for example, has caused problems in both camps. The tail of *Archaeopteryx* is fundamentally different from that of all extant birds. Up in the air it would generate considerable amounts of both lift and drag but the structure of the bony support makes it difficult to control these forces. In both scenarios it is not at all clear how flight could have evolved through stepwise evolution in which each step increased the fitness of the new generation. For example, in running animals the generation of lift would reduce the maximum speed achieved through the pushing force of the hind legs (formula 1 race cars use spoilers to counteract lift). A stepwise theory requires a function, not necessarily flight, which could provide increased fitness through a gradual improvement of features also required (in a later stage) by the flight apparatus. I proposed an alternative scenario that does just that (Videler 2000).

In my view *Archaeopteryx* was a marine shorebird, capable of running across water and mudflats in search for food. Running over water is a fairly widespread technique in the animal kingdom and does not require a miracle to be performed. Basilisk lizards run over water supporting the body weight by repeatedly hitting the water surface with their feet. Could *Archaeopteryx* have used a similar technique?

5.4 *The Jesus-Christ dinosaur hypothesis*

Central American lizards of the genus *Basiliscus* resemble miniature dinosaurs. One of their common names is Jesus-Christ lizards referring to their ability to run across water (Deventer 1983). All four species are bipedal runners. The smaller individuals run across water to escape from predators and to exploit new feeding areas (Fig. 5.3(a)). Adults vary in weight between 200 and 600 g. Males reach maximum lengths of 1 m, females grow up to 0.6 m and weigh 300 g maximum. Three-quarters of their length is occupied by the tail. The lizards have lengthy hind limbs with long, slightly flattened, toes. The toes rotate when the feet hit the water. Lateral fringes along the toes increase the area after rotation (Laerm 1974). Maximum running speeds of

Fig. 5.3 The Jesus-Christ lizard (a) and *Archaeopteryx* (b) running at the water surface. The drawing of the lizard is based on a photograph by Stephen Dalton which appeared as a cover picture (Alexander 1992). The relative dimensions of *Archaeopteryx* are taken from the Berlin specimen (from Videler, 2000).

2.3 m s^{-1} have been reported (Rand and Marx 1967). Weight support while running on water has nothing to do with surface tension but is achieved dynamically by slapping the water surface. During each step, three phases can be distinguished. The flat foot hits the water surface in the slap phase. Subsequently, it pushes down creating an air cavity in the water during the stroke phase. The foot withdraws quickly before the air cavity collapses in the retraction phase. The size of the feet and the length of the legs are important parameters determining the effectiveness of this form of locomotion. The water running ability is size dependent; a 200 g lizard can barely support its weight. There is no correlation between size and speed but larger adults cannot run as far as the juveniles. Stride frequencies vary between 5 and 10 Hz; there is always one foot in the water and hence step periods vary between 0.1 and 0.05 s respectively. The feet of heavy adult males sink too deeply when striking the water and the retraction speed is not fast enough to be out before the air cavity collapses. In conclusion, the capacity to run over water depends on the body mass, the stride frequency and on the speed of the slapping feet.

The ancestors of *Archaeopteryx* were, according to my hypothesis, Jesus-Christ dinosaurs exploiting the running over water trick to escape from predators and to travel between islands in the coral lagoons of Central Europe 150 million years ago (Fig. 5.3(b)). These ancestors were already feathered and could probably float on water. Initially, they used the fore limbs to keep equilibrium while running over water. Gradually the arms evolved into wings generating lift to increase the distance covered at higher body weights. At first, both thrust and weight support were provided by the slapping feet, later the wings gradually took over some weight support. In this scenario every little step towards increased lift generation by the wings had a positive effect on the fitness of the animals, offering better escape performance and a wider feeding range. Natural selection was working to optimize the lift generating properties of the wings. The wings of *Archaeopteryx* (as we know them from the fossil evidence) were adapted to generate lift without flapping at the low speeds the feet could provide, but could not yet be used to produce thrust.

The fossil remains of the London, Berlin, Munich, and Eichstätt specimen show that the feet can be flexed over more than 90° which is sufficient to enable the movements required to slap the water surface. *Archaeopteryx* might have possessed fringes but no fossilized remains of that soft tissue have been recognized so far. Anatomical details of the hind limbs show no obvious reasons why *Archaeopteryx* would not be able to perform the foot folding action executed by the Basilisk lizards. Running over water requires both strength and extreme movements of the ankle joint. *Archaeopteryx* has these features in common with the Basilisk lizard.

The wings of *Archaeopteryx* are studied in detail by Rietschel (1985); his reconstruction of the right wing is shown in Fig. 5.4. It is exceptionally large for a bird of that size and it would require a lot of power to flap such heavily built wings. The hand part occupies about 40% of the length of the

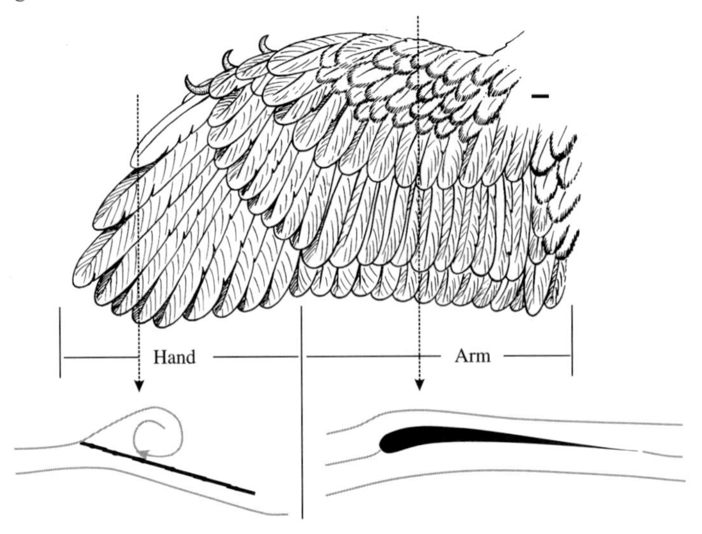

Fig. 5.4 Rietschel's (1985) reconstruction of the right wing of *Archaeopteryx* in ventral view (with the author's kind permission; scale bar is 1 cm). Cross sections through the hand and arm part of the wing are shown with expected lift generating flow patterns (from Videler 2000).

entire wing. The arm part was well cambered. Cross sections through the arm would have shown a rounded leading edge, a large cord, and a sharp trailing edge. The arm wing obviously generated lift conventionally, even at high angles of attack. The hand part consists of 12 primaries with narrow vanes forming the leading edges of the feathers. The leading edge of the hand wing is sharp and wedged backwards, the primary shafts are slightly curved to the rear. The hand wings are in fact swept-back delta wings with sharp leading edges. At low speeds and high angles of attack delayed dynamic stall creates leading edge vortices above that type of wings (see Chapters 1 and 4), a configuration especially adapted to cope with angles of attack as high as 60–70° at low velocities. *Archaeopteryx* could not flap its wings because it lacked the apparatus to do that, but it could presumably spread and close its hand wing as a fan regulating the amount of lift required for each speed. This concept provides a clear role for the three claws on the fingers. Aircraft designed to generate lift with attached flow over the proximal wing parts as well as with separated leading edge vortices over sharp-edged swept-back distal wing sections, often require a device to separate the two flow patterns. Wing fences, and saw-toothed leading edges (Fig. 5.5a) are installed to generate vortices that clear up the boundary layer and separate one type of flow from the next. The fingernails, situated between the arm and hand part of the wing, are leading edge protuberances possibly serving, as in aircraft, to control the formation of leading edge vortices (Ashill *et al.* 1995; Barnard and Philpott 1997; Lowson and Riley 1995; Videler 2000). Leading edge claws of bats (Fig. 5.5b) and some pterosaurs (Fig. 5.5c), and the alula of

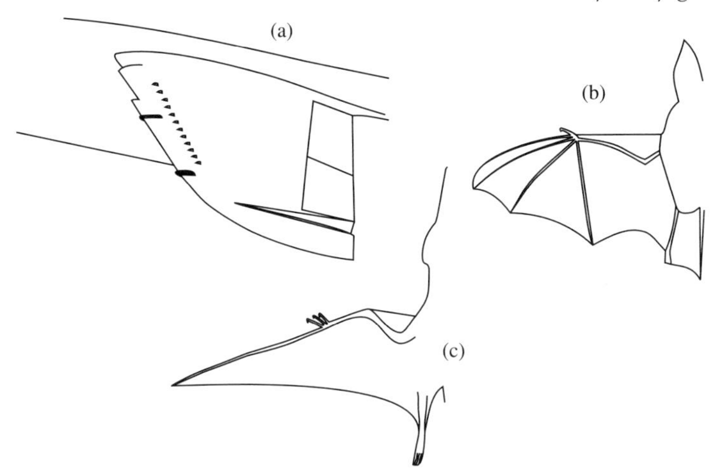

Fig. 5.5 Leading edge devices on wings. (a) Wing fences and saw teeth on the wing of the Harrier, a British Aerospace fighter plane. (b) The first digit of a bat wing. (c) A pteranodon's three free digits.

birds could have the same function. The claws of *Archaeopteryx* probably performed that task in a dynamically optimal way. Each of the three claws could be used according to the amount of spreading of the hand wing, or even in quick succession if needed. The claws that were not in use could be kept withdrawn between the feathers. This function would explain why the claws do not show any wear and why a bi-stable joint (Fig. 5.2(c)) formed the connection between the last digit and the rest of the finger. Due to this feature the nails could be in two resting positions: while extended and while retracted.

The tail, kept in a straight position backwards, could generate varying amounts of lift and drag to control and stabilize the run across the water. At the low speeds used for running slight movements would not have had disastrous effects on the control but could have helped to keep the animal well balanced above its thrusting feet beating the water surface.

New ideas about *Archaeopteryx* are commonly treated with extreme reservation because the fossils have an almost sacred status (Witmer 2002). An article in the journal *Archaeopteryx* (Ma *et al.* 2002) challenged my idea, which was published in the same journal in 2000, and tried to prove that it was wrong. The following paragraph reflects the reply.

5.5 *How could* Archaeopteryx *run over water?*

A quantitative assessment balancing the forces involved is required to answer this question. A quantitative biomechanical model by Glasheen and McMahon (1996*a*,*b*) explains the running over water behaviour of Basilisk lizards. Their model offers allometric equations showing the size-dependence of the important parameters. These are used to make estimates for *Archaeopteryx*. The model is explained in Box 5.1. The approach offers the opportunity to

Box 5.1 ***Summary of Glasheen and McMahon's (1996a,b) model for the Basilisk lizard,*** **Basiliscus basiliscus**

Dimensions of Basilisk lizards over an almost 100-fold range in body mass are used to construct scaling rules for:
The leg length L_L:

$$L_L = 0.1704 M_b^{0.338} \quad \text{(m)} \tag{5.1.1}$$

The foot length L_F:

$$L_F = 0.1321 M_b^{0.302} \quad \text{(m)} \tag{5.1.2}$$

The body mass M_b is expressed in kg (the multiplication factors differ from those of Glasheen and McMahon (1996*b*) because they specified mass in grams in the allometric equations).

The *maximum slap impulse* was estimated using:

$$\text{maximum slap impulse} = \left(\tfrac{4}{3}\right) \rho_{water} r_{eff}^3 u_{peak} \quad \text{(Ns)} \tag{5.1.3}$$

where ρ_{water} is the density of fresh water ($\rho_{freshwater}$) in case of the lizards and that of sea water ($\rho_{seawater}$) when the model is applied to *Archaeopteryx* (Table 5.2). The radius of a disk that would generate the same slap impulse as a physical model of a Basilisk foot is indicated as r_{eff}. The value of r_{eff} can be obtained from the allometric relationship:

$$r_{eff} = 0.0249 M_b^{0.252} \quad \text{(m)} \tag{5.1.4}$$

The model for the *maximum stroke impulse* is given as:

$$\text{maximum stroke impulse} = 0.5 C_D S \rho_{water} (u_{rms}^2 + g L_L)(L_L / u_{rms}) \quad \text{(Ns)} \tag{5.1.5}$$

The product of the drag coefficient C_D and surface S of the model feet and the body mass of the lizards was significantly related as:

$$C_D S = 0.0014 M_b^{0.517} \quad (\text{m}^2) \tag{5.1.6}$$

The model assumes a constant downward velocity u_{rms} of 2.5 m s^{-1} during the stroke to a depth equal to the leg length L_L. The gravitational acceleration g is fixed at 9.81 m s^{-2}.

The *minimum needed vertical impulse* is the product of the body weight and the period between steps (T_{step}):

$$\text{minimum needed vertical impulse} = M_b g T_{step} \quad \text{(Ns)} \tag{5.1.7}$$

predict the impulses that could have been produced by *Archaeopteryx* slapping the water surface with its feet. The minimum impulse required by the body weight for each step must be smaller than the sum of the vertical impulses produced during the slap and stroke phases. The total lift force from the gliding wings and tail reduced the body weight of *Archaeopteryx*. The lift force and the associated induced drag were estimated using a modification of the mass flux model of Chapter 4. The mass of air deflected downwards by the wing and tail of the running animal is taken into account. The equations describing that model are in Box 5.2. The thrust impulse produced by the feet during a slightly oblique stroke should exceed the drag impulse estimates for the duration of each step. The resulting impulse balances will answer the question if it was physically possible for *Archaeopteryx* to run over water. (Table 5.1 gives the physical quantities used in the various models.)

Box 5.2 ***Gliding over water: the approximate mass flux model used for* Archaeopteryx**

The running animal deflects air down with wings and tail as shown in Fig. 5.6; it generates lift, uses ground effect to enhance that lift and suffers drag induced by the lift generation as well as parasite and profile drag.

Lift force L

The mass of air involved (M_a) equals the mass of air in a cylinder with a diameter equal to the wing span b and a length equal to the product of the running velocity u and time t.

$$M_a = \rho_{air}\pi\left(\tfrac{1}{2}b\right)^2 ut \quad \text{(kg)} \qquad (5.2.1)$$

where ρ_{air} represents the air density (Table 5.1). Over time t, air is accelerated downward reaching velocity w. The downward momentum imparted to the air is

$$M_a w = \rho_{air}\pi\left(\tfrac{1}{2}b\right)^2 utw \quad \text{(Ns)} \qquad (5.2.2)$$

This expression is divided by t to get the rate of change of the momentum, which equals the aerodynamic reaction force on the animal in upward direction:

$$L = \rho_{air}\pi\left(\tfrac{1}{2}b\right)^2 uw \quad \text{(N)} \qquad (5.2.3)$$

The unknown variable is w. Figure 5.6 shows that w is related to ε and u by:

$$w = u \tan\varepsilon \quad (\text{m s}^{-1}) \qquad (5.2.4)$$

Combining equations (5.2.3) and (5.2.4) turns L into the form:

$$L = \rho_{air}\pi\left(\tfrac{1}{2}b\right)^2 u^2 \tan\varepsilon \quad \text{(N)} \tag{5.2.5}$$

Ground effect G

The gliding wings moved at a short enough distance over the water surface to benefit from the ground effect G. This can create a lift enhancement without drag increase of:

$$G = 2L/AR \quad \text{(N)} \tag{5.2.6}$$

where AR is the aspect ratio of the wings equal to the span b squared divided by the surface area of the wings (Anderson and Eberhardt 2001).

Induced power P_{ind} and induced drag D_{ind}

The induced power for the generation of lift, equals the kinetic energy imparted in downward direction to the added mass of air divided by t:

$$P_{ind} = 0.5M_a w^2 t^{-1} \quad \text{(W)} \tag{5.2.7}$$

Using (5.2.1) removes t and gives:

$$P_{ind} = 0.5\rho_{air}\pi\left(\tfrac{1}{2}b\right)^2 w^2 u \quad \text{(W)} \tag{5.2.8}$$

The induced power also equals the induced drag times the velocity u:

$$P_{ind} = D_{ind}u \quad \text{(W)} \tag{5.2.9}$$

Hence the expression for the induced drag is:

$$D_{ind} = 0.5\rho_{air}\pi\left(\tfrac{1}{2}b\right)^2 w^2 \quad \text{(N)} \tag{5.2.10}$$

The downwash angle ε is the only unknown variable in the aerodynamic model equations.

Profile and parasite drag $D_{par+prof}$

Profile drag is the drag on the wings and tail when kept in such a position that the air would not be deflected downward. Parasite drag is associated with pressure and friction on the rest of the body during the movement through the air. Profile and parasite drag are difficult to assess. They are in the order of:

$$D_{par+prof} = 0.5\rho_{air}AC_{par+prof}\,u^2 \quad \text{(N)} \tag{5.2.11}$$

where A is the frontal area while running without generating lift and $C_{par+prof}$ the fraction of the area effectively blocking the air flow.

Table 5.1 Values of the physical parameters used in the model approaches.

Gravitational acceleration (g)	9.81	m s^{-2}
Air density (ρ_{air})	1.23	kg m^{-3}
Sea water density ($\rho_{seawater}$)	1024	kg m^{-3}
Fresh water density ($\rho_{freshwater}$)	998	kg m^{-3}

Table 5.2 Relevant dimensions of the Berlin *Archaeopteryx* used in the calculations.

Body mass (M_b)	0.25 Kg
Weight	2.45 N
Span (B)	0.6 m
Wing area	0.06 m^2
Aspect ratio (AR)	6
Length upper leg (femur)	0.054 m
Length lower leg (tibiatarsus)	0.071 m
Length foot (L_F)	0.076 m
Distance hip to heel (L_L)	0.125 m

5.5.1 Calculations of the interaction between feet and water

Measurements of the length of the skeletal elements of the legs are taken directly from the Berlin fossil (Fig. 5.1). The right leg as it appears on the slab (it could have been the animal's left but there is still dispute about how it fossilized) is almost completely preserved and measurements can be taken from it without correction because it is situated in one plane over its entire length. Yalden (1984) made educated guesses of the mass range of *Archaeopteryx*. A value of 0.25 kg seems to be a reasonably accurate estimate for the Berlin specimen. The relevant dimensions of the Berlin *Archaeopteryx* are given in Table 5.2.

The size of the legs and feet of *Archaeopteryx* are compared with those of equally sized Basilisk lizards using the allometric equations of Glasheen and McMahon (Box 5.1). The leg and foot lengths of Basilisk lizards scaled close to isometry over an almost 100-fold range in body mass. *Archaeopteryx* has a shorter foot length but the total leg length is longer.

The Basilisk model divides each step into the three phases mentioned before: (1) slap-, (2) stroke-, and (3) protraction-phase. The minimum needed vertical impulse (in Ns) is the product of the body weight and the step duration. There is always one foot in the water and each stride takes two step periods. We accept these conditions for *Archaeopteryx* assuming that there is no gliding phase without a foot in the water. The slap is assumed to be directed vertically. The maximum slap impulse is proportional to the density of the water, the velocity of the slap, and the third power of the radius of a disk that would generate the same slap impulse as a physical model of a Basilisk foot, tested by Glasheen and McMahon. The slap velocity of the Basilisks was extremely variable and not a function of body size.

Glasheen and McMahon used an upper limit of 3.75 m s^{-1}. That is 1.5 times the root mean square value of the downward velocities measured during the stroke. There is no obvious reason to assume that these values could not be used for *Archaeopteryx*. The maximum stroke impulse is the product of the average drag force on the descending foot and the time over which that drag force is applied. The model assumes a close to vertical downward stroke at a constant velocity to a depth equal to the length of the leg. The foot is kept orthogonal to the direction of travel throughout the entire stroke. The stroke is at some small angle β backwards. The vertical and horizontal components of the maximum stroke impulse are found by multiplying it with cosine β and sine β respectively. The sum of the vertical components of the maximum stroke impulse and the maximum slap impulse must be greater than the minimum needed vertical impulse to make running over water possible. The horizontal components of the stroke impulse must be larger than the impulse needed to overcome the drag at the running speed.

The potential lift generating capacity of *Archaeopteryx* is not taken into account in the results presented in Table 5.3. Glasheen and McMahon found that stride frequencies between 5 and 10 Hz were used by Basilisk lizards with no clear dependence on body mass. These are therefore used here as well. The maximum slap impulse of *Archaeopteryx* is a fraction higher due to the fact that it was slapping seawater, which has a higher density than fresh water. The longer legs of *Archaeopteryx* mainly cause the higher maximum stroke impulse. With a stride frequency of 10 Hz (step duration of 0.05 s) the total maximum impulse exceeds the minimum impulse needed for both animals. The impulse surplus of *Archaeopteryx* is larger than that of the Basilisk lizard of the same mass, again due to the difference in leg length. With a stride frequency of only 5 Hz the impulse surplus is negative for both *Archaeopteryx* and the Basilisk which means that they are unable to run at the water surface. The model predicts that a Basilisk lizard of 0.25 kg needs

Table 5.3 A comparison of the water running abilities of *Archaeopteryx* and a Basilisk lizard.

	Berlin specimen		Basilisk lizard		
Length foot (L_F)	0.076	0.076	0.087	0.087	m
Distance hip to heel (L_L)	0.125	0.125	0.107	0.107	m
Max. slap impulse	0.028	0.028	0.027	0.027	Ns
Max. stroke impulse	0.130	0.130	0.106	0.106	Ns
Total maximum impulse	0.158	0.158	0.133	0.133	Ns
Stride frequency	10	5	10	5	Hz
T_{step}	0.05	0.10	0.05	0.1	s
Minimum needed impulse	0.123	0.245	0.123	0.245	Ns
Surplus impulse	29.0	−35.5	8.6	−45.7	%

Parameter values used for both animals are: $M_b = 0.25$ kg; $u_{rms} = 2.5$ m s^{-1}; $u_{peak} = 3.75$ m s^{-1}; $r_{eff} = 0.018$ m; $C_D S = 0.00068$ m^2. The table compares stride frequencies of 10 and 5 Hz for each animal. The lift generating capacity of *Archaeopteryx* is not taken into account.

a minimum stride frequency of 9.2 Hz to perform the trick. *Archaeopteryx* has the possibility to generate lift which reduces the required impulse.

5.5.2 Estimates of aerodynamic forces generated by the running *Archaeopteryx*

The large wings and the feathered tail of *Archaeopteryx* generated lift by deflecting oncoming air down during runs over water. The interactions between the air and the wings and the long sideways-feathered tail must have been rather complex. The model simplifies that complexity by assuming that the affected air can be represented by a cylinder of air with a uniform flow. Figure 5.6 shows how the animal is running at a uniform speed and how it deflects a cylinder of air downwards due to the presence of the stretched wings and the spread tail. The diameter of the cylinder is estimated to be approximately equal to the wing span. The approximate mass flux method described in principle in Chapter 4 (Box 4.1) can be adapted to estimate the amount of lifting force the wings and tail of *Archaeopteryx* could generate during a run at a uniform speed. The maximum bipedal running speed of $2\ \mathrm{m\,s^{-1}}$ of an *Archaeopteryx* sized dinosaur estimated by Alexander (1976) is used in the calculations. Box 5.2 shows the equations used in the calculations. The volume of air involved per unit time is equal to the cross sectional area of the cylinder times the running velocity. Multiplication with the air density gives the affected air mass per unit time, which is multiplied by the downward velocity reached to find the force applied on the air. The running animal pushes that air down over the downwash angle ε with its wings and tail to generate lift. The lift force is the reaction force on the animal from the

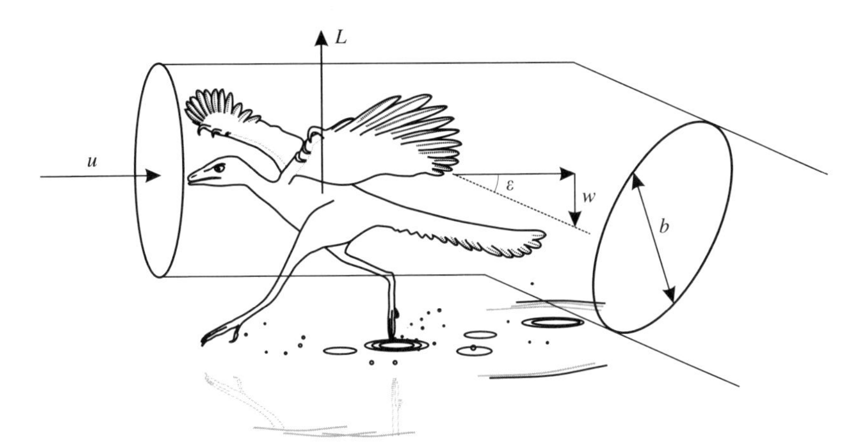

Fig. 5.6 Artist impression of *Archaeopteryx* running over water at speed u. Relative to the animal, the flow of air has velocity u. The air is deflected downward over an angle ε by the combined action of the wings in gliding position, the body and the feathered tail. A cylinder of air with a diameter b equal to the wingspan is affected. The air is accelerated downward to reach a velocity w. The aerodynamic reaction force L is directed upward. See text for further explanation.

force used to accelerate the air down. This action will induce a drag force on the animal in the opposite running direction. The induced drag times the running velocity is the induced power. The induced power is also equal to the kinetic energy per unit time imparted in downward direction to the air. The induced drag can now be found by dividing the induced power by the running velocity. The drag on the animal consists of induced drag associated with the generation of lift, of profile drag on the wings and the tail, and of parasite drag on the head, the body, and the legs. Profile drag is the drag on the wings and tail when kept in such a position that the air would not be deflected downward. Parasite drag is associated with pressure and friction on the rest of the body during the movement through the air. Profile and parasite drag are difficult to assess. They are proportional to the frontal area while running without generating lift and to a drag coefficient which is the fraction of the area effectively blocking the air flow.

The model integrates the interactions between animal and air by using reasonable estimates for the direction and velocity of the resulting airflow behind the animal. Estimates of the lift forces therefore include the integrated effects of wings, body, and tail, ignoring the possibility that the arm wing and the hand wing generate lift and drag in different ways. The wings are close enough to the water surface to make use of lift enhancing ground effects (see Chapter 4). The magnitude of the ground effect is proportional to the lift and inversely proportional to the aspect ratio (span squared divided by the surface area) of the wings (Anderson and Eberhardt 2001) (Fig. 5.7).

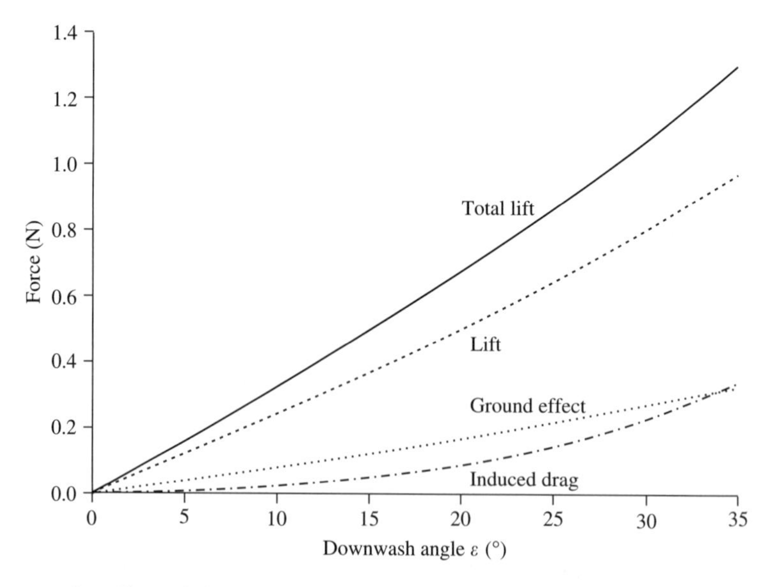

Fig. 5.7 The effect of the downwash angle on lift and drag forces experienced by *Archaeopteryx* running over water at 2 m s^{-1}. Force calculations are based on Sunada and Ellington's (2000) mass flux model approach. See text for further explanation.

The results of the calculations of the aerodynamic forces for values of downwash angles ε between 0° and 35° are shown in Fig. 5.7. The total lift force generated by deflecting a cylinder shaped virtual mass of air, enhanced by the ground effect gradually increases with the downwash angle. The induced drag on *Archaeopteryx* running at 2 m s^{-1} increases also but remains small compared to the values lifting the animal. Figure 5.8 shows how the lift forces at the range of downwash angles determine the net impulse surplus and hence the ability to run at the surface for three stride frequencies. We already saw that with a step period of 0.05 s *Archaeopteryx* could run on water even without the help of a lifting force. A step period of 0.1 s requires slightly less than 0.16 Ns of impulse due to lift forces from the wings and the tail. That amount would be generated according to our model with downwash angles exceeding 26°. The consequences of the average step period value of 0.07 s are also indicated in Fig. 5.8, showing that downwash angles larger than about 11° are required in that case.

The sum of the profile and parasite drag is estimated to be in the order of 0.006 N using a frontal area estimate of 0.009 m^2 and a coefficient of 0.25 and thus is considered small enough to be further ignored. The forward directed impulse that must be produced during each step to overcome the induced drag of 0.165 N at ε = 26° is about 0.017 Ns. That value is reached with a backward stroke angle β of only 7.5° and leaves 0.129 Ns of the 0.130 Ns for the vertical component of the stroke impulse showing

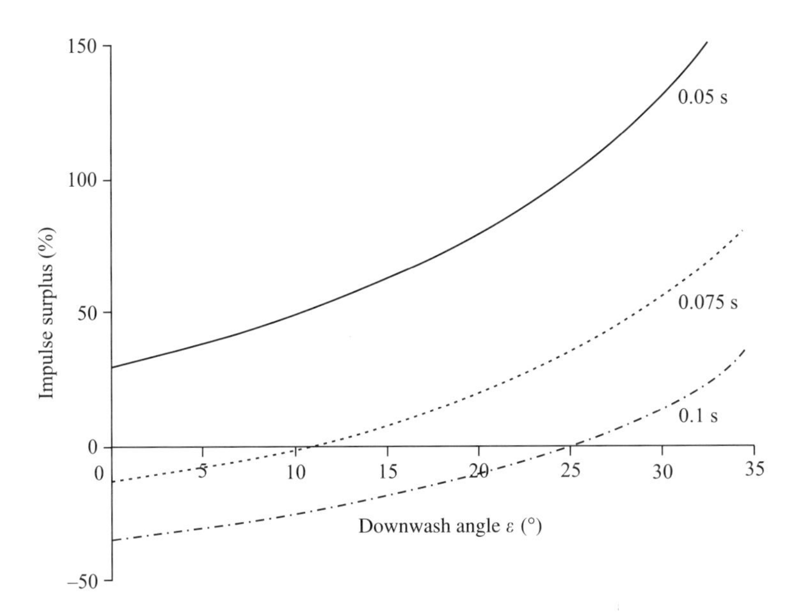

Fig. 5.8 The impulse surplus a 0.25 kg *Archaeopteryx* would be able to generate while running at 2 m s^{-1} across the water surface using the technique of the Basilisk lizard and generating lift force by deflecting air downwards with steady wings and tail. Model predictions for three step durations are shown.

that surplus percentages are hardly affected by such small backward stroke angles.

The analysis shows that a 0.25 kg *Archaeopteryx* could run over water if it could use stride frequencies as high as 10 Hz or else could generate lift.

5.5.3 Ecology and evolution

Many aquatic bird species run over water while beating the wings prior to take-off. It is generally accepted that *Archaeopteryx* was not able to beat the wings forcefully and therefore had to keep them in gliding position during runs across the water. The Western Grebes also keep their wings static while running over water performing characteristic courtship display. Figure 5.9 is a drawing made of one picture of a film fragment of the BBC series 'The life of birds' by D. Attenborough. The film shows the rushing ceremony performed by two birds in perfect harmony. The bodies, weighing up to 1200 g, are lifted completely out of the water by the slapping feet. The wings do not give the impression to contribute much to the lifting forces because the trailing edge wing feathers flutter during the run. The legs of grebes are set far back, a position ideal for under water propulsion. Three front toes

Fig. 5.9 The Western Grebe (*Aechmophorus occidentalis*) running over water during the rushing ceremony of its courtship display. Drawing based on a film frame taken from the BBC series 'The life of birds' by David Attenborough.

have broad independent lobes; the hind toe has a small lobe (Del Hoyo *et al.* 1992). The film fragment shows that there is always one foot in the water. These grebes manage to keep bodies above the water which are more than four times the weight of *Archaeopteryx*. Unfortunately no biomechanical analysis of this remarkable behaviour has been published yet. There is also no indication of the importance of the lobes on the toes during the runs at the surface. Anyway, none of the *Archaeopteryx* fossils show signs of the presence of lobes. The grebe case shows that there are more biomechanical solutions to the problem. *Archaeopteryx* could have used the Basilisk or the Grebe technique or even used a completely different method to run at the surface.

The ecological advantages of running over water for *Archaeopteryx* could have been the possibility to escape from terrestrial predators and the exploitation of distant feeding grounds around islands in a shallow tropical sea. All the fossils have been found in marine offshore sediments. Some specimens reveal that the animals must have died close to where they fossilized. The bodies of the Berlin, the London, the Eichstätt, the Solnhofen, the Solnhofen Aktien-Verein, and the tenth specimen (Mayr *et al.* 2005) were very much intact which excludes the possibility of long-distance transport from a terrestrial place of death. The wings of the London and Berlin individuals are stretched as in a gliding position. Running over water using gliding wings and tail to generate some lift was obviously not without risk but the model calculations show that *Archaeopteryx* could do it.

Elzanowski (2002), inspired by the shape of the teeth, assumes that soft bodied arthropods made up a substantial part of the diet of *Archaeopteryx*. Dr. M. Kölbl-Ebert (Jura-Museum, Eichstätt) brought the presence of large water skaters, *Chresmoda obscura* to my notice. These large insects occurred in the area where *Archaeopteryx* lived, and may be a likely candidate for such a food source. Their exoskeleton was probably rather thin to reduce the body weight allowing surface tension forces to carry the approximately 5 cm long animals. The picture emerges of *Archaeopteryx* as a highly specialized shorebird, feeding on water skaters on shallow waters and wet mudflats. The ability to run across the water surface fits in this scenario.

My hypothesis paints a picture of a bird-like animal that could run over water over large distances at a speed of at least 2 $m\,s^{-1}$. Its wings were perfectly suitable to generate lift to take care of much of its body weight but could not generate thrust. Such a creature could glide but not fly actively. If *Archaeopteryx* was going to evolve into a flying animal it would have to loose weight and it would have to develop the morphological requirements to beat its wings. That would include a proper sternum preferably with a crest to accomodate the supracoracoideus muscle and a tendon system to take care of the wing's movement during the upstroke. There are three *Archaeopteryx* skeletons found for which the position in the layers of sediment in Solnhofen is known. The London specimen was found deepest and is the largest one. The seventh fossil (the Munich or Solnhofer Aktien

Verein specimen) was found in a layer 14.6 m higher up. It was considered a new species, *Archaeopteryx bavarica*, because its size was about 73% that of the London remains and it probably had a calcified sternum indicated in Fig. 5.10. The Maxberg *Archaeopteryx* from an in-between layer 8.5 m deeper down was of an intermediate size. These facts (Wellnhofer 1993) do not prove but support the view that *Archaeopteryx* was in the process of evolving into a flying bird. Younger fossils from the same site should provide the answer to the question how a flying *Archaeopteryx* was built. However, by the time *Archaeopteryx* could fly it would no longer be running between islands and the chances to get killed and drowned during one of these trips would be reduced and also the chances to find a fossil of such a rare event would be remote. Recent discoveries show that soon after *Archaeopteryx* left Solnhofen the world was full of more or less flying feathered dinosaurs.

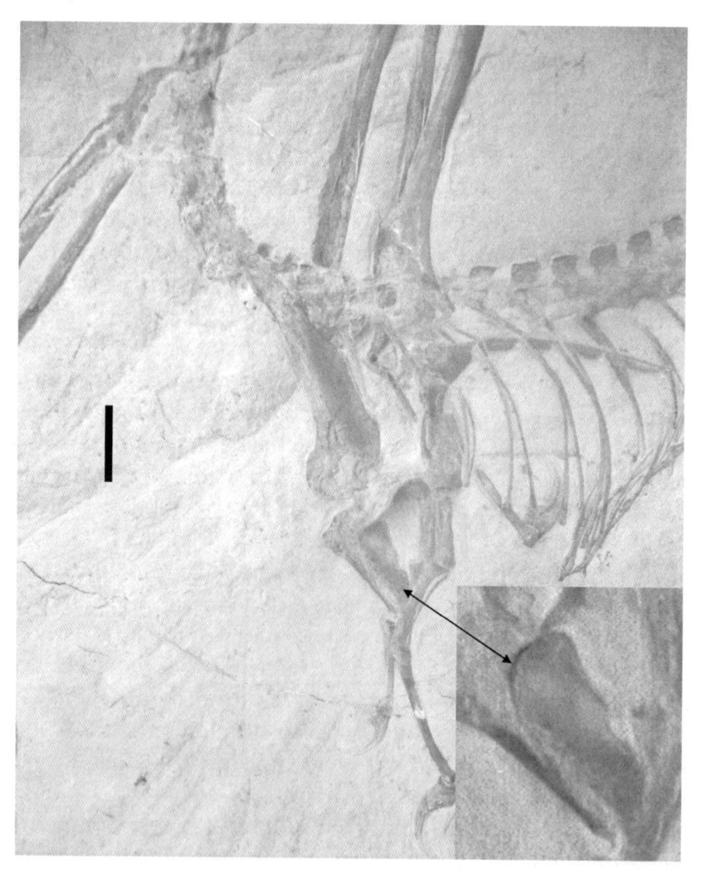

Fig. 5.10 The sternum of the Solnhofen Aktiën Verein (Munich) *Archaeopteryx* (scale bar is 1 cm).

5.6 *Other fossils with characteristics possibly related to the beginning of flight*

From the time of *Archaeopteryx* until the end of the Cretaceous period rapid adaptive radiation of feathered dinosaurs took place all over the world. In places as far apart as Spain and China witnesses of this process were recently unearthed. Many workers in the field try to use the data to construct phylogenetic trees. However, the diversity is so large that almost every new fossil gets its own branch. The only reasonable conclusion so far seems to be that flight evolved along more lines of evolution. Furthermore, interpretation of the flight-related features suffers from the fact that we are still ignorant about the aerodynamic details of bird flight as demonstrated in Chapter 4. There are most likely many ways to get airborne and several evolved in different bursts during that long innovation period between 150 and 65 million years ago. The mass extinction event at the end of the Cretaceous puts an end to several bird lineages and marked the beginning a new period of rapid radiation starting during the Palaeocene and continuing during the Eocene up to 35 million years ago. It could well be that the mass extinction event wiped out some of the alternative pathways to flight and restricted further evolution to the pattern that we find in modern birds.

No potential ancestors of *Archaeopteryx* have been discovered in the Solnhofen lithographic limestone. But theropod dinosaurs found elsewhere in the world look more or less like the picture we have in mind. Rich fossil beds in North-eastern China, for example, contained two specimens of a small theropod dinosaur with an extremely long tail, *Sinusauropterix prima* (Chen *et al.* 1998). The skeletons of *Sinusauropterix* reveal a close kinship with *Compsognathus*. The largest was 68 cm long and had 64 vertebrae in the tail. The animals had long hind legs and a relatively large skull with a toothed beak. The front legs look short and stubby. These fossils were considered extremely interesting because the skin of *Sinusauropterix* was covered with a thick layer of feather-like structures from head to tail. There are multi-branched thick strands, several centimeters long, which are much thicker than the hair of mammals of similar size. The shaft-like strands were probably hollow. The impression arises of simple feathers each with a shaft and numerous loose filamentous barbs. There are no signs of aerodynamic characteristics. The thickness of the layer suggests heat retention as the most probable function. There is no consensus on the identity of these structures as feathers. Some authors believe that the structures are collagenous fibres remaining after the decay of the dermis (Lingham-Soliar 2003). Ageing of the sediments where *Sinusauropterix* was found is still a large problem. They are estimated to be of mixed Jurassic-Cretaceous character and could be younger than the Solnhofen limestone. The same authors also discovered a bird-like animal with symmetrical feathers, *Protarchaeopteryx*, probably also a feathered dinosaur with 30 caudal vertebrae. Several specimens of *Caudipteryx* from the same region clearly had feathers but possibly were secondarily flightless birds or not birds at all but belonging to another group of dinosaurs the oviraptorosaurs (Zhou and Hou 2002). Another fossil, showing an interesting feature related to flight in a non-flying dinosaur was discovered in 1996 in Patagonia (Novas and Puerta 1997). *Unenlagia comahuensis* was a 2 m long theropod which lived much later than *Archaeopteryx* during the Upper Cretaceous. Some vertebrae, most of the pelvic girdle, bones of the hind legs and most interestingly for our current interest, the left scapula and an incomplete humerus were found. The shoulder joint is well preserved and reveals the degrees of freedom of the humerus. As in birds the shoulder joint of *Unenlagia* was facing laterally, offering the possibility to lift the humerus and hence the arm upwards. This animal must have been able to make up and down movements with its forelimbs. Why it would need to do that remains a mystery. There are no signs of feathers found so we have no proof for the idea that the arms were actually feathered wings that could provide forces by beating the air.

The rich possibly late Jurassic—early Cretaceous fossil beds in China contain many birds found in lacustrine sediments. Again, until there is clear evidence of ancestral relationships we should not study each new species as a step in a phylogenetic range but try to see the structures as adaptations producing evolutionary fitness in a certain ecological niche. Flight-related features of a few striking examples are used here to create some feeling for the diversity of designs in early birds.

Xu *et al.* (2003) describe a most remarkable little dinosaur from the famous Jiufotang formation in Western Liaoning (China). *Microraptor gui* is about 77 cm long including a long sideways feathered tail occupying half that length. This guy has four wings, two formed by the arms and two by the legs. Both front and rear wings are shaped by pennaceous primary and secondary feathers some with weakly asymmetric vanes and have preserved wing coverts. The skeleton of the trunk has a flat bony sternum; fused scapula and coracoid and a sturdy rib cage. The fossa of the shoulder joint faces laterally. The front wings are supported by the humerus, the ulna and a thin radius, unfused metacarpals, and three fingers ending in claws. Some well-organized feathers are associated with the first digit and are interpreted as a precursor of the alula. The usual hind limb skeleton supports the hind wing leaving only the clawed toes free. The authors believe that this animal was arboreal and capable of gliding. Surprisingly, the existence of a four-winged dinosaur was predicted by C. William Beebe (1915) on the basis of presumed early developing flight feathers on the upper part of the hind leg of a juvenile dove. Beebe imagined the existence of a proavis with four wings as a 'tetrapteryx' stage in the evolution of flight and here it is.

Another long-tailed bird from the same area was obviously eating the seeds preserved in its stomach. *Jeholornis prima* (Zhou and Zhang 2002) was a tall bird of approximately 70 cm from head to tail tip. The tail occupied about half that length with 22 articulating caudal vertebrae. There is a short sternum with un-fused lateral trabecula indicating a large origin for main flight muscles; the humeri have large deltoid crests. Scapula and coracoid are not fused. The furcla is robust boomerang shaped. There is no triosseal canal

(Zhou, personal communication). The authors believe that it was a good flyer and arborial with large clawed feet suitable for perching on branches.

Confuciusornis sanctus (Hou *et al.* 1995, 1996; Martin *et al.* 1998) discovered in 1995 in China most likely could fly. Large numbers have been preserved. They must have been about 25 cm tall. The number of tail vertebrae is reduced to 4–5 and there is a pygostyle. The animals had beaks without teeth. An unkeeled sternum and a boomerang-shaped furcula were present. The humeri had enormously expanded deltoid crests. Large males and small females have been found side by side. The males had long ornamental tail feathers. Each wing had three unreduced fingers and an enlarged phalanx of the first digit with a strong claw. The species must have lived in flocks near fresh water lakes in a tropical forest. Their lifestyle is compared to that of parrots.

Liaoningornis longidigitris found in the same deposit as *Confuciusornis*, is the first ancient bird found with keeled sternum attached to coracoids. It had sharp curved claws and its small size probably improved flight conditions. *Protopteryx fengningensis*, a starling sized bird again from the same formation, showed even more features of modern birds. It had a pygostyle, elongated coracoids, strut-like clavicula and a sternum with a crest and lateral processes. The hand skeleton is longer than the forearm and an alula is attached to the alular digit. It provides the first fossil evidence for the existence of a triosseal canal (Zhang and Zhou 2000).

Probably the best early flier was *Eoalulavis hoyasi* found in Spain (Sanz *et al.* 1996). It was a small wader with a wing span of 17 cm. Its name gives away that it was considered the first bird with an alula. Another derived feature is the presence of a triosseal canal. So the first birds with all the basic structures of the modern flight apparatus appeared 105 million years ago, 45 million years after *Archaeopteryx*.

This conclusion is true if we ignore the claim of Chatterjee (1997) who believes that he found fossil remains of an equally sophisticated bird in a 225-million-year-old deposit in Texas. He named that presumed mother of all birds *Protoavis texensis*. If this claim is true it predates the evolution of birds by 75 million years to the end of the Permian and rules out a dinosaur ancestry for birds. *Protoavis* had a large keeled sternum, a triosseal canal and impressions of shafts of feathers on the arm skeleton. Comments of experts are extremely sceptical. Chatterjee's drawings and text in his book *The Rise of Birds*, devoted to *Protoavis* are very compelling but support for this exciting claim awaits real evidence.

After the mass extinction event 65 million years ago most groups of Mesozoic birds had disappeared and the modern orders of non-passerine flying birds started to radiate as descendants of a few surviving groups. The ancestors of ducks, loons, plovers, tube-noses, and maybe some other groups are believed to have survived the crisis, but hard evidence is scarce. By the start of the Eocene the basic structure of the bird flight apparatus was more or less fixed and remained unchanged ever since although we have to keep in mind that every single species shows its own deviations from this basic pattern.

Following the evolution of flight into that kind of detail is not possible at this stage. The most numerous group of living birds, the passerines, started to emerge and diversify during the mid-Tertiary between 10 and 30 million years ago. Fossil records of the early development are scarce (Feduccia 1999).

The basic pattern of modern birds includes a keeled sternum kept at a fixed distance to the shoulder joint by the coracoidal bones; a triosseal canal acting as a pulley for the supracoracoideus, the muscle responsible for raising the wing and an alula at the leading edge between the arm wing and the handwing replacing the protruding claws of Mesozoic birds.

5.7 *Summary and conclusions*

The most common opinion is that flight of birds started to evolve during the Jurassic period some 150 million years ago. Fossils of small theropod dinosaurs with feathers are found from that period. The most challenging series is unearthed from the limestone deposits of the Solnhofen area in Bavaria, Germany. *Archaeopteryx* (10 skeletons and 1 feather) kept palaeontologists busy for more than a century and is still at the heart of the debate regarding the beginning of flight. Therefore, the main part of this chapter concentrates on this group of animals which lived along the flat open coasts of a tropical back-reef area off the Tethys Sea. *Archaeopteryx* had large well-designed wings and a long sideways feathered tail. This apparatus was perfectly adapted to allow gliding at low speeds; flapping was probably not possible due to the lack of hardware to allow that.

Since the discovery of the first *Archaeopteryx* fossils more than a century ago, conflicting scenarios, the arborial and cursorial, were used to describe how flight might have evolved in these animals. The arguments pro and con are still being exchanged. Knowledge about the environment in which *Archaeopteryx* lived is hardly taken into account despite lots of fossil evidence. Some of the arguments violate the law of parsimony. Therefore a new hypothesis is offered explaining most of the peculiar anatomical features of *Archaeopteryx* and giving it an ecological niche in the environment in which it lived and died. It is depicted as a shorebird capable to run over water and mudflats in search of food and to escape from predators. The technique used by the Basilisk lizard could have generated most of the lifting forces required; the wings and tail in gliding position helped to carry the animal during the running action. A quantitative assessment using two models showed that *Archaeopteryx* could do this. The model developed for Basilisk lizards shows the impulse per step generated for stride frequencies between 5 and 10 Hz. Using a realistic frequency of 5 Hz lift from the wings and tail was needed to keep the animal at the surface during runs at 2 m s^{-1}. At that speed lift and induced drag forces are calculated for a range of downwash angles using an adapted version of the approximate mass flux model described in Chapter 2. Additional lift from ground effect is taken into account. This approach shows that *Archaeopteryx* could run over water. Similar behaviour of extant birds is discussed. Abundant water skaters found fossilized in the same deposits as *Archaeopteryx* are suggested as a possible food source that could be exploited with the proposed specialization. Some of the fossils form a time series probably indicating a development in the direction of flapping flight.

Other fossils of feathered dinosaurs are found abundantly in China but are also known from other regions all over the world. These remains of Mesozoic bird-like animals reveal the existence of parallel lines of evolution of flight-related characters. Several examples of the Mesozoic radiation are given and the flight-related structures are discussed. The mass extinction event 65 million years ago brought an end to the recognized large groups of Cretaceous birds. Only a few minor groups of birds capable of flapping flight survived and were ancestral to the modern lineages of birds rapidly radiating during the beginning of the Tertiary.

6 *Bird flight modes*

6.1 *Introduction*

A bout of flight behaviour starts with the take-off event followed by a passage through the air over a certain distance and ends with a landing manoeuvre. By analysing repeated flights under controlled conditions the choice of flight strategies of a bird can be detected. Being the pilot, a bird must take decisions regarding altitude, speed, acceleration, deceleration, the use of the wings during flapping or gliding, and the actions of the tail, to mention just a few important items. Our studies of free flight of common kestrels under controlled conditions give some indication of the complexity of this behaviour.

Perhaps the most characteristic feature of avian flight is the cyclic wing beat during powered flight. Wings not only flap up and down but also fold, stretch, rotate, and change speed. The description of these movements can be made with respect to the body, in relation to the air around the bird or in an earthbound frame of reference. Kinematical analysis concentrates on changes of position as a function of time without reference to forces. The kinematics of animal flight must be studied three-dimensional because the movements are made in all three directions in an x–y–z orthogonal frame of reference. Wing beat cycles have been studied under standard conditions and with some experimentally controlled changes in search for stereotypic aspects. Detailed analysis of a single wing beat of a small bird emphasizes the existence of unsteady effects. We try to find out if there are general rules for flapping flight which are obeyed by most species.

Hummingbirds are able to hover on the spot in still air for prolonged periods; some small passerines can do that too but only briefly. Studies unravel the details of this flight technique and show some limits of hummingbird performance.

Approximately one dozen species is able to maintain a fixed position over the ground by flying against the wind at the speed of the wind. The various species of kestrel are most commonly known for this windhovering behaviour. The stable position in an earthbound frame of reference of hovering and windhovering birds offers the possibility to film their flights in the wild with a (high-speed) camera mounted in fixed position. The latter is a condition required for an accurate quantitative analysis of the movements.

Repeatedly interrupted flapping is common among birds. During the interruptions birds either fold their wings or spread them as during a glide. Some species use both modes. Saving energy is a common explanation for these intermittent flight techniques. That is also the case for formation flight of large birds, but we must find out if there is real evidence to support these clarifications.

Various forms of gliding are of course the ultimate flight strategies to remain airborne in a cheap way. At least four different techniques used by birds to profit from favourable atmospheric conditions have been described and shall be briefly explained.

Manoeuvres involving sharp turns are daily practise for birds but experiments aimed at gaining some understanding of how it is actually done are not easily designed. There have been many attempts to measure the speed of birds in the wild. The reliability of these data is often not easy to judge. Even the most sophisticated methods to measure flying speed in the field are error prone. The aim here is to obtain a realistic feeling about the range of speeds flown by birds of different groups and sizes.

6.2 *The flight plan*

A simple flight starts with the take-off action followed by acceleration to some cruising speed and ends with a landing manoeuvre. We first concentrate on characteristic features of such standard flights before focussing on the movements of wings, body, and tail during the cruising speed phase.

In the field it is virtually impossible to measure the characteristics of an entire flight from take-off to landing accurately. Unpredictable conditions and the complexity of the required instrumentation are the main precluding factors. We therefore designed an experimental set-up, which allowed us to measure unrestricted flight behaviour under controlled conditions (Videler *et al.* 1988*a,b*). Two wild adult common kestrels one female ('Kes') and one male ('Jowie') were trapped, kept, and trained using falconry techniques (Glasier 1982). The birds were exercised daily after working hours in the 142 m long, 3.4 m wide, and 2.4 m high straight corridor of the Biological Centre in Haren (The Netherlands). Each bird flew up and down between the gloves of two falconers (volunteering graduate students). Small bits of minced mouse were randomly offered at the gloves after 80% of the landings. A range of experiments were conducted over a period of about half a year before the birds were released back to the wild. In one experiment we measured the flight strategies from start to landing without and with weight carried by the birds. The average body masses during captivity were 190 and 160 g for Kes and Jowie respectively. These figures are about 20–30 g lower than the mass at capture. At this level, the birds were in good condition and keen to fly. There was one flight session for each bird per day. The birds were allowed to fly up and down as long as they were keen. We changed the weight of the animals by making them fly with leaden instead of leather anklets weighing either 0.3 N (mass 31 g) or 0.6 N (61 g) per pair. These weights represent the range of weights of prey items of common kestrels in the wild (Masman *et al.* 1986). Under the same loading conditions we made them cover two distances to see if the birds used different strategies for shorter and longer flights. One session was devoted to one distance of either 50 or 125 m and to one type of loading. The number of flights varied between 50 and 156. Kes flew 94 and Jowie 78 times on average. The number of flights did not seem to be related with the body mass at the start of the session. Only very small pieces of mouse were offered at the glove to get as many flights as possible.

During the experiments, flight data were recorded on a computer connected to a quartz clock for precise time keeping. Electronic switches in the gloved left hand of the falconers clocked landing and take-off times with an accuracy of 0.01 s. The number of flights, the duration of each crossing, and the flight direction could be deduced from these data. At four positions along the track, infrared-light-sensitive cells, mounted on the floor under infrared lights in the ceiling, recorded the instant of passing of a bird. These registration points subdivided the 50 and 125 m flights into segments of 10 and 25 m respectively (see Fig. 6.1: Kes, session 1 and 3). The birds always changed from flapping flight to gliding on approaching the end of the track. The instant of this change was hand-clocked for each flight and stored in memory. These data provided accurate estimates of gliding times. A more detailed registration of the starting and landing parts of the 125 m flights was obtained by concentrating the positions of the infrared-light-sensitive cells at one end of the corridor 10 m apart (Fig. 6.1: Kes, session 5). Markings on the sidewall of the corridor made it possible to obtain an estimate of the flight altitude. Velocities were calculated by dividing the distance between two registration points by the time taken to cover that distance. For each flight, cruising speed was estimated to be the speed between two central registration points (between 20 and 30 m, and 50 and 75 m, for 50, and 125 m flights respectively).

6.2.1 Strategies for different distances and weights

Flight strategy data were obtained from 13 sessions for each bird. Each bird flew both distances during at least two sessions without added mass and with weights of 0.3 and 0.6 N. Kes flew 1226 recorded flights over a total distance of 100 km: 42.5 km without added weight, 34 km with 0.3 N and 23.5 km carrying 0.6 N. Jowie was less laborious with 1017 flights over 85 km: 29 km without added weight, 33 km with 0.3 N and 23 km carrying 0.6 N.

The flight behaviour of the kestrels was extremely stereotyped. Flapping flight was used to descend from the falconer's hand (at a height of about 1.8 m) and to cross about two-thirds of the way to the other falconer. This was followed by a glide that ended with a short swoop up to the glove. The flight altitude was usually stable at some height between 0.3 and 0.8 m during each flight. Differences in flight path altitude did not effect the flight durations.

There were differences between the birds. Without added load or with 0.3 N, Kes flew about 0.3–0.4 $m\,s^{-1}$ faster than Jowie. Over a distance of 125 m, with a load of 0.6 N, this speed difference vanished completely.

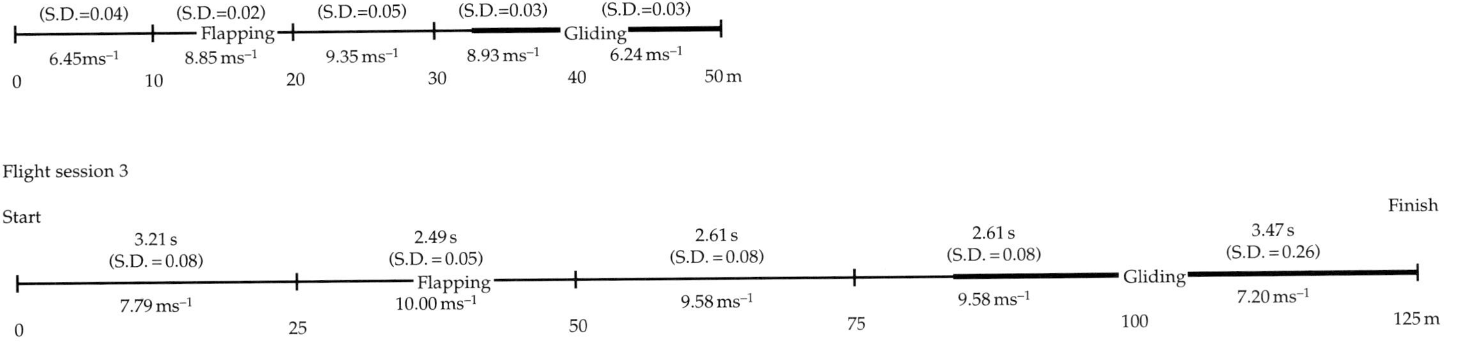

Fig. 6.1 Examples of averaged data, collected during three flight sessions with a trained female common kestrel (Kes) flying in a windless coridor. Session 1: average body mass 192 g; 156 flights. Session 3: average body mass 193 g; 56 flights. Session 5: average body mass 190 g; 92 flights, registration points are concentrated in one end of the corridor (Videler *et al.* 1988*a*).

The average unloaded cruising speeds during the 125 m flights were 9.71 m s^{-1} for Kes and 9.45 m s^{-1} for Jowie. The average speed of both birds over 125 m was 1 m s^{-1} faster than over 50 m. Cruising speed was of course relatively less influenced by starting and landing and was 0.5 m s^{-1} faster during the 125 m flights. This consistent result suggests that both birds deliberately choose a lower speed to cover a shorter distance.

Both Kes and Jowie on average decreased their flight speed with added weight. Figure 6.2 visualizes this point. Addition of 0.3 N reduced the cruising speed by an average of 0.4 m s^{-1}. The added weight seemed to affect Kes more strongly than Jowie. Despite the general trend the average velocities of Kes during two 125 m sessions with 0.3 N added weight were higher instead of lower than during sessions without loading. The standard deviations of the average and cruising speeds of the first session for Kes with an added weight of 0.6 N were notably higher than all the other standard deviations measured. A detailed look at the 96 flights during this session shows a peculiar change of behaviour after flight 38. The standard deviations show the usual order of magnitude, if the data for the first 38 flights are separated from the later ones. Kes obviously started to fly very fast, even faster than during the session without added weight. The durations of the last 58 flights were similar to those of the other 50 m/0.6 N sessions of Kes and Jowie. These data seem to suggest that birds indeed decide which flight strategy they choose given the conditions at the start. Sometimes it turns out not to be the best for some unknown reason and flight behaviour is changed instantly.

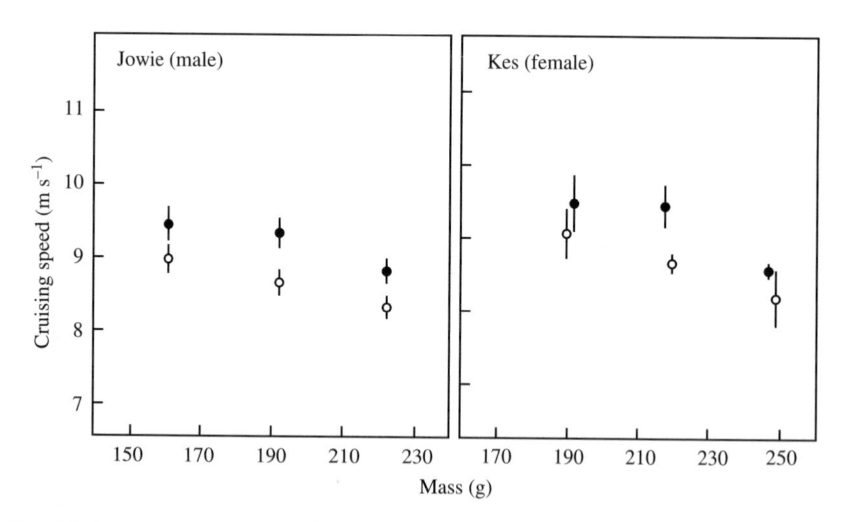

Fig. 6.2 Average cruising speed (with standard deviations) of a male and a female common kestrel during 125 m (filled symbols) and 50 m (open symbols) flights in a windless corridor plotted against total weight. Leaden instead of leather anklets were used to obtain the higher weight classes (Videler *et al.* 1988*a*).

6.2.2 Take-off

Both kestrels always flapped their wings at take-off; height loss over the first few metres was minimal and cruising speeds were reached after about 20 m. We did not measure accelerations or the forces exerted on the gloves during departure. Bonser and Rayner (1996) and Bonser *et al.* (1999) studied take-off performance of the European starling (average mass 76 g, 0.75 N). Starlings keep the wings folded during the initial phase of leaping while a push-off force is generated by the legs; accelerations between 14 and 48 $\mathrm{m\,s^{-2}}$ were measured. The peak reaction force was about 1.2 N and the push-off angle about 70° for branches varying in thickness from 4.5 mm to 20 mm. Starlings do not adjust the push-off force or angle with the thickness of the perches. The energy lost due to deformation of the thinnest perch was 0.015 J which could have raised the bird 2 cm vertically. Maximum leap forces during take-off vary among species being four times the body weight for starling and only 2.3 times the body weight for pigeons with a maximum acceleration of 15.6 $\mathrm{m\,s^{-2}}$ (Heppner and Anderson 1985).

Tobalske *et al.* (2004) expected that the very small hind legs of hummingbirds would restrict the thrust force at take-off. They measured the take-off performance of the rufous hummingbird with an average body mass of 3.2 g. Take-off angles varied between 65% and 78%. An instrumented perch measured push-off forces of 0.047 N (1.6 times the body weight). High-speed films showed that the animals used wing beat frequencies at escape take-offs of 53 Hz giving the animal a peak acceleration approaching 40 $\mathrm{m\,s^{-2}}$. It is difficult to separate the forces from the wings in interaction with the air from those of the legs. The wings unfold and complete an upstroke during the 122 ms of the leg thrust duration. The total acceleration force peaked during escapes at 0.055 N (1.8 times the body weight). The leap forces as multiples of body weight are smaller than those of starlings and pigeons and the shorter legs might be responsible for some fraction of the difference.

6.2.3 Gliding to a halt

Accurate measurements of landing events are rare. Available data reveal different strategies among and within species.

Both the take-off and landing forces of starlings increase linearly with body mass: a 60 g starling lands with an impact of 0.6 N and takes off with twice that force; an 80 g bird lands with 1.8 N and starts on average using 2.3 N. Take-off forces of starlings are 45% higher than landing forces. Take-off and landing angles are similar and close to 70° (Bonser and Rayner 1996).

In the field trained kestrels always land on the falconer's glove against the wind. The touch down is without a noticeable impact (gloves are recommended because they take a firm grip with their sharp talons after landing). Figure 6.3 shows Kes a fraction of a second before landing on the glove. The arm wings serve as airbrakes and are kept almost perpendicular to the flight direction; detached flow on the upper part of the left arm wing lifts the

Fig. 6.3 A trained common kestrel landing on the glove of the falconer after a free flight in the field. Following common practise the bird is fitted with leather anklets and a piece of nylon rope threaded through the eyelets with a knot on one side. See text for an aerodynamic interpretation of the event.

coverts. The bird approaches without losing altitude keeping the feet at the height of the glove. The alula is stretched and lifted up probably inducing a leading edge vortex on top of the swept-back hand wing with its sharp leading edge to produce the required lift at speeds approaching zero. Birds landing on a branch will commonly use the same technique: brake to reduce speed to almost zero and simultaneously generate enough lift to end just above the branch.

The landing behaviour of the kestrels in the windless corridor is different; there the birds decelerated while gliding with stretched wings at the same flight level to a point at about 5 m in front of the falconer where the final swoop to the glove starts. At the end of the glide they sweep down and straight up to land on the glove without any noticeable impact. The forward kinetic energy is wasted in the swoop.

The glides near the end of the flights in the corridor were usually straight and uninterrupted. The duration of the glides decreased with weight increase from about 36% down to about 10% of the total flight duration. Decelerations were estimated using the data of flights of Kes's session 5 in the direction where the registration points were concentrated at the end of the track (Fig. 6.1). Four flights with uninterrupted glides at one altitude over slightly more than 30 m distance were selected for detailed analysis. In still air, without height loss, lift equals weight and drag equals mass times deceleration. The analysis showed deceleration values varying between 0.77 and

0.91 m s^{-2}, with an average of 0.84 m s^{-2}. From these deceleration values, lift/drag or glide ratios of 12.8 : 1 and 10.8 : 1 (11.7 : 1 on average) can be calculated. Glide ratios shall be further discussed in Section 6.7.

6.3 *Cruising flight characteristics*

Precise quantitative kinematic descriptions of wing beats of free flying birds are rare. The reason is that it is very difficult to film at least one complete wing beat cycle with a camera in fixed position. A fixed camera position is required to obtain pictures showing the movements in an earthbound frame of reference. The bird needs to pass close enough to get a sufficient detailed picture. The animal must be in focus and the frame rate of the camera should be high enough to take a sufficient number of frames per cycle.

Kinematic studies are easier if birds can be persuaded to fly steadily at various speeds in a wind tunnel. Drawbacks are the noisy conditions, the confined space which could influence the kinematics and the fact that the birds cannot choose their preferred speed. Therefore insights in wing beat kinematics extracted from studies of voluntary flight in still air are treated first.

6.3.1 Wing beat kinematics

Marey (1890) was towards the end of the nineteenth century way ahead of his time when he used three electromagnetically synchronized cameras in fixed perpendicular positions to take time series at 50 frames s^{-1} of pictures of freely flying birds. He gave the pictures to an artist who made a wax sculpture of the shape of the bird at each time interval. Bronze casts were made using the lost wax technique. Series of bird statues accurately visualized the wing beat cycles (Fig. 6.4).

Marey summarized his findings in eight kinematic rules:

1. The downstroke lifts the body and increases the speed; the upstroke also generates lift but decreases the velocity.
2. The wing tip describes an elliptic trajectory with an oblique major axis from back to front.
3. The direction of the movement of the trajectory is such that the wing tip goes forward and downward, and upward on the way back.
4. The wing is extended and almost flat during the downstroke.
5. During the upstroke the surface of the wing is inclined with respect to the flight direction, the underside faces forward.
6. The duration of the downstroke is generally longer than that of the upstroke.

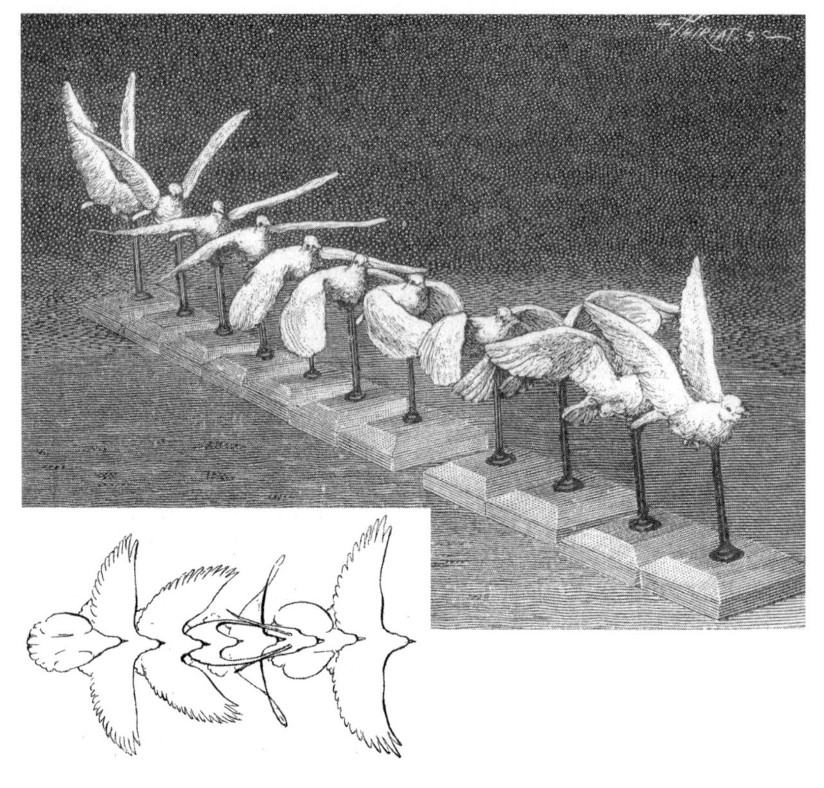

Fig. 6.4 Bronze statues of one wing beat cycle of a pigeon based on film pictures taken simultaneously from three perpendicular sides. Inset shows drawings made from the top view (Marey 1890).

7. In flight the wing is only rigid during the downstroke and partly folded during the upstroke. Folding of the wing during the upstroke becomes less evident in bigger birds.
8. During the upstroke the primaries rotate around their longitudinal axis. These feathers leave slits to let the air pass freely.

Marey does not give details of how he exactly extracted these rules from his films of pigeons and gulls and it is therefore not easy to judge how generally valid they are.

We studied wing beat patterns shown by the kestrels Kes and Jowie in the windless corridor. During these experiments (Videler *et al.* 1988*b*) the flying birds were filmed in side view and simultaneously from below. The 16 mm high-speed camera used was designed to stop the film completely before each picture was taken. In our case it made these still pictures at a rate of about 200 frames s^{-1}. It was mounted horizontally in a fixed position in a side passage of the corridor. The optical axis of the lens was perpendicular to the longitudinal axis of the corridor. A 2.35 m long and 1.5 m wide mirror in the middle of the corridor, tilted at 45° to the horizontal plane, faced the camera. A bird flying along at about 1 m over the centre of the mirror appeared on the film in side and bottom views. We used indoor plants as obstacles to guide the birds over the centre of the mirror at the right height. The camera was triggered with infrared sensitive cells by the approaching bird, so that it was running at constant speed before the bird entered the field of view. Slightly more than one complete flapping cycle was recorded during each crossing. The distances between the falconers and the camera position were large enough to make sure that the birds crossed the mirror at steady cruising speeds. Each frame of film contained a direct lateral image and a mirror image of the ventral side of the bird. We defined a three-dimensional frame of reference with the origin at the point where the camera's optical axis penetrated the mirror. The x-axis ran from the origin parallel to the longitudinal axis of the corridor. The flight direction of the birds was roughly along the x-axis. The vertical axis in the lateral image was the z-axis. The ventral image contained the origin, the x-axis, and the horizontal y-axis coinciding with the optical axis of the camera. Our set-up was four times as fast as that of Marey but we had only two images. However, we could analyse the cyclic events more precisely using computer aided digitizing techniques and we manipulated the weight of the birds by hanging 31 or 61 g lead on their feet. The lateral and ventral images of the beak, the base and the point of the tail as well as the two wing tips in each frame were digitized. The surface areas of the projections on the horizontal plane of the wings and of the tail were measured from the ventral image of each frame. The inclination angle of the bird was defined in the x–z plane as the angle between the x-axis and a line through the point of the beak and the end of the tail. To measure the inclination angle of the tail the line through the tail tip and tail base was used. The wing stroke angle was also determined in the x–z plane using the tip of the wing facing the camera. The point with the highest z-value was shifted along the x-axis to the x-position of the image with the lowest z-value. The wing stroke angle was then calculated as the angle between the x-axis and the line through the extreme wing tip points. All angles were corrected for the difference between the beak direction and the x-axis in the x–z plane to obtain angles with respect to the flight direction of the bird. Flapping flight is a cyclic motion where the period T equals one complete upstroke and downstroke. Periods are the units for comparisons. Fourier analysis was used to fit harmonic functions through the digitized data points; details of the numerical method are shown in Box 6.1.

Figure 6.5 shows the outlines of the lateral and ventral view of the male Jowie (body mass 162 g) flying at cruising speed without and with 61 g added load. The average frame rate was 198 frames s^{-1}. Every fifth frame is drawn. Without load, the average velocity is 8.1 m s^{-1} and the wing beat frequency 5.9 Hz. The added weight reduces the speed to 7.1 m s^{-1} and the wing beat frequency is slightly increased to 6.2 Hz. Note that these frequencies are within the range shown in Table 6.1.

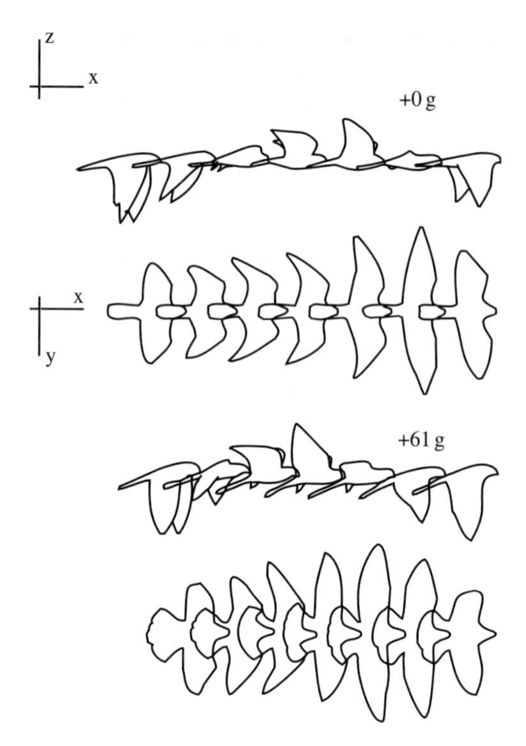

Fig. 6.5 Outlines of the lateral and the ventral view of a 162 g male kestrel at cruising speed without and with 61 g added load (Videler *et al.* 1988*b*).

Box 6.1 ***Fourier analysis of wing beat cycles of kestrels***

In an orthogonal frame of reference is x defined as the flight direction, z is the vertical, and y the lateral direction.

The sequences of z-coordinates of the wing tips, showing the largest amplitudes, are used to get the best approximation of the period of the movement T by fitting a harmonic function (standard statistical programme packages can do that). Displacements in the x-, y-, and z-direction of the beak, the tip of the tail, and the wing tips are approximated with least square fitting methods by harmonic functions in the form:

$$f(t) = a_0 + b_0(t - t_c) + \sum_{j=1}^{3} \left(a_j \cos \frac{2j\pi t}{T} + b_j \sin \frac{2j\pi t}{T} \right) \qquad (6.1.1)$$

where t_c is the time point of the central frame of the sequence, halfway between the first and the last frame. The first two terms represent a straight line motion at constant speed, the others are Fourier terms describing the harmonic motion. a_0 is the average position and b_0 represents the average speed. Three Fourier terms are sufficient because higher frequencies drown in the noise which was usually about ±8 mm. Figure 6.6 shows the actual displacements and the fitted functions in the x- (a), y- (b), and z-directions (c) of the beak, the tip of the tail, and the wing tips. The left wing tip was away from the camera and disappeared behind the rest of the bird during parts of the sequence. (Only the right wing tip was used for the comparisons, assuming symmetrical movements.)

The surface areas of the projections on the horizontal plane of tail and wings were also analysed as periodic functions with time period T (d). They are approximated by:

$$s(t) = a_0 + \sum_{j=1}^{3} \left(a_j \cos \frac{2j\pi t}{T} + b_j \sin \frac{2j\pi t}{T} \right) \qquad (6.1.2)$$

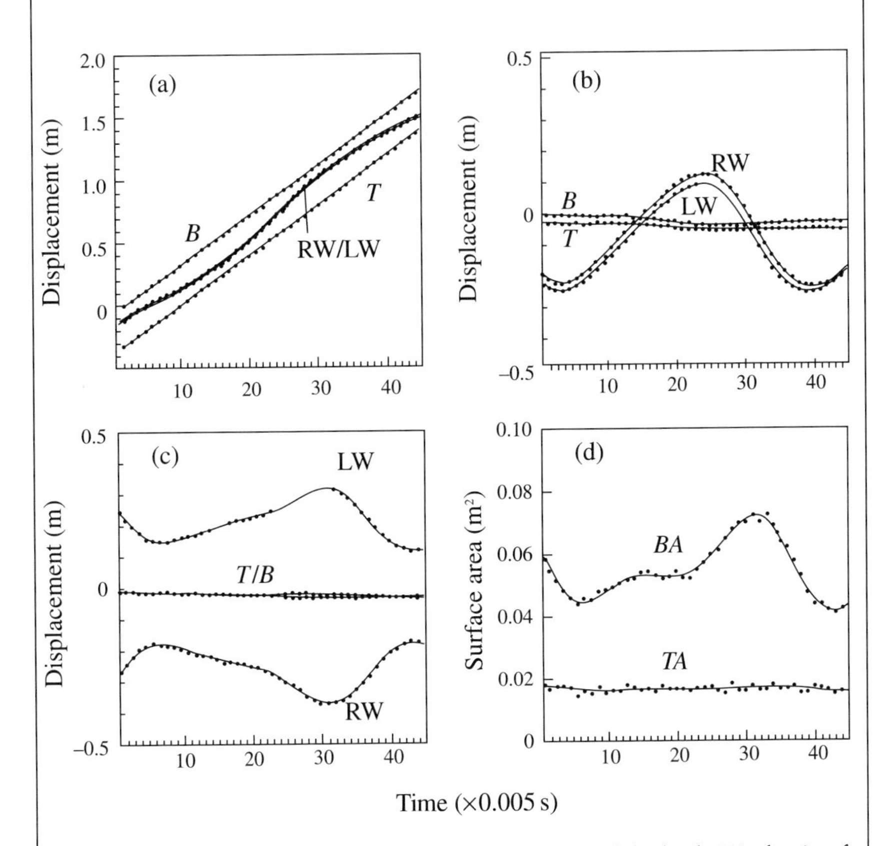

Fig. 6.6 (a)–(c) The position, digitized every 1/200 s, of the beak (B), the tip of the tail (T), and the left and right wing tips (LW, RW) of a male kestrel flying indoors at cruising speed over a mirror. Digitized points and fitted functions in the x- (a), the z- (b), and the y- (c) directions are drawn. (d) The projections on the horizontal plane of the total surface area of the bird (BA) and of the tail area (TA), with their fitted functions (Videler *et al.* 1988*b*).

The average surface area equals a_0 and the other terms describe the periodic changes as the sum of three sets of Fourier frequencies. (Higher frequencies are not included because their contributions remain within the limits of the measuring fault of ± 4 cm^2.) Figure 6.6(d) gives measurements and functions of the total area of the bird (BA) and the tail area (TA).

The functions describing the displacements of the beak in the x-direction were used to estimate the average velocity (b_0 in these equations). The durations of the up anddownstroke (T_u and T_d) were estimated from the durations between the instants when the wing tip reached the highest (t_h) and lowest (t_l) position along the z-axis. The wing beat frequency is:

$$f = \frac{1}{T_u + T_d} = \frac{1}{T} \qquad (6.1.3)$$

The inclination of the stroke plane ϕ was defined in the x–z frame of reference to be:

$$\phi = \arctan \frac{[W_z(t_h) - W_z(t_l)] - [B_z(t_h) - B_z(t_l)]}{[W_x(t_h) - W_x(t_l)] - [B_x(t_h) - B_x(t_l)]} \qquad (6.1.4)$$

where $W_z(t)$, $W_x(t)$, and $B_z(t)$, $B_x(t)$ are displacement functions of the wing tip and the beak along the z- and x-axes. The subscripts h and l indicate the wing tip positions with the highest and lowest z-values respectively. This gives an accurate estimate of the inclination of the stroke relative to the mean path of motion of the beak.

In the x–z frame of reference angles between the x-axis and the lines through the beak and tail tip and tail tip and tail base represent the total angle of inclination (β) and the inclination of the tail (β_t), respectively. To obtain the angles of inclination relative to the flight direction, these angles were corrected for non-level flight by adding the angle δ between the flight path of the beak and the x-axis.

$$\delta = \arctan \frac{B_{0z}}{B_{0x}} \qquad (6.1.5)$$

where B_{0z} and B_{0x} are the beak velocities in the z- and x-directions.

Figure 6.7 exemplifies the harmonic functions during one period T of the displacement, velocities, and accelerations of the beak, tip of the tail, and the right wing tip of Jowie flying with 31 g. The velocity functions are obtained by differentiating the displacements and the accelerations by differentiating the velocities (both with respect to time).

Table 6.1 Kinematic parameters of one female (Kes) and one male (Jowie) common kestrel during loaded and unloaded indoor flights (Videler *et al.* 1988*b*).

Bird		Kes			Jowie		
Body mass	(kg)	0.198	0.189	0.189	0.162	0.162	0.162
Added weight	(N)	0	0.3	0.6	0	0.3	0.6
n		6	4	3	2	3	2
Velocity	($m\,s^{-1}$)	8.1	8.4	7.7	8.1	7.8	7.1
Wing beat period T	(s)	0.18	0.18	0.16	0.17	0.17	0.16
Frequency $1/T$	(Hz)	5.5	5.5	6.2	5.9	5.9	6.2
Upstroke duration T_u	(s)	0.10	0.10	0.09	0.10	0.09	0.08
Downstroke duration T_d	(s)	0.08	0.08	0.07	0.07	0.08	0.08
Downstroke ratio T_d/T		0.43	0.43	0.46	0.43	0.45	0.48
Total inclination angle β	°	3	7	9	3	7	11
Tail inclination angle β_t	°	14	21	22	12	19	23
Wing stroke angle ϕ	°	91	87	91	86	84	80
Projected wing area average	(m^2)	0.043	0.044	0.048	0.040	0.043	0.045
Projected wingspan average	(m)	0.48	0.47	0.49	0.48	0.52	0.55
Projected tail area average	(m^2)	0.008	0.010	0.014	0.008	0.009	0.019

Maximum wingspan: Kes 0.72 m, Jowie 0.70 m. Projections are measured of the dorsal view.

The downstroke took $0.42T$ and hence was shorter than the duration of the upstroke. The displacements in the x-direction were about 1.3 m in one period of 0.170 s accounting for an average velocity of 7.7 $m\,s^{-1}$. Head and tail moved along straight lines in the x- and y-plots and showed oscillations of about 1 cm around the average position in the z-direction. The wing tip moved away from the body during the beginning of the downstroke to a distance of 0.33 m from the body axis at $0.2T$ and approached to 0.15 m at $0.6T$ in the first part of the upstroke. The total excursion of the wing tip in the z-direction (the wing beat amplitude) was about 0.35 m and slightly larger downwards from the average position than upwards. The velocity in the y-direction and accelerations in the x- and y-directions of beak and tail were zero on average. The small fluctuations were noise and had no physical meaning. In vertical direction, fluctuations of the velocities and accelerations of beak and tail tip were in counter phase, indicating small oscillations with period $0.5T$ of the body axis in the x–z plane. We did not see signs of the speed increase of the bird during the downstroke and the decrease during the upstroke predicted by Marey (1890). The wing tip velocity at the beginning of the downstroke reached a maximum value of about 10 $m\,s^{-1}$. During the first half of the downstroke and last part of the upstroke the wing tip velocity in the flight direction was greater than the average speed. The speed was lower during the rest of the period, with a minimum of 5 $m\,s^{-1}$ at $0.6T$. The bird pulled the wing tip towards the body (y-direction) just before the end of the downstroke at 5 $m\,s^{-1}$, which was faster than the movement away from the body near the start of the downstroke, which reached 3 $m\,s^{-1}$. The maximum speed in the z-direction during the downstroke was 8 $m\,s^{-1}$ which was faster than the upstroke maximum

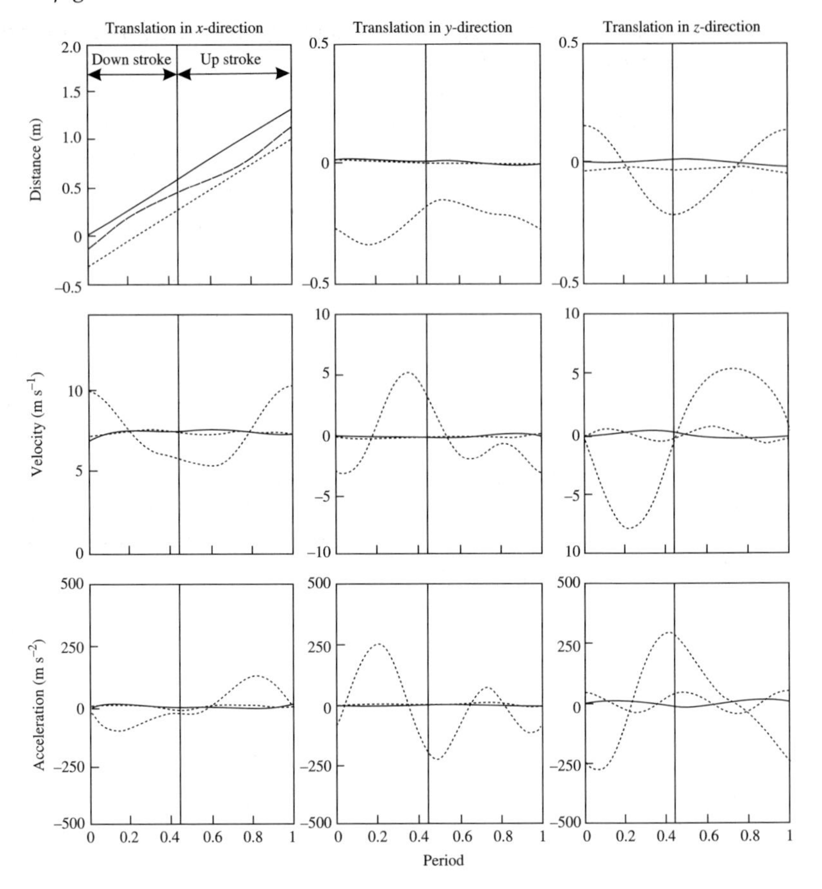

Fig. 6.7 The harmonic functions during one period T of the displacements, velocities, and accelerations of the beak (solid line), the tip of the tail (dashed line), and the right wing tip (broken line) of a male kestrel carrying 31 g (Videler *et al.* 1988*b*).

of 5.5 $m\,s^{-1}$. Wing tip accelerations in that direction reached two extremes, one of -250 $m\,s^{-2}$ and one of $+250$ $m\,s^{-2}$ during the downstroke. One was the downward acceleration at the beginning and the other the deceleration of the downward movement near the end.

The general results of the detailed analyses of 13 sequences of Kes and 7 for Jowie presented in Table 6.1 also witness the consistency of the kinematic patterns. Jowie flew slightly slower than Kes and his flight speed decreased with an increase of added weight, consistent with the results from the flight strategy experiments (Videler *et al.* 1988*a*). The average velocity of Kes during the 0.3 N added-weight experiments reached 8.4 $m\,s^{-1}$, an increase compared to the unloaded average.

Wing beat frequencies in both birds reach the highest values under maximal load. Downstroke ratios showed the highest values at the highest load but they were less than 0.5. The shorter wing beat cycle under load was mainly caused by a faster upstroke, implying that the relative importance of the downstroke increased.

The average vertical wing tip displacements during unloaded flights were 166 and 167 mm for Kes and Jowie respectively. These amplitudes under 0.6 N added-weight conditions were 190 and 184 mm. The average wing tip displacement in the fore-aft or x-direction of Kes carrying 61 g was 49 mm compared with 69 mm unloaded. In Jowie's case, these figures were 59 and 78 mm.

Table 6.1 shows that the wing stroke angles of Kes do not seem to be affected by increased weight. She beats her wings slightly more vertically than Jowie, whose wing stroke angles decreased by a few degrees with increased weight. Jowie tends to increase his wingspan on average with increased load; Kes does not seem to bother. Wing areas increase slightly in both birds when loaded. The total and the tail inclinations increase with added mass. The projected average tail area also increases substantially. The birds probably use the tail as a delta wing, at high angles of attack with leading edge vortices to create extra lift and consequently considerable drag (Hoerner and Borst 1975). The drag penalty decreases the velocity despite the increased wing beat frequencies.

The maximum wingspan measured by stretching the wings manually without undue strain was 0.72 m for Kes and 0.70 m for Jowie. These values were never reached during flight. The average wingspan during the wing beat cycles was between 60% and 80% of the maximum values. The largest span occurred in the middle of the downstroke where it was 92% of the maximum in the most extreme case. The average projected wingspan values in Table 6.1 show a clear increase with increased weight for Jowie but not for Kes.

6.3.2 Details of a single wing beat

The kestrel studies provided general kinematic data of birds in free flight. The spatial resolution however was not good enough to show details of the movements of the parts of the wings during upstroke and downstroke. We therefore look in some detail at wind tunnel studies by Bilo (1971, 1972) providing quantitative details on the wing beat of a small passerine. A house sparrow was trained to fly in a wind tunnel. The wind speed used during the experiments was 8.15 m s^{-1}. Stereo photographs taken at 520 frames s^{-1} with a camera in fixed position were used for the kinematic analyses of the upper surface of the right wing (Fig. 6.8). The flapping frequency was about 22 Hz. The results are described in two articles, a big one of 72 pages dedicated to the downstroke (Bilo 1971) and one of 11 pages covering the upstroke (Bilo 1972). Eight years later Bilo used the same data to emphasize the unsteady kinematics of the sparrow's wing beat in an article written in English (Bilo 1980). From the stereo pictures 3-D reconstructions could be made using an earth (camera) bound, a body bound, and a wing bound orthogonal frame of reference. This technique allowed calculations of the

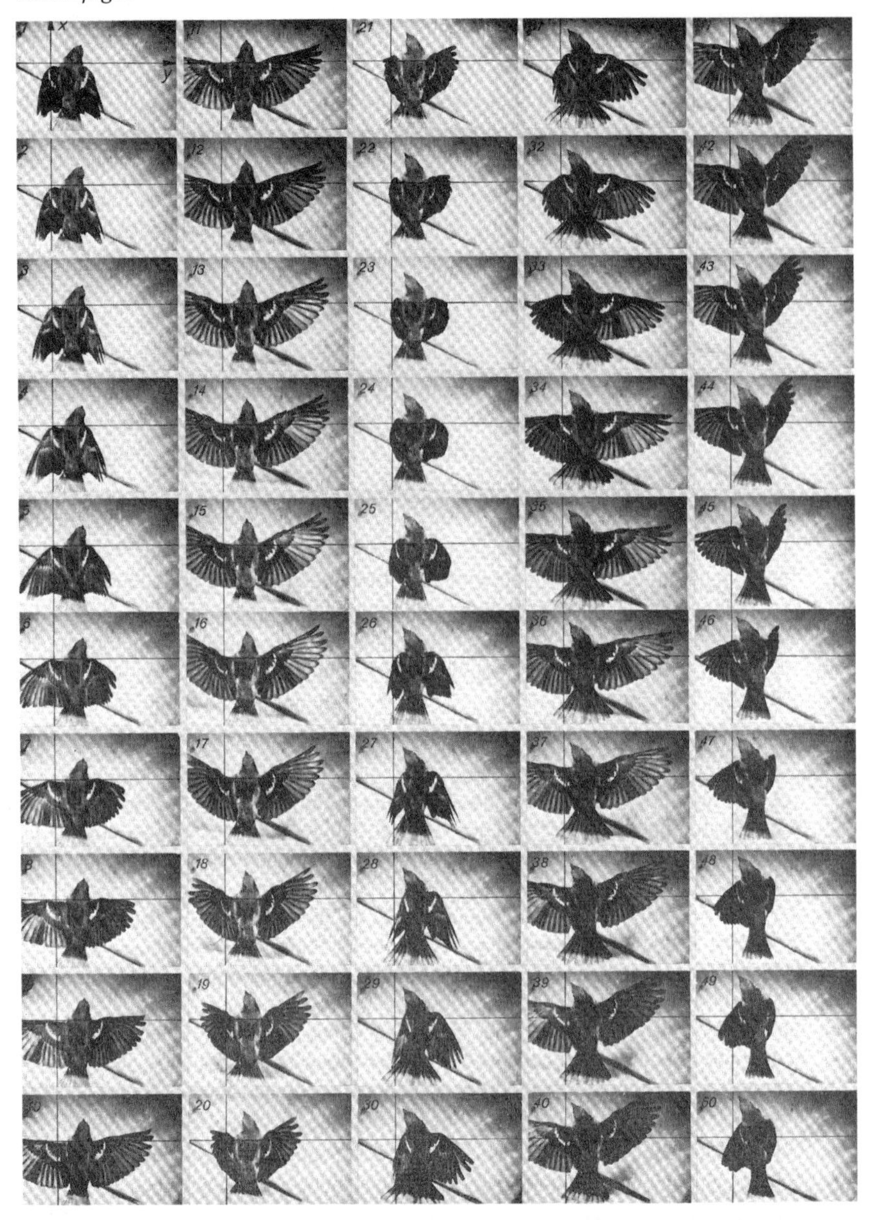

Fig. 6.8 A house sparrow flying in a wind tunnel. The left pictures of a stereo film are shown. The time interval between the pictures is 1.92 ms (Bilo 1971, reprinted with kind permission (© 1971 Springer Verlag)).

movements in space, of the changes in shape and of the angles of attack of the air on the wing. One downstroke was analysed quantitatively in great detail (Fig. 6.8, frames 30–47). The duration of that downstroke (T_d) was 33 ms. Relative to the body the hand wing including the wing tip moves gradually downward and slowly forward until about $0.7T_d$. At the start, the vertical angle of the wing tip is about 40° (at 0° the wing tip would be straight above the bird) and the horizontal at about 70° (at 90° the wing is stretched out straight with respect to the main body axis in a horizontal plane). The hand wing starts to fold backward during the last $0.3T_d$. The wrist follows a similar downward forward trajectory until $0.4T_d$, when it abruptly starts to move forward at a faster rate and slows down the vertical movement. This means that the hand wing sweeps further down to an angle of 170° while the wrist stops the vertical movement at about 140°. At the end of the downstroke the wrist has moved forward to about 125° while the wing tip is stretched straight out at 90°. This movement implies in a wing bound frame of reference, that at the wrist the angle between the arm and hand changes back and forth continuously. In the plane of the wing the angle swept is about 30°, whereas a vertical movement of the hand goes from about 15° above the horizontal to just fewer than 30° below. We have to keep in mind that the movements we see on the film are the result of muscle activities moving the skeleton inside the wing and reactive forces from the air on the feathers forming the outer shape of the wing. The interaction is obviously highly dynamic. The sudden deceleration of the downward movement and the violent forward acceleration of the wrist coincide with high frequency rotating and twisting oscillations around the leading edge of the hand wing. During the middle of the downstroke (between 0.35 and $0.7T_d$) the angle of attack on the hand wing oscillates with amplitudes of between 5° and 10°. The frequency of these oscillations is about 260 Hz, which is about 12 times the flapping frequency. The angle of attack of the arm wing only changes gradually from values around −5° at the start up to about +10° near the end. The torsion of the wing is measured from the change of angle of attack on the wing from the body outward. During the second half of this downstroke the torsion of the hand wing starts to twist rhythmically again with a frequency of about 260 Hz. The fastest rotational velocity of these twisting movements of the hand wing is measured at more than 5700° s^{-1}. Bilo also measured the profile curvature of the upper surface of the wing during the downstroke. The results of these measurements reveal that even the camber of the wing at every distance from the body outward changes in time during the course of the downstroke.

The analysed upstroke took eight frames (Fig. 6.8, frames 22–29) or 15 ms, which is less than one-third of the stroke duration and much faster than the upstroke taking 33 ms. The arm wing rotates upward around the shoulder joint, bringing the wrist up with the hand wing trailing behind in an almost vertical position. During this movement the arm is folded close to the body with the underarm (radius and ulna) almost parallel to the longitudinal body axis. The strongly adducted arm is rotated upward around the shoulder joint. Figure 6.8 shows that this movement is almost completed in frame 26. The angle of attack on the arm wing during that first part of the upstroke remains about constant at around 20°. Between frame 26 and 29 the arm rotates forwards around an axis parallel to the leading edge (pronates) and the angle of attack decreases to slightly negative values at the beginning of the downstroke.

At the start of the upstroke, the hand wing is angled downward around the wrist joint. From that position, the hand wing is rotated backward and upward (Fig. 6.8, frames 24–29). The wing tip describes a circle segment of about 150°. During this movement, the hand wing closes at first (frames 24–27) and opens after passing the horizontal position (28–29). During the 150° rotation of the hand wing the individual primary feathers rotate around their shaft and the hand wing opens as a Venetian blind (frames 27–29). During the final phase of the upstroke the entire wing stretches without any further upward movement. The wing is fully stretched and spread during the early stages of the downstroke.

6.3.3 Wing beat kinematic rules?

Flight kinematics can be highly stereotypical although the possible variations are extremely large. Individual birds adapt the way they fly to the circumstances. If these do not change we see individuals repeating the same strategy. What exactly the flight pattern is may be different for different individuals belonging to the same species. Among species there are as many styles as there are species. For example, wing beat frequencies vary from extremely high values of up to about 70 Hz for hummingbirds to the lowest measured figure of less than 2 Hz for the Goliath Heron (Rayner 1988). Passerines beat their wings at frequencies up to 22 Hz. In the field individual birds can change their wing beat frequency considerably, moreover they also can control the relative duration of the downstroke (Oehme and Kitzler 1974). A Eurasian collared dove (wing beat frequencies between 3.9 and 6.1 Hz) varied the relative downstroke period between 0.14 and 0.55 times the wing beat cycle. The measured frequencies of rooks varied between 3.2 and 4.1 Hz and the downstroke ratios between 0.33 and 0.72. Our kestrels in the corridor increased the wing beat frequencies with added weight. These higher frequencies were due to faster upstrokes. Both birds kept the downstroke duration approximately constant. Not only the timing but also the shape of wings and tail changes during wing beat cycles.

Marey's kinematic rules are not generally applicable; we therefore must continue to investigate wing beat characteristics of many more species under controlled conditions to discover the details of various styles. It might be the case that we find that every species has a style of its own and that there are as many individual differences in wing beat kinematics as there are differences in walking styles of individual human beings.

6.4 *Hovering*

The house sparrow was flying in the wind tunnel at a fixed position with respect to the earth because it flew at the speed of the wind generated by the tunnel. Flying at one spot without opposing wind is much more difficult. Like most small passerines a house sparrow does it only very briefly usually when approaching the nest or catching an insect. Flying in one position without wind is called hovering flight, only hummingbirds varying in mass between 2 to 8 g can do it for up to several minutes. Sunbirds, weighing between 10 and 20 g, are more reluctant to hover but will do it to reach flowers that cannot be probed by perching. The wings of hummingbirds are usually hardly visible; those of hovering sunbirds beat slower and can be seen as a blurred disk. The genuinely hovering birds can change position sideways or backwards. Hovering is relatively easy to film with a high-speed camera in fixed position. It is therefore surprising to find that there are only very few quantitative kinematic analyses available in the literature.

In Chapter 2, Section 2.2.3 Stolpe and Zimmer's (1939) description of the morphology of the hummingbird flight apparatus was used to illustrate the differences with the general bird wing design. Here their kinematic results, obtained by making films at up to 1500 frames s^{-1} of hovering behaviour from the lateral and dorsal side, describe the principle of hovering flight. The two species investigated, the black Jacobin and the glittering-bellied emerald were animals kept in captivity in the Berlin zoo.

A schematic representation of the hovering wing activity is given in Fig. 6.9. The body is usually in a rather vertical position with the longitudinal axis at about 40–50° to the horizontal. The wing tips sweep a figure of eight.

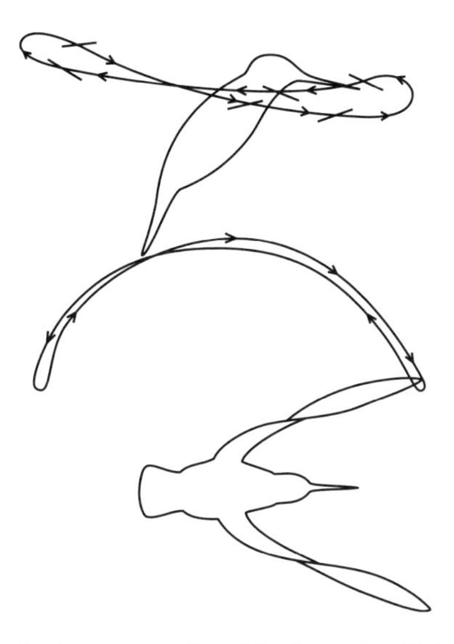

Fig. 6.9 Wing tip path during one stroke of the hovering flight of a hummingbird in side view (upper panel) and top view (lower panel). The direction of the movement is indicated. The side view shows also the angles of wing cross sections near the wing tip (based on Stolpe and Zimmer 1939).

The forward (down) stroke starts high back and moves to a lower position in front of the bird. The back (up) stroke is nearly horizontal. The angle swept is about 130°, reaching equally far forward as backward. The hand wing is rotated almost upside down during the backstroke and the anterior edge of the hand wing is the leading edge throughout the entire beat cycle. Most of the hand wing has a positive angle of attack during both strokes. Wing beat frequencies were between 36 and 39 Hz for the glittering-bellied emerald and 27 to 30 Hz for the black Jacobin. The tracks of the wing tip in side and top view are figures of eight. Maximum wing tip velocities of the wing tip of the black Jacobin of 20 m s^{-1} were reached at about 0.6 times the half stroke period. The downstroke ratio is approximately 0.46. The wing joints allow the hand wings to rotate around the longitudinal wing axis over about 150° near the tip of the hand wing. These rotations at the changeover for each half stroke are fast, reaching values up to 10.204° s^{-1}. The precise nature of the interaction between hovering hummingbird wings and air is not known and awaits quantitative flow visualization.

Maximum performance of ruby-throated hummingbirds was tested experimentally by Chai *et al.* (1997). Hovering performance of two males and four females was filmed at ambient temperatures of 5°C and 25°C and under unloaded and loaded conditions. The loading procedure was simple and neat as follows. A 76 cm long thread with beads evenly distributed along its length was curled up on a feeding table. One end was tied to a flexible 0.2 g rubber loop. The bird could bear this loop as a necklace. It had to lift up the thread of weights from the table to hover within a screened enclosure. The weight carried could be determined from the number of beads lifted from the table. Chains of 3 and 4 g total weight were used. Both males and females lifted around 80% of their body weight up in the air. The duration of the weight-lifting exercises were in the order of 1 s. Stroke amplitudes and wing beat frequencies were used as measures for performance. The results were rather different for individual birds but some trends emerged. Under maximum load the temperature difference did not seem to have effect. The males of slightly less than 3 g used higher frequencies (56–58 Hz) than the about 1 g heavier females (49–52 Hz). The stroke amplitudes were under load roughly the same for all birds, varying between 175° and 190°. The unloaded performance showed much more variation. The four females used low frequencies of around 42 Hz and amplitudes between 145° and 155° at the highest temperature. The scatter among them was much higher at 5°C but they all used higher frequencies and lower amplitudes. The unloaded trend is not clear but maximum hovering performance is shown by both sexes to be independent of temperature.

Extreme hovering kinematics was performed by ruby-throated hummingbirds hovering in a mixture of oxygen and helium with a density down to one-third of normal air (Chai and Dudley 1999). Females with body masses of about 4 g beat the 48 mm long wings at 47 Hz on average. Young males with body masses of 3.9 g and intermediate 45 mm wing length used 53 Hz. Adult males with an average mass of 3.6 g and 42 mm long wings

Table 6.2 Load-lifting capacities and morphological and kinematic variables of males of four hummingbird species.

	Blue-throated hummingbird ($n = 2$)	Magnificent hummingbird ($n = 3$)	Black-chinned hummingbird ($n = 5$)	Rufous hummingbird ($n = 1$)
Mass (g)	8.4 ± 0.3	7.4 ± 0.2	3.0 ± 0.2	3.3
Maximum load (% of body mass)	174 ± 21	190 ± 14	104 ± 8	88
Duration (s)	0.43 ± 0.01	0.48 ± 0.12	0.64 ± 0.09	0.65
Wing length (mm)	85 ± 2	79 ± 3	47 ± 1	42
Frequency loaded/ unloaded (Hz)	$31 \pm 0/23 \pm 2$	$32 \pm 1/24 \pm 1$	$60 \pm 3/51 \pm 4$	62/52
Amplitude loaded/ unloaded (°swept)	$185 \pm 1/151 \pm 7$	$188 \pm 2/150 \pm 6$	$162 \pm 5/126 \pm 6$	185/163

n = Number of individuals tested; averages ±SD are calculated from average values of individuals (Chai and Millard 1997).

reached the highest frequencies of 60 Hz just prior to failure in air with a density of 0.57 times that of normal air. Under these extreme conditions sweep angles of all groups approached 180°. It looks as if mass and wing dimensions determine maximum wing beat frequencies of ruby-throated hummingbirds.

The load-lifting capacities during hovering of recently captured males of four other hummingbird species were tested by Chai and Millard (1997) using the heavy necklace method described above. Table 6.2 shows that the two largest species have by far the largest carrying capacities up to values around twice the body mass for the 7.4 g magnificent hummingbird. The data show again that hummingbirds increase both frequency and amplitude to cope with loads. The sweeping angle often even exceeds 180°, which implies that the wing tips overlap at the extremes of the stroke. The maximum frequencies of the large birds reach about half the values of the smaller ones. The frequency increase of the two large species is about 35%; that of the smaller birds just under 20%. No such size difference seems to exist for the increase in amplitude.

6.5 *Windhovering*

To avoid confusion, the term 'hovering' should be strictly reserved for flight at a fixed position in still air. Flying *sur place* against the wind is a different technique, which has been termed 'windhovering' (Videler *et al.* 1983). Windhover is an ancient common English name for the kestrel, a species well-known for this behaviour. It is their main method to prey on small mammals, lizards, and insects. Several other raptors including ospreys, some terns, the long-tailed jaeger, and pied kingfishers also specialize in this mode of flight. Windhovering is used for hunting and fishing. By flying upwind at the speed of the wind, the birds can keep the eyes in a fixed position relative to the ground. This enables them to detect moving prey on the ground or in the water. Their fixed position in the air makes it possible to study the kinematics of windhovering quantitatively in the field. The main problem is to film the behaviour with a camera in a fixed position. The bird has to be close enough to show details of its movements and it has to be in focus. Other requirements for proper analyses of what happens up there are knowledge of the mass and dimensions of the bird on the film and preferably, information about the instantaneous wind speeds as closely as possible near the bird. We mounted a high-speed camera fitted with a strong telephoto lens (600 or 850 mm) on the roof rack of a vehicle together with a wind velocity transducer on the top of a 4 m long pole. The hunting behaviour of the birds was studied before filming attempts were made. There is usually a fixed pattern and a few preferred hunting areas with abundant prey. Detailed knowledge of the hunting habits of individual birds made it not too difficult to be close enough to attempt filming. The camera was directed at the head of the windhovering bird and as soon as it was in focus, the camera was secured in fixed position and started at a frame rate of 100 or 200 frames s^{-1}. A reference grid was built in the camera and movements seen on the film are movements relative to the grid and to the ground. Filming was stopped as soon as the bird made a strike or moved away. The wind speed could be recorded on the side strip of the film by a flashing light emitting diode (LED) mounted inside the camera. Another LED marked the other side of the film every 0.01 s driven by a quartz clock for precise time keeping and calibration. We used this method to make films of common kestrels, rough-legged buzzards, greater kestrels, black-winged kites, scissor-tailed kites, and pied kingfishers.

Insight into the windhovering technique also requires knowledge about the morphometric characteristics of the bird appearing on the film. We managed to catch the raptors by bal-chatri (Cavé 1968). This small wire cage containing a mouse or some other small lively rodent is attached to the ground near the hunting bird. Nylon loops on top of the cage snares the feet of the raptor at the strike attempt. The pied kingfishers appearing on the films were caught using mist nets.

There is a correlation between average wind speed and the duration of windhovering bouts in the kestrel (Fig. 6.10). At wind speeds lower than about 3 m $s^{-1}$and higher than 13 m s^{-1} the duration was about half that during intermediate velocities. Kestrels cannot windhover at wind speeds below about 2 m s^{-1}. Under these conditions, they look for prey from a perched position. At low speeds the birds must generate predominantly lifting force. Kestrels keep their body in an almost vertical position and wings and tail are spread maximally to do that. The wing beat amplitude is high. Contrastingly, during the highest wind speed conditions the body is horizontal, the tail is completely closed, and the wings are narrow and look slender. The wing beat amplitude is low and the frequency fast. The bird is obviously trying to generate the large amounts of thrust needed to

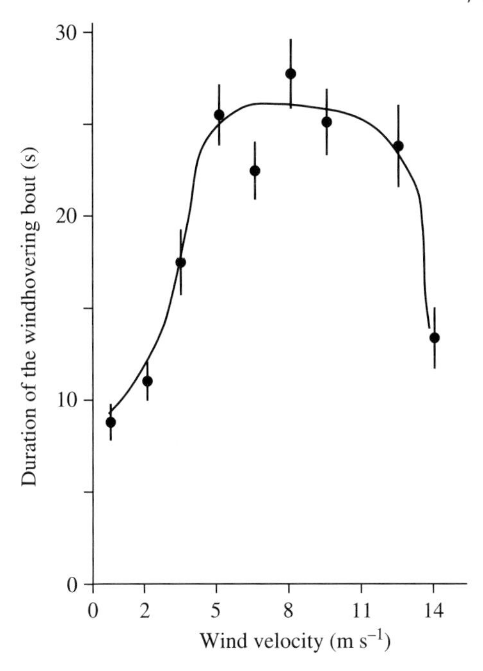

Fig. 6.10 The duration of windhovering bouts plotted against the average wind speed. The data points result from 794 measurements of the behaviour of one female during October 1979. The data were collected in a Dutch polder where the countryside is flat and open without woodland or high obstacles that could cause serious distortion of the wind flow (Videler *et al.* 1983).

counteract the high wind speed drag. In none of the birds filmed could we find correlations between the kinematics and the wind speed variations on a small time scale. Not even in the case of the pied kingfishers where we measured the instantaneous wind speed within a distance of a few metres of the windhovering bird.

Windhovering birds usually manage to keep the head in an extremely stable position in mid-air. Kestrels restrict lateral and vertical displacements of the head to plus or minus 6 mm with respect to the average position in gusting winds with speeds varying between 4.8 and 7.5 $m\,s^{-1}$ (Fig. 6.11). The displacements are not correlated with the wing beat cycle. The head stability of pied kingfishers can be even greater than that of kestrels, being normally less than 4 to 5 mm around the average. In order to keep the head in such a stable position the birds must react rapidly to changing wind conditions. The flexibility of the neck is used to dampen the movements of the body. The centre of gravity and the centre of forces generated by the wings do not necessarily coincide. Birds can control the position of these centres using the wings, the legs, head, and tail. The average position of the wings can be shifted up and down and fore and aft. The legs are usually tucked up and hidden under the feathers of the belly behind the thickest part

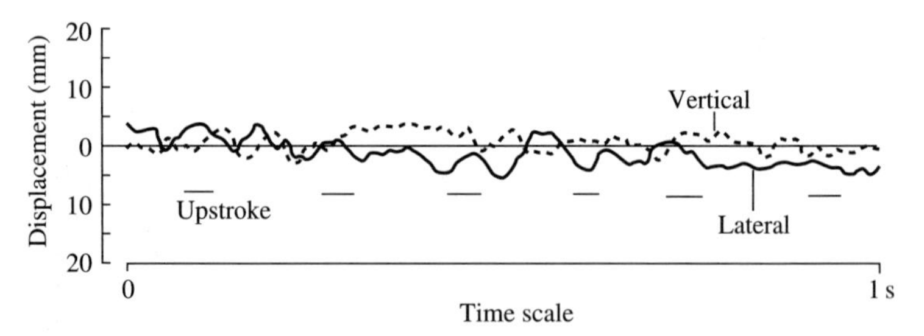

Fig. 6.11 The vertical and lateral displacements of the point of the beak of a windhovering kestrel relative to nominal position as a function of time. The wing beat cycle is indicated by the durations of the upstroke (Videler *et al.* 1983).

of the body reducing drag to a minimum. But they can also be stretched out and lowered down completely as well as moved to the front and back. Tails can spread and close, move up, down, and sideways. Figure 6.12(a) shows an example of a wing beat cycle during windhovering of a male kestrel (body mass 207 g, wingspan 76 cm) in frontal view. Filming speed was 100 frames s^{-1}. The wing beat shown lasts 0.15 s at a wing beat frequency of 6.7 Hz. Upstroke and downstroke use equal amounts of time in this case. During the bout of windhovering from which this cycle was taken, the average wind velocity was 5 m s^{-1}, varying between 3 and 8 m s^{-1}, and the average downstroke and upstroke periods were 0.09 and 0.07 s respectively. The average frequency is about 6 Hz and the maximum wing tip amplitude fluctuates around 25 cm. The bird flies with stretched wings rotated backwards at the start of the upstroke and forward towards the end of the upstroke. Figure 6.12(b) gives one windhovering cycle under similar conditions in side view. Here the upstroke takes 0.07 s and the downstroke 0.09 s, the wing tip beats obliquely forward at an angle of about 30° with respect to the vertical. The eyes are kept in a stable position with respect to the earth (indicated by the cross), the tail moves slightly up at the beginning of the upstroke.

The kinematic variation under steady conditions can be large even for a single bird. The two series of drawings of wing beat cycles of a windhovering pied kingfisher in Fig. 6.13 illustrate this. There is large variation in the use of the tail and in the execution of the upstroke. The duration of both wing beat cycles is 0.1 s. Despite the high frame rate it is difficult to measure when the downstroke ends and the upstroke begins in both cases. The wing tips are still going down after the wrist already changed direction. Results did not show a consistent relationship between wind velocity and windhovering kinematics. Insight into what really happens during these different interactions between the bird and the air awaits quantitative flow visualization under controlled conditions in a wind tunnel. So far such an extremely complicated experiment has not been feasible.

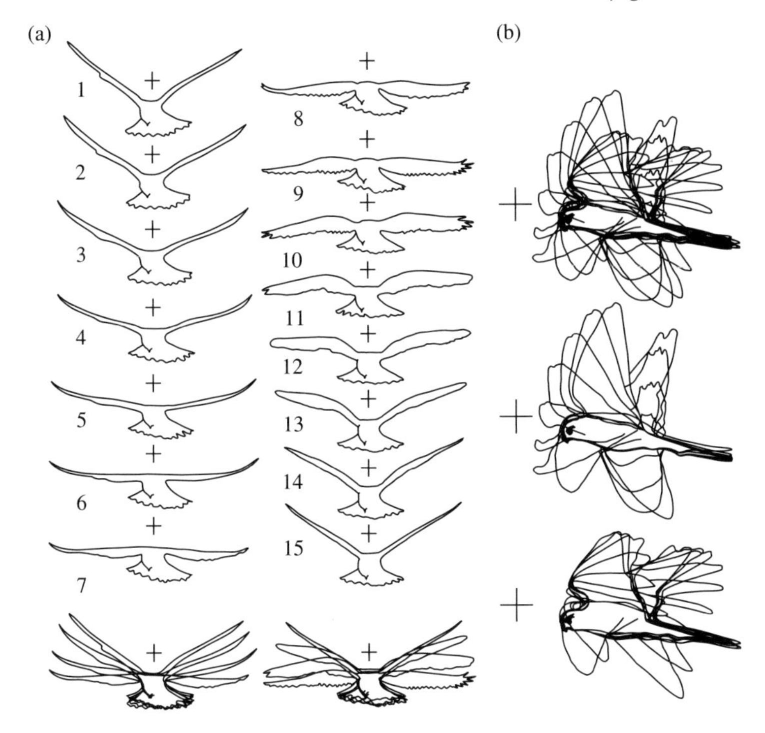

Fig. 6.12 (a) The downstroke (left) and the upstroke (right) of a windhovering kestrel in frontal view. Time interval between the frames is 0.01 s. The plus drawn above each frame is a fixed point in an earthbound frame of reference and is used to show the movements of the stroke halves in the lower panels. (b) The movements of the wings also filmed at 100 frames s^{-1} in side view. The upper panel shows the wings during the entire stroke; the panel in the middle has only the downstroke and the bottom panel only the upstroke (Videler 1997).

6.6 *Techniques to reduce the energetic demands of flapping flight*

Flapping flight is energetically expensive (see Chapter 9). Natural selection has favoured a variety of ways to economize on flight costs in birds. An overview of the techniques that have been described so far and the evidence there is about the effects is given below.

6.6.1 Intermittent flight

Many birds try to save some energy by reducing the amount of flapping of the wings. This is commonly done by punctuating flapping with short bouts without wing movements; either with the wings stretched in gliding position or folded against the body. The first option is intermittent gliding flight, the latter flap bounding flight. Flap bounding birds fly rhythmically up and down along an almost sinusoidal path. They increase speed by flapping the

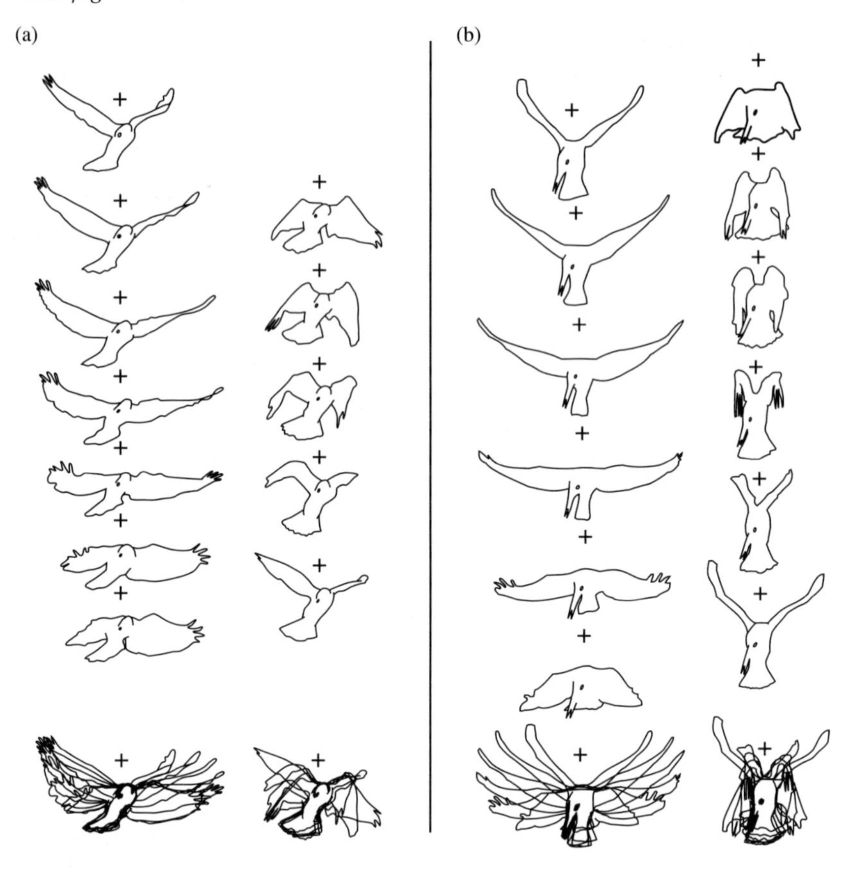

Fig. 6.13 Two wing beat cycles of a windhovering pied kingfisher illustrating individual variation. The animal is a male, body mass 75 g, wingspan 46 cm, distance between the eye and the tip of the beak is 7 cm. Films were taken at 200 frames s^{-1} with a camera in fixed position, every other frame is shown. The plus symbols above the pictures represent a fixed point in the earthbound frame of reference. In each case the downstroke is drawn on the left and the upstroke on the right. The pictures at the bottom show the drawings in the column above in overlap. Both cycles last 0.1 s (wing beat frequency 10 Hz); Wind speeds were measured at the flying altitude a few metres away from the bird. (a) The average wind speed was 3.4 m s^{-1}. The wings are brought straight up in semi-folded position; the tail spreads and folds rhythmically between 70% and 100% of the maximum width, reaching the highest values during the second half of the downstroke. (b) The average wind speed was 2.6 m s^{-1}. During the upstroke the wings are strongly folded and move back and upward close to the body. Rhythmic spreading of the tail is extreme, varying between 40% and 100% and reaching the highest values at the end of the downstroke (Videler 1997).

wings during the lower parts of the track and keep the wings folded close to the body while flying as a bullet along the upper half of the path. The birds are supposed to save energy because the drag is lower during the periods when the wings are folded. This would be especially advantageous at high average velocities because drag forces increase with the velocity squared. Estimates for the amount of energy that can be saved reach values of 35% for a chaffinch (Rayner 1977). However, the explanation given is not entirely satisfactory. Small birds use the technique also at low speeds and even during hovering. The fixed gear hypothesis (Rayner 1985*c*) suggests that the intermittent rest periods help to optimize the power output of the muscles by allowing a higher, close to optimal, frequency during the flapping period to reach a certain average speed.

During intermittent gliding flight the active phase can be a steep climb followed by a gradual gliding descend with extended wings or the flight can be level and the bird decelerates during the glide without loosing altitude. While using the first option the animal gains potential energy during the climb and loses it again during the subsequent glide. Models predict that the minimum cost of transport of a starling, for example, could be reduced by about 11% if the bird would choose the optimal climb and glide strategy (Rayner 1985*c*).

Tobalske *et al.* (1999) studied the kinematics of flap-bounding flight in the zebra finch over a wide range of speeds in a wind tunnel. As an example of this intermittent flight strategy the collected data will be used here. The birds (weighing 13.2 g on average; wingspan about 169 mm) were filmed at 300 frames s^{-1} during flights at speeds varying from 0 to 14 $m\,s^{-1}$. Flap-bounding was used by the birds at all speeds. Figure 6.14 illustrates the behaviour at 12 $m\,s^{-1}$. The zebra finch stretches the wings during the downstroke and brings them back up close to the body. The periodic flapping is clearly indicated by the recording of the up and down movement of the wing tips. The bounding bout starts when the wings are folded close to the body during an upstroke. Note that the wings are held more tightly against the body than during the mid-upstroke. The bout lasts slightly more than 100 ms before flapping recommences with an upstroke. The altitude is increasing and continues to increase when the bounding bout starts to reach a maximum height after about two-third of the bounding period. The angle between the body and the horizontal also starts to increase towards the end of the flapping phase and has reached the highest value when bounding starts. In the second half of the bounding bout the angle reduces.

During the flapping phases wing beat frequencies increased and wing tip amplitudes and the body angle decreased with increasing flight speed. The angular velocity was highest during hovering, decreased to a minimum at 8 $m\,s^{-1}$ and slightly increased with further increase in flight speed. The birds obviously adapted their wing beat kinematics for every speed they were forced to fly at. The percentage of time spent flapping was about 88% during hovering and decreased gradually to reach a value of approximately 55% at 14 $m\,s^{-1}$.

During bounds vertical and horizontal forces could be calculated using measurements of the acceleration following Newton's second law wherein force is mass times acceleration (see Chapter 1). Here too the body angle decreased with flight speed with a clear effect on lift and drag. Lift force

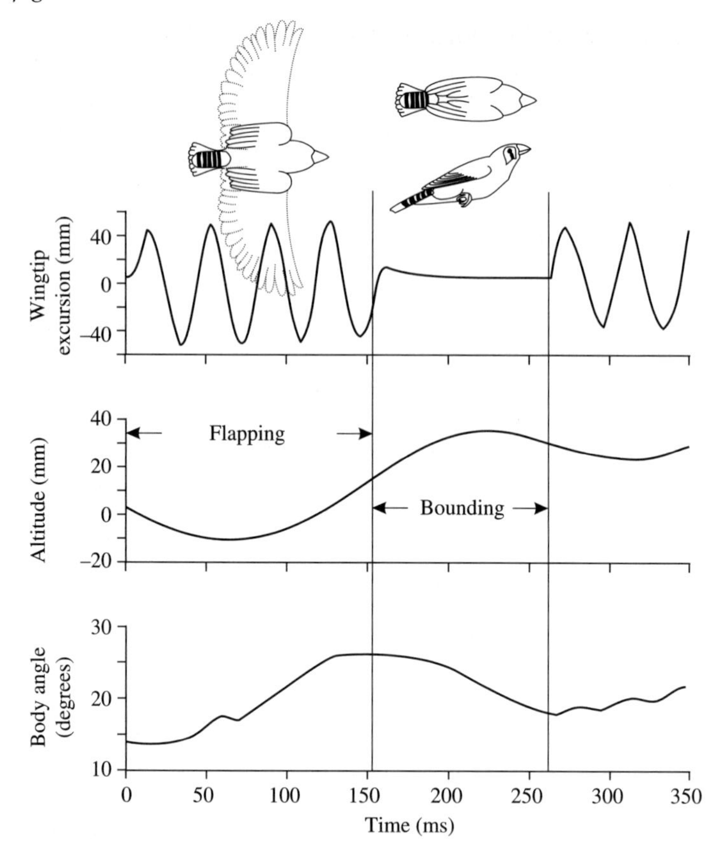

Fig. 6.14 Timing of a flap-bounding cycle in relation to the wing movements, changes in altitude, and body angle relative to horizontal in a zebra finch flying at 12 $m\,s^{-1}$ in a wind tunnel. The dorsal view above the flapping period shows the wing contour at mid-downstroke (thin line) and at mid-upstroke (bold line); the wings during the bound are fully flexed. The side view picture has a body angle of 25° representing the attitude during the first half of the bounding period (based on Tobalske *et al.* 1999, with kind permission of The Company of Biologists).

obtained from the body during bounds increased to a maximum of 15.9% of the body weight at 10 $m\,s^{-1}$; body drag also reached a maximum at that speed. Zebra finches were obviously changing the bounding technique and the aerodynamic function according to flight speed. An increase of body lift was the target at slow speeds and drag reduction probably the objective at the highest speeds. This study suggests that energy saving is indeed the rationale behind flap-bounding flight in birds.

Woodpeckers in flight can be easily recognized because of their typical flap-bounding flight. Tobalske (1996) studied the behaviour in the field of six American species varying a factor 10 in body mass (downy woodpecker, 27 g; red-naped sapsucker, 47 g; hairy woodpecker, 70 g; Lewis' woodpecker, 107 g; northern flicker 148 g; pileated woodpecker, 262 g). All species showed flap-bounding flight. Glide-flapping and glide-bounding (during a bound wings are stretched quickly and repeatedly) were observed as a manoeuvre prior to landing. During the bounding phase of the flap-bounding cycles the wings are kept close to the body. The wingspan of the northern flicker, for example, was 5–10 cm during the bound compared to 10–17 cm during mid-upstroke. Wing beat frequencies and bounding phase durations were negatively correlated with mass, whereas flight speed and percentage of time flapping increased with mass. Flapping percentages were highly variable among these woodpeckers with values ranging between 30% and 93% of the flight time.

Common magpies (average mass 158 g and wingspan 57 cm) were flown in a wind tunnel at a speed range of 4–14 m s^{-1} (Tobalske and Dial 1996). The birds used flap-gliding at speeds up to 8 m s^{-1}. At higher speeds the percentage of bounds during the non-flapping phases increased up to 60%. Flapping beat frequencies did not increase with speed but the body angle, the stroke planes, and the tail spread decreased with speed increase. The birds decelerated usually during the non-flapping phases, only occasionally losing altitude. Magpies too obviously changed the emphasis of the flight technique from weight support to thrust production. In the same wind tunnel study rock doves (average mass 316 g and wingspan 62 cm) flew at the higher speed range of 6–20 m s^{-1}. The pigeons did not flap-bound but progressively flexed their wings closer to the body during glides at higher speeds. At a given speed budgerigars, starlings, and magpies decrease the percentage of bounds relative to glides with increasing body mass. This could explain why pigeons, weighing twice as much as magpies, do not show bounding behaviour. Barn swallows tested in a wind tunnel at speeds between 4 and 14 m s^{-1} started to flap glide at high speeds using short pauses in the middle of the upstroke. The pause durations varied between 10 and 25 ms. Barn swallows also showed a high degree of kinematic plasticity (Park *et al.* 2001).

Intermittent flight also occurs during hovering and windhovering. Zebra finches, hummingbirds, and sun birds perform flap bounding by folding the wings briefly during hovering. Kestrels glide intermittently during wind-hovering bouts with wings stretched and are capable to keep the head in a fixed position with respect to the ground during the glide. We (Videler *et al.* 1983) found that during windhovering in gusty winds with speeds varying between 4.8 and 7.5 m s^{-1}, gliding bouts lasted on average 0.3 s. The head was kept in position by stretching the neck while the body was gliding backwards. Kestrels can stretch their neck over a distance of about 4 cm. The maximum gliding time allowed by this distance matched the duration of the average gliding bouts.

6.6.2 Formation flight

Aircraft and bird wings generate wing tip or trailing vortices in flight as illustrated in Fig. 1.6. Behind the wings the trailing vortices are situated slightly inward from the wing tip. There is down-wash more inwardly and up-wash on the outside. Airplanes flying in formation make use of the up-wash by positioning a wing of the trailing plane with the tip in the up-wash region. Under optimal conditions the rear plane can save 15% fuel. Flapping wings of birds produce lift and thrust. The lift production oscillates but the time average will show up-wash behind the wing tips and birds flying behind in a V-shaped formation can make use of that. Condition is that the ratio of up and down flapping speed to forward speed is small. This is the case for large birds and therefore formation flight is restricted to that category (Hummel 1995). Unfortunately this is theory and real aerodynamic measurements on birds in formation do not exist. But there are measurements which provide circumstantial evidence that formation flight by large birds is more than a way to stay in contact with each other in the air and to communicate. Fifty-four skeins of pink-footed geese were photographed from below and distances among the birds were measured by Cutts and Speakman (1994). The spacing between the wing tips was on average 17 cm plus or minus 2 cm. That narrow range indicates that the birds were seeking position with respect to the wings in front with great precision. The authors claim that it is too far out because the theoretical optimal position was 6 cm inward of the wing tip of the leading bird. The problem is, however, that we have no idea about the error in the theory which is developed for aircraft when it is applied to flapping birds. According to the same theory the spacing to the rear should have been 50 cm but it was 150 cm plus or minus 100 cm. Once again, we need to know more about the flow patterns behind large birds before we can be sure about the function of formation flight. Pelicans are known for their tendency to operate as groups. Weimerskirch *et al.* (2001) compared heart rates and wing beat frequencies of great white pelicans flying solitary and in formation. When in formation birds had a heart rate that was 11.4–14.5% lower than when flying alone under the same conditions. Average wing beat frequencies decreased from the leader in a formation to a minimum value for the bird in fourth position. Saving energy is an attractive reason for this complex behaviour and what we know about it is rather convincing but detailed evidence is required.

6.7 *Gliding*

In still air, birds can save energy by gliding while losing either altitude or speed. Wind conditions offer more complicated possibilities to glide over prolonged periods.

6.7.1 Hanggliding

Obstacles such as hills, cliffs, dikes, hedgerows, or even waves obstructing horizontal winds force the airflow in upward direction. Many birds make use of such up-draughts to stay aloft but it is not easy to measure the conditions in these rising currents of air. In the Netherlands kestrels were observed to hang, almost without wing-flapping, in fixed positions over a sea dike (Videler and Groenewold 1991). We measured the wind speed, the vertical wind angle, and the horizontal direction from the ground up to 9 m above the dike at 0.5 m intervals and recorded the positions where kestrels were hanging more than 90% of the hunting time. The kestrels preferred a position about 6.5 plus or minus 1.5 m over the windward slope with sea winds blowing at 8.7 plus or minus 1.5 m s^{-1} almost perpendicular to the longitudinal dike axis. In the preferred position the upward wind angle ranged between 6° and 7° with respect to horizontal. The lowest value requires a glide ratio of 9.5 which is safely below the range of values of 10.8–12.8 found for the kestrels gliding in the corridor experiments in Section 6.2.3. The kestrels in the field obviously must take the gusty wind conditions into account and choose to hang in a position where they can perform the trick easily with some space for corrective manoeuvring.

6.7.2 Dynamic soaring

Out in the middle of the ocean there are at least three wind-related phenomena that can be used by birds to glide over considerable distances without spending energy in flapping flight. Winds experience drag at the water surface which causes the wind velocities to be lower near the surface. The wind gradient is pronounced under the high wind conditions occurring in the Southern oceans. In a strong wind (Beaufort scale 7) winds blowing at a speed of about 15 m s^{-1} at 20 m altitude will be reduced in speed to about 10 m s^{-1} at about 1 m above the surface. Albatrosses and other large seabirds are believed to make use of speed reductions in the wind gradient although actual measurements have never been made. The required behaviour is termed dynamic soaring or gliding. The principle has been understood since Lord Rayleigh's article in *Nature* in April 1883. The airspeed, which is the velocity between a gliding bird and the air about it, can be used by a bird to gain elevation. It uses the kinetic energy of $\frac{1}{2}mv^2$ to do so (see Chapter 1). In the equation m is the mass of the bird and v the airspeed. An example worked out in Table 6.3 uses more or less realistic figures to explain the dynamics of the dynamic soaring technique. A wandering albatross repeatedly glides down wind and sweeps up against it. Starting point is a bird gliding at an altitude of 20 m to leeward along with a 15 m s^{-1} wind. Let us assume that the airspeed is about 12 m s^{-1} which is probably close to the minimum sinking speed of a wandering albatross (Pennycuick 1989). At that airspeed the albatross with an assumed glide ratio of 20 sinks minimally at a rate of about 0.6 m s^{-1}. The speed over the ground, the sum of the wind speed, and the airspeed, is 27 m s^{-1}. While sinking, the albatross will find itself flying in gradually slower winds and the airspeed increases even if the ground speed would remain the same. Of course, if the bird chooses to dive down faster than at its minimum sinking speed the airspeed near the surface of the sea will be

Table 6.3 Calculated airspeeds during dynamic soaring of the wandering albatross.

Height above sea level (m)	Wind gradient ($m\,s^{-1}$)	Leeward down Airspeed ($m\,s^{-1}$)	Windward up Airspeed ($m\,s^{-1}$)
20	15	12.0	36.8
15	15	12.4	37.7
10	14	13.0	38.4
5	13	14.1	38.6
1	10	16.5	37.2
0.1	7	20.0	34.0

The following assumptions are used for calculation: Best glide ratio, 20 : 1; minimum sinking speed, 12 $m\,s^{-1}$; minimum sink speed, 0.6 $m\,s^{-1}$; wind speed at 20 m is 15 $m\,s^{-1}$; constant groundspeed during descend of 27 $m\,s^{-1}$; gravitational acceleration is 9.8 $m\,s^{-2}$; there is no drag; during the ascend velocity is sacrificed for altitude with for every m altitude increase $V_2 = \sqrt{(V_1^2 - 2g)}$ (Wilson 1975); velocity gradient calculated following Sutton (1953) with $h_0 = 0.001$. Speed data based on Pennycuick (1989).

even higher, but the gain from the effect of the wind gradient will be the same. To keep the example simple we assume that the bird keeps the ground speed at 27 $m\,s^{-1}$. Just above the surface the bird turns windward and, if it does not lose too much speed in the turn, faces the wind at the groundspeed of 27 $m\,s^{-1}$ it had just before moving to windward. The airspeed is now the groundspeed in the windward direction plus the speed of the wind facing the bird. At 0.1 m above the surface the wind velocity in the gradient is assumed to be only 7 $m\,s^{-1}$. So, the bird starts the climb at the airspeed of 34 $m\,s^{-1}$. While climbing, speed and hence kinetic energy are sacrificed for potential energy (height). The potential energy gained by an elevation of 1 m is the mass of the bird m times the gravitational acceleration g. It is equal to the difference between the kinetic energies before and after the 1 m climb $(\frac{1}{2}mv_1^2 - \frac{1}{2}mv_2^2)$. Mass cancels out in this equation and we can calculate the velocity sacrifice because v_2 is the only unknown factor (Wilson 1975). The sacrifice is small and more than compensated by the fact that the bird climbs up the wind gradient into areas with higher wind speeds. This implies that although the airspeed decreases due to the climb it also increases due to the higher velocities it meets at the higher altitudes. This cyclic extraction of kinetic energy from the wind velocity gradient could go on for ever but we have to keep in mind that drag losses are not taken into account in this example and the system will be less effective than suggested. Also it is important to note that real proof that it actually happens does not exist (Pennycuick 2002).

6.7.3 Sweeping flight and gust soaring

The other air movements that can be used by oceanic birds are the updraught caused by the rising of waves. With a 6 $m\,s^{-1}$ wind and waves with a steepness of 1 m height to 12 m width, the up-current near the wave is 1.65 m s^{-1}. Sliding along just above and on the windward side of the crests of long steady waves is a trick performed by many species including albatrosses, fulmars, sea gulls, and pelicans. Birds usually use the upward wind velocity to increase their speed. I saw pelicans using the big surf in Baja California to travel along the Pacific coast without beating the wings. The glide along the wave crest parallel to the coast ends when the waves break, the birds use the high speed attained at that moment to sweep up, exchanging kinetic energy for potential energy (as shown above for the albatrosses), and glide down against the wind to pick up a new wave further out in the ocean. There the birds turn 90° and start to glide along the wave crest again with the wind from the side and approaching the coast with the wave. Wilson (1975) termed this behaviour sweeping flight.

Pennycuick (2002) described another way to make use of high waves. Not above but behind the wave to leeward there is a region where flow separated by the sharp crest creates calm. Albatrosses are observed to fly there. They extract energy by diving up from that region into the severe wind above the wave. The typical behaviour includes a belly-to-wind roll used to extract as much energy as possible from the gust. The pulse of kinetic energy due to the large difference in wind speed is enough to allow pullups to the heights actually observed in albatrosses.

6.7.4 Thermal soaring

Hot air rises, as we all know, and over land temperature differences due to differential heating of the bottom by the sun occurs easily but there can be temperature driven rising air at sea also underneath clouds. Upward air velocities in thermals can reach values up to 5 m s^{-1}. Under ideal conditions the positions of the thermals are given away by cumulus clouds and can reach hundreds of metres high. Streets of thermals often develop over flat planes. Many large bird species with broad slotted wings make use of thermals to gain height by circling in tight curves as close as possible around the centre. The common strategy of the birds is to circle upward in a thermal and glide down to the next and gain height again by circling up. Vultures, eagles, kites, buzzards, storks, pelicans, and secretary birds commonly use thermal soaring during migration or while searching for prey. Pennycuick (1971*a*) followed thermal soaring birds in a motor glider across the Serengeti, measuring their sinking and forward speeds. Most observations could be done on the African white-backed vulture. Its flight characteristics can be used as an example of a bird well adapted to this behaviour. It has a glide ratio of 15 : 1 (at 13 m s^{-1}), a minimum gliding speed of 9 m s^{-1}; a minimum sink speed of 0.76 m s^{-1} (at a minimum sinking speed of 10 m s^{-1}). Pennycuick's motor glider (a Schleicher ASK-14) had a glide ratio of 28 : 1 (at 26 m s^{-1}) and the same minimum sink speed but at twice the minimum sinking speed of the vulture. A typical airliner has a glide ratio of 16 : 1 and the best man made glider can travel 60 m for every metre descend. The space shuttle tumbles down 1 km every 4 km glide distance. Albatrosses with maximally 23 : 1 are the birds with the best glide ratio (Anderson and Eberhardt 2001). This comparison shows in fact that the vultures have a surprisingly bad glide ratio. They are obviously not adapted to fast gliding over large distances but are possibly configured for slow tight turning to make the best use of the up-draught in thermals. Birds with narrow slender wings have a higher turn radius and have difficulties to remain airborne in thermals with small radii. However, slotted broad wings are also the required attributes allowing steep take-off angles from the ground.

Pennycuick (1972) suggests that there is another way to make use of the presence of thermals to travel for free across large distances. It is possible to make a cross-country flight from thermal to thermal without wasting time in circling. If the thermals are abundant enough, linear soaring can be used to travel fast. Each time the gliding flight along an approximately straight track reaches a thermal the speed is reduced (vultures use their legs as airbrakes) to gain as much height as possible while passing through the rising air. (Pennycuick 1971*b*).

6.8 *Manoeuvring*

A flying bird can move around three axes through its centre of gravity. Rotation about the longitudinal axis through the centre of the body from beak to tail is called a roll. It can be controlled by lift forces generated by the wings. The further away from the centre of gravity the stronger the lever effect on the roll control will be. Lift forces on the hand wings are the dominant forces. Birds with long pointed V-shaped tails can also use the tail for roll control. Pitch is a rotation around an axis parallel to the length of the stretched wings. Head down and tail up and the other way around. Pronation and supination of the wings and changes of the angle between the longitudinal axis of the tail and the body in the median plane control pitch rotation. The third axis runs vertically through a horizontally flying bird and rotation around that axis is called yaw. It can be induced by differences in drag or thrust between the wings. In principle a bird can turn using yaw but that is not the common way to change direction. Usually a roll is made in the direction of the turn and stopped when the wings are at a certain angle with the horizon: the bird is banking. During the bank the lift force is no longer vertical and has a horizontal component turning the bird.

Asymmetric aerodynamic forces generated by the wings and the tail are required during flight manoeuvres. The asymmetry can either come from drag or from lift forces. During gliding flight the left and right wings can be stretched to different extents at the elbow and/or wrist, and they can be asymmetrically supinated or pronated providing differences in the angles of attack. Birds could even change the camber or the appearance of parts of the wings differently on the right and left. The tail can also contribute significantly to manoeuvring forces by tilting in one direction using various degrees of spreading. Flapping offers even more possibilities to generate large asymmetric forces both during upstroke and downstroke or both using lateral asymmetries in the velocities or accelerations. Warrick and Dial (1998) induced pigeons to fly turns around a barrier to a perch next to a bird of the opposite sex. The birds were fitted with infrared light reflecting markers and filmed with four high-speed cameras. The short distance flights were slow at about 3 $m\,s^{-1}$ and the birds used banking with flapping wings to get around the barrier. Bank angles were in the order of 30° with accelerations of 600 $rad\,s^{-2}$ ($3450°\,s^{-2}$). The roll starting the bank was initiated by asymmetric downstroke velocities between the wings; one wing was beating faster than the other early in the downstroke, the difference being about 17% of the maximum. The pigeons did not use differences in angles of attack or surface areas between the wings. The rolling movement was stopped either at the end of the same downstroke or at the beginning of the next upstroke again by asymmetric velocities of the wings but now in opposite direction. Pigeons are obviously using large alternating and opposing forces during upstroke and downstroke instead of subtle changes of the aerodynamically important features of the wings.

6.9 *Accurate measurements of speed*

Birds travel through the air at a variety of speeds, using different flight modes and under diverse conditions. They can fly horizontally or up and down, at largely different altitudes and under all possible wind and weather conditions, by day and night. The purpose of a flight will also affect the chosen speed. Travelling velocities during long distance migrations differ usually from speeds used for short foraging trips.

Chapter 4, Section 4.2 showed that basic considerations about power requirements for flight predicted the existence of two optimal speeds for birds. There is a most economical speed where the amount of energy used per unit flight time is minimal (the minimum power speed) and a slightly higher speed where the smallest amount of energy is required to cover the flight distance (the maximum range speed). Whether these alternative choices really exist for a bird depends on the shape of the curve of the flight power against speed. U-shaped power curves predict the two optimal speeds, deviating curve shapes usually do not.

We are interested in real characteristic speeds used by birds in the field. These are difficult to measure. Accurate methods are rare and complex. Air speed is displacement over time with respect to the air surrounding the bird. We saw in Section 6.7.2 that ground speed, displacement in time in an earthbound frame of reference, includes the effect of the wind on the displacement of the bird. Vertical displacements and altitude also make airspeed estimates difficult. Airspeed measurements require knowledge of the wind speed at the flying altitude as close as possible near the bird and at the time the bird is flying there. It is therefore not surprising that published speed records vary tremendously even within a single species. Bruderer and Boldt (2001) presented the airspeeds of 139 Palaearctic species adjusted to represent flight at sea level. They use their own long-term radar measurements as starting point and annotated these with reliable estimates of others to reach realistic assumptions. They have been using tracking radar since 1968. The tracking range starts at 100 m distance and a single chaffinch in tail-on view can be followed up to 4.5 km away. Winds were consequently measured near the ground with anemometers and high up by tracking pilot balloons at regular intervals. This set-up is not without error but it is currently the most accurate method available. Measured average airspeeds vary approximately between 6 and 23 $m\,s^{-1}$ (22 and 83 $km\,h^{-1}$). Speeds do not seem to increase with body mass. The smallest birds measured, the goldcrest of 6 g, flew at speeds varying from 6 to 12 $m\,s^{-1}$ (average 9 $m\,s^{-1}$). Pelicans and vultures with about 10 kg at the other side of the scale were flying 15 $m\,s^{-1}$ on average. The vast majority of the species measured flew at speeds between 6 and 16 $m\,s^{-1}$. As expected, differences within a single species were large. The green sandpiper's record of 22.7 $m\,s^{-1}$ was measured during a slightly (1.9 $m\,s^{-1}$) descending flight, its lowest velocity of 5.8 $m\,s^{-1}$ was clocked just after release while gradually ascending. Kestrels flew at about 8 $m\,s^{-1}$ in our windless corridor but were flying between 12 and 13 $m\,s^{-1}$ during migration. Very high speeds were observed for mallard (17.6–24.4 $m\,s^{-1}$, average 21.4 $m\,s^{-1}$) and goosander (21.2 $m\,s^{-1}$). Swifts were flying at disappointingly slow velocities. The variation found for the common swift was from 6.4 $m\,s^{-1}$ during roosting flights to 11 $m\,s^{-1}$ while migrating. The maximum speed recorded was 17 $m\,s^{-1}$ (61 $km\,h^{-1}$). Flights of the alpine swift varied between 8 and 20 $m\,s^{-1}$. The handbook of the birds of the world volume 5 (Del Hoyo *et al*, 1999) mentions a horizontal speed of 170 $km\,h^{-1}$ (47 $m\,s^{-1}$) for the white-throated needletail. No indication is given about the method used to estimate the speed record. Diving speeds of peregrine falcons have been frequently overrated. Peter and Kestenholz (1998) measured a maximum of 51 $m\,s^{-1}$ (184 $km\,h^{-1}$) at the end of a 334 m stoop.

6.10 *Summary and conclusions*

Birds are pilot and aircraft in one. They have to take decisions about the take-off procedure, the subsequent flight direction, altitude, wing, and tail kinematics and speed and must choose the right procedure to land. Given the same conditions, birds tend to use exactly the same flight strategy, although there are small differences among individuals. Slight but consistent changes in speed and kinematics can be induced by changing the flight distance and body weight under otherwise identical conditions. Feet usually push-off during take-off in a standard way without taking the thickness of perches into account. The forces used can amount to multiples of the body weight. Take-off angles of 70° are common. Some species including hummingbirds flap their wings during take-off. Birds can land on a perch and landing procedures require the speed to decrease to zero without loss of altitude. We only just start to understand how birds solve that problem.

The study of wing beat kinematics during flapping flight demands high-speed pictures preferably in three dimensions. Marey (1890) was the first to successfully make these. We studied changes in wing shape during downstroke and upstroke and modifications of beat frequencies and amplitudes during flight with and without carrying weights under otherwise standard conditions in kestrels. Focussing on a single wing beat of a small bird with a very high time resolution reveals high rates of changes of the kinematic parameters. Fast accelerations, frequent rotations, and shape changes turn wing beats into aerodynamically highly unsteady events. Wing beat kinematics are usually stereotypical for individual birds and even within a species but the variation is large among the various groups and little of that variation has been studied.

Hovering is flying at one position without wind. Many small birds can do that for very brief periods only. Hummingbirds have morphological and kinematic adaptations that allow them to hover for prolonged periods. They are able to turn the large hand wing around at the end of the downstroke and make the upstroke with the hand wing in upside down position. Maximum performance during hovering was investigated by making the birds lift weights and by flying them in low density air. Hummingbirds are capable of performing wing beat frequencies of 60 Hz and of lifting twice their body mass while flying on the spot.

Windhovering is a very different technique performed by a small number of specialists. These birds manage to fly against the wind at exactly the speed of the wind. The head is kept in a precisely fixed position over the ground or the water surface to facilitate the detection of moving prey. In gusty winds aero-acrobatic movements of the wings, the tail, and the feet are often required to keep the head steady.

Flapping flight is a highly demanding form of locomotion and we expect that natural selection favoured the evolution of energy saving techniques. Birds show a variety of flight styles that could represent such techniques but it is not easy to prove that they do. During intermittent flights the flapping movements of the wings are repeatedly stopped briefly. Bounding birds fold the wings close to the body during these pauses; intermittent gliding birds keep the wings stretched. Precise measurements of bounding flights of zebra finches at a range of speeds give insight in the kinematics of this behaviour and suggest that bounding offers increased lift during slow flights and reduced drag during fast flights. This is consistent with the observation showing that larger birds glide intermittently during slow flights and bound while going fast. Anyway, series of short motionless periods will probably reduce flight costs if compared to continuous flapping, but direct proof is not available yet. Saving energy by flying in close formation is well established for aircraft but not for birds. However, there is now circumstantial evidence that birds fly in particular formations not only to keep in close contact but also to save energy—though necessary proof is required to establish this fact.

Atmospheric conditions can offer the opportunity to fly without beating the wings. Gliding in up-draughts caused by obstruction of the wind is very common and practised by many average-sized species. Kestrels use it to hang motionless over dikes in the Netherlands in positions where winds of about $9\ \mathrm{m\,s^{-1}}$ blow 6–7° upwards. Over the open southern oceans gradients of strong winds with decreasing velocities towards the surface make it possible to gain kinetic energy. Albatrosses could use this possibility by dynamic soaring, but it is not sure that they do. The action includes fast glides downwind into slower and slower winds towards the surface and 180° turns into the wind at high speed. Kinetic energy is exchanged for potential energy by increasing altitude once again making use of the wind velocity gradient but now in opposite direction. The same trick to gain height at the expense of speed is used by birds in interaction with large waves. The up-draught over the wave makes it possible to glide along the wave and increase speed. Thermals where air moves upward due to temperature differences in the atmosphere often develop over land. A range of large birds with broad slotted wings use thermals to travel over considerable distances for almost free. The common wing shape is probably required for vertical take-off from the ground.

Manoeuvring of aircraft or flying animals is based on rotations around three major axes through the centre of gravity: yaw around the vertical axis, pitch around the axis through the stretched wings, and roll around the axis running from head to tail. Accurate measurements of a standardized manoeuvre of pigeons provide insight in the kinematic events.

Airspeed estimates of free flying birds are very hard to make. Speed over the ground can be measured with some accuracy but wind speed estimates near the bird are hard to obtain. The most reliable data including many species of widely diverse groups show that the differences are small and not dominated by size. The majority of species is able to fly at speeds between 6 and $16\ \mathrm{m\,s^{-1}}$. The highest speeds are reached during long dives of specialized birds of prey.

7 *The bird flight engine*

7.1 *Introduction*

The functional anatomy and biomechanics of the flight apparatus is of great interest if one wants to know how birds fly. The internal proceedings form a challenging problem. A large group of scientists in the USA made a huge concerted effort to solve it during the last 15 years or so. Their names appear in the list of references related to this chapter.

A major step forward was the *in situ* visualization of the movements of the skeletal elements of a bird in flight using X-ray films at high speed. Simultaneous electromyograms (EMGs) of a range of flight muscles were recorded during various manoeuvres including steady level flight. The timing patterns of muscle onset and offset provide insight into the relation between muscle recruitment and the wing and tail movements but not in the forces exerted by the muscles because the relationship between the timing of the EMG and force production is not straightforward.

Direct measurements of the force produced could only be made for the pectoralis muscle. A clever technique was developed using the insertion point on the humerus as a strain gauge. Pectoralis force production and length changes in the muscle were monitored in a flying bird offering the possibility to calculate the work done by the muscle during each wing beat. Simultaneous high-speed films or video showed the wing beat frequencies used, making it possible to derive figures for the mechanical power generated by the pectoralis.

The main upstroke muscle, the supracoracoideus also received attention but direct force measurements could not be made. Timing and function of this muscle are now reasonably well understood.

Muscles involved in moving the tail and the tail feathers are also organized in a complex configuration. EMGs throw some light on the recruitment patterns during walking and different flight modes. To find out how the combined action precisely moves the tail will require more research.

Not all the skeletal actions during flight are directly related to movements of the wings. Some have a respiratory function in a rather complex way.

Box 7.1 gives a summary of the structure and function of vertebrate muscles to provide the basic knowledge required to understand the main motor of a bird.

Box 7.1 ***Brief summary of cross-striated muscle structure and function***

Figure 7.1 provides a schematic overview of a vertebrate muscle fibre. Vertebrate cross-striated skeletal muscles consist of multinucleate fibres surrounded by a membrane (sarcolemma). A muscle fibre contains mitochondria and is densely packed with myofibrils. These are in close contact with sarcoplasmatic reticulum and are filled with bundles of thick and thin filaments. In longitudinal direction myofibrils consist of a sequence of identical units called sarcomeres. Sarcomeres are a few microns long. Within a sarcomere the thin filaments are attached to both ends and interdigitate with the thick filaments in the middle. The cross-striated appearance under a microscope is caused by the regular arrangement of the thick and thin filaments in the A and I-bands of each sarcomere. The overlap between thick and thin filaments increases when the muscle shortens during contraction, decreasing the distance between the Z-lines separating the sarcomeres. A transverse tubule system, open to the world outside the fibre through holes in the sarcolemma, often coincides with the position of the Z-lines (based on Woledge *et al.* 1985).

During shortening of muscle the thin filaments slide along the thick ones. The active protein of the thick filaments is myosin and the thin ones consist of the protein actin.

Shortening is caused by cyclically moving connections between myosin and actin. Parts of the myosin molecules, the cross bridges, stick out and

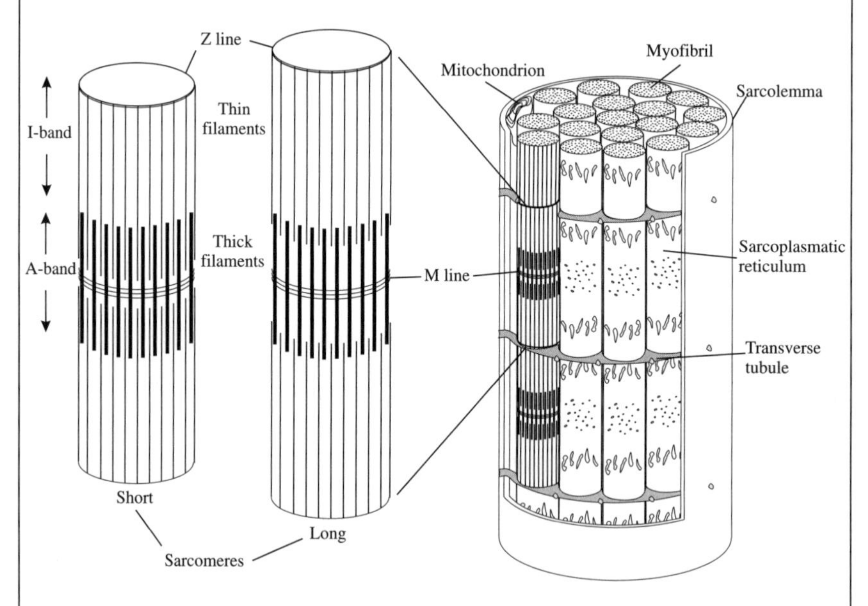

Fig. 7.1 The structure and nomenclature of a vertebrate muscle fibre. The shortening action of a sarcomere is illustrated on the left. See text of Box 7.1 for more detailed information.

interact with the thin filaments moving these along. The breakdown of ATP (adenosine triphosphate) into ADP (adenosine diphosphate) provides the energy for the movement. Each cross bridge consists of a head and a thin neck connecting it to the rest of the myosin. Under resting conditions the head is attached to the actin molecule. The neck is stiff and acts as a lever arm.

A cartoon of the cross bridge cycle is drawn in Fig. 7.2; the cycle is represented by four stages (a)–(d). Stage (a) shows one myosin cross bridge in a non-activated muscle. The head is attached to the actin molecule on the right. The cycle is initiated when ATP attaches to its binding site on the cross bridge at stage (b). ATP binding induces the cross bridge to dissociate from the actin and to change the angle with the rest of the myosin. Stage (c) depicts cleavage of ATP into ADP.Pi (the third inorganic phosphate Pi loosely connected to ADP). Without ATP at the binding site the head reconnects to the actin. The next event is the release of the inorganic phosphate Pi from the cross bridge at stage (d). The dissociation rearranges the structure instantly, causing a swing of the lever arm over an angle of 70° and a shift of the actin over a distance of about 11 nm.

Only two stages of the myosin cross bridge cycle have been visualized so far. The crystal structure of a cross bridge from chicken skeletal muscle represents the end of the power stroke (stage (a)). Combined protein crystallographic studies of parts of cross bridges from other studies show the stage where the head is detached from the actin (stage (b)). The behaviour of the structures during the dynamic conversion between the stages is still enigmatic (Holmes 1998).

The cross bridge cycle shown in Fig. 7.2 illustrates the events during shortening of the muscle. We must speculate that a similar mechanism is used to create tension during isometric or excentric contractions.

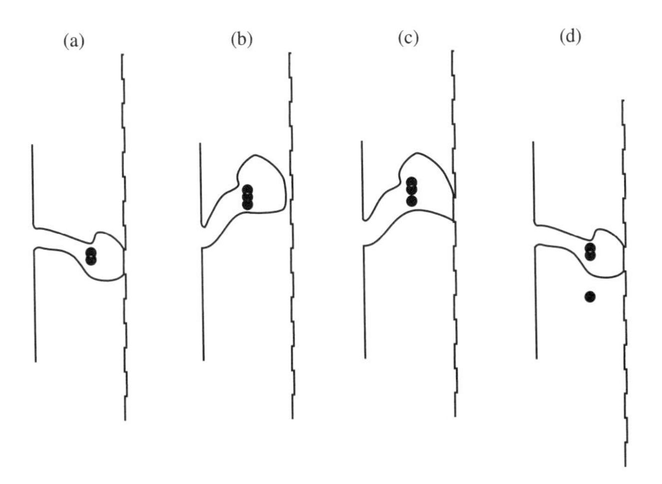

Fig. 7.2 Schematic representation of a partly hypothetical cross bridge cycle in 4 stages (a)–(d). See text of Box 7.1 for a detailed explanation.

7.2 *A glimpse under a starling's cowling*

The anatomical descriptions in Chapter 2 provide insight into the structure of the most important parts of the flight apparatus. Now we want to know how it works. Jenkins *et al.* (1988), in a study that was a classic from the moment it appeared in print, managed to make X-ray films at 200 frames s^{-1} of a European starling in dorsal and ventral view flying in a wind tunnel at speeds between 9 and 20 m s^{-1}. The radiographic films showed the internal movements of the skeleton in three dimensions during full wing beat cycles. Figure 7.3 illustrates the position of the skeletal elements at four instants in dorsal and lateral view. At the start of the upstroke–downstroke transition the humerus is kept almost parallel to the main body axis in the dorsal view and moves upwards in the lateral view to reach an angle of 55°–60° with respect to the mean path of motion before starting to move down after the extension of elbow and wrist. The humerus moves down over an angle of 110° in a plane perpendicular to the mean path of motion. During the downstroke the hand skeleton is kept parallel to that plane. At the end of

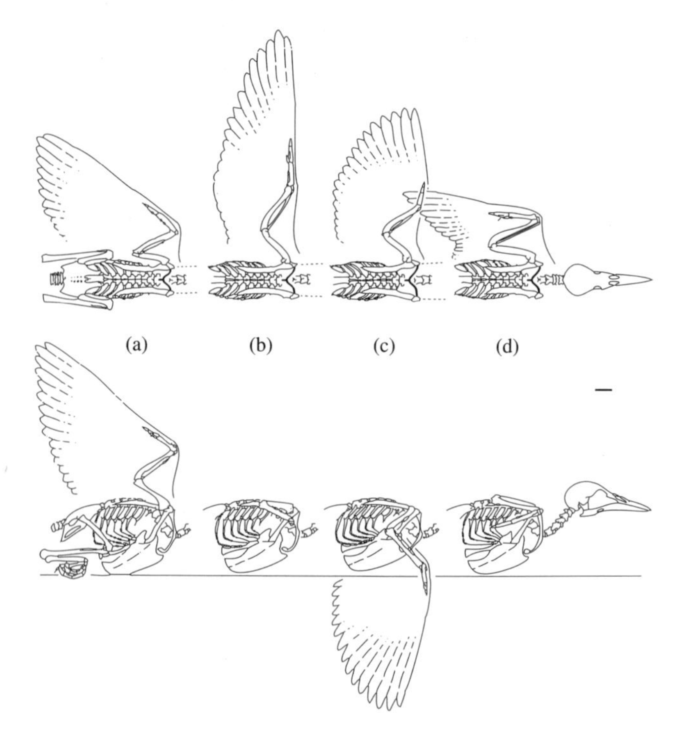

Fig. 7.3 The skeleton of the European starling visualized using X-ray films during flight in a wind tunnel. Top panels show the dorsal and bottom panels the side view at the end of the upstroke (a), the mid-downstroke (b), the end of the downstroke (c), and the mid-upstroke (d). The scale bar is 1 cm. The lateral series shows the movement of the sternum relative to the horizontal plane indicated by the line. (Reprinted with permission of Jenkins *et al.* 1988. © 1988 AAAS.)

the downstroke the finger tips are in the forward-most position just before elbow and wrist flex and the humerus starts to elevate at the start of the upstroke. The wings are brought up in semi-folded position.

The X-ray films showed for the first time the dynamic behaviour of the wishbone during the wing beat cycle. The distance between the wishbone (furcula) heads widens during the downstroke and decreases during the upstroke covering a distance of about 6 mm which is about half the resting distance (Fig. 7.4). Spreading starts just before the upstroke–downstroke transition when the humerus rotates forward and continues during the downstroke. Each wishbone head is attached to the distal end of the coracoid and the end of the scapula. The shoulder joint in starlings is formed by the coracoid and the scapula. This implies that the humerus at the shoulder joint moves over the same distance as the furcula heads. The coracoid does not bend sideways but rotates and slides at the joint with the sternum. The scapula moves along by sliding sideways while the head follows the furcula head and the posterior end remains in position. During the downstroke, when the furcula widens, the sternum goes up and tilts. The tilting motion is caused by the rear part which rises more than the front part (Fig. 7.3, bottom). The force required to widen the furcula heads varies between 0.6 and 0.8 N. The furcula acts as a spring releasing part of the force during the upstroke while closing the gap again. The muscle most likely responsible for the widening of the furcula is the sternocoracoides. It is a short muscle connecting the sternum with the ventral half of the coracoid. Stimulation of the pectoralis does not result in widening of the furcula heads. The function of this spring action of the furcula is not at all clear. It might help to bring the

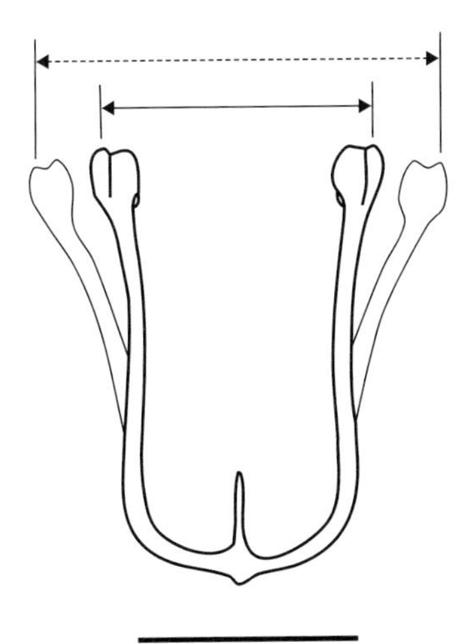

Fig. 7.4 Frontal views of the furcula (wishbone) of the European starling. The distance between the heads varies between 18.7 mm during the downstroke and 12.3 mm during the upstroke. The scale bar is 1 cm (Reprinted with permission of Jenkins *et al.* 1988. © 1988 AAAS.)

wing up but most likely has a respiratory function by helping to inflate and deflate the interclavicular and thoracic air sacs. The timing between wing beat cycles and respiration was subject of detailed studies, some results are discussed in Section 7.8.

7.3 *Muscle activity*

Understanding how the flight engine works requires knowledge of how the motors generate force, work, and power. Muscles are the motors and length change is the dominant action. Box 7.1 provides a brief overview of how cross-striated vertebrate muscles are built and how they are supposed to work. Chapter 2 showed that wing movements are generated and controlled by large numbers of muscles. Each muscle extends between two points or areas of attachment on bony elements either across a joint or between two unconnected bones. Tendon plays a role in most muscle–bone connections. Shortening of muscles may decrease the angle between the articulating bones or bring the elements together. Box 7.1 shows how thin and thick filaments in the sarcomeres slide along each other to shorten the muscle. During isometric contractions muscles generate force without length change. Muscles can also be activated while being stretched. Lengthening, induced by an outside force, can be passive and the outside force has to overcome the internal resistance of the muscle. A much larger force is required to lengthen a muscle while it actively resists to being stretched.

Electrical activity can be measured inside active muscles. EMG records the pattern of bursts of electric activity. The nature of this activity is not completely clear but the phenomenon can be used to obtain an idea of the recruitment patterns of muscles during the wing beat cycle. The relationship between the timing of the EMG and the timing of the force development in the muscle was studied by Goslow and Dial (1990) using the pectoralis of anaesthetized starlings. An isometric twitch contraction recorded after a single stimulus of the appropriate nerve showed a single maximum force peak 30 ms after the stimulus and a gradual decline during slightly more than 80 ms after that. The EMG signal started at the instant of stimulation and lasted only 5 ms. The force peak was reached after the EMG stopped. (Exactly the same time delay to peak force was found in fish muscle by Wardle in 1985 illustrated in Videler 1993, figure 7.10.) Using supramaximal stimulation of the nerve at a stimulus frequency of 125 Hz under the same isometric conditions resulted in a continuous force exerted by the muscle. This force was reached 65 ms after the stimulus and the start of the EMG activity and continued for 25 ms after the EMG stopped. The total time of force increase and decline during the twitch experiment lasted 110 ms. A normal wing beat cycle duration of a starling is only 75 ms. Maximum isometric measurements *in vitro* obviously do not reflect what is actually happening during dynamic length changes in the muscle *in vivo*. But nevertheless, these data indicate that the EMG starts at the instant of muscle stimulation by the nerve and that force development lasts at least as long as an EMG is recorded.

Chapter 2 mentioned that bird wings contain 45 muscles and that muscle activity during flight of only 18 of these muscles has been seriously studied using EMG techniques. In the European starling the EMG activities of 11 muscles sampled from 16 birds flying at wind tunnel speeds of 9–20 $\mathrm{m\,s^{-1}}$ revealed consistent cyclic patterns (Dial *et al.* 1991). Figure 7.5 indicates the position of nine of these muscles in a lateral and dorsal view. Only a brief description of origin and insertion is given below; more details of the usually

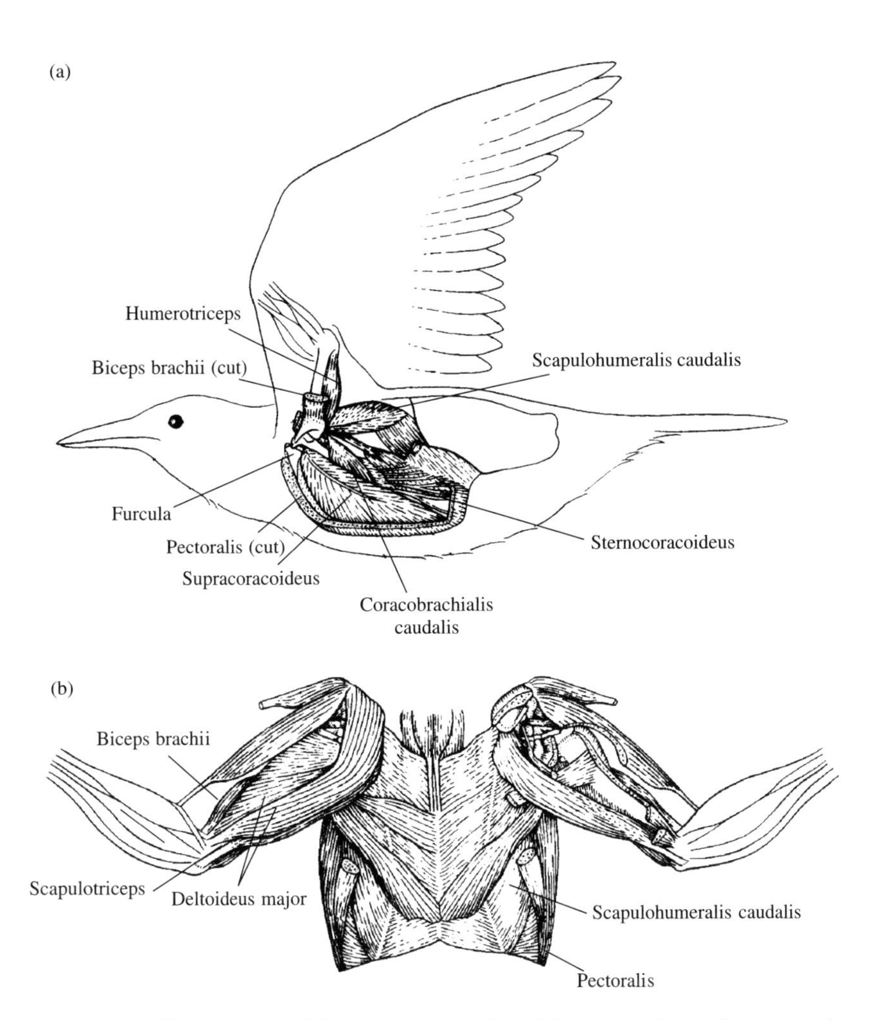

Fig. 7.5 Shoulder anatomy of the European starling. Names are given of nine muscles for which the EMG timing is shown in Fig. 7.6. (a) Lateral view of the muscles. Some superficial muscles have been cut to expose the deeper layer. The position of the furcula head is indicated. (b) Dorsal view (Dial *et al.* 1991, with kind permission of The Company of Biologists).

very complex muscle anatomy can be found in Vanden Berge (1979) and George and Berger (1966). Mechanical reasons for the excessive complexity are virtually unknown so far. It is good to realize at this point that bird flight cannot be regarded as fully understood until a functional explanation is found for every detail of the flight apparatus. Timing of muscle activity is a first step towards that distant goal.

The following functional interpretation attempt is based on a rough idea of the main connections made by each muscle in combination with the cyclic activity pattern derived from EMG recordings. Origins and insertions of the nine shoulder muscles are treated first followed by their EMG activity during a wing beat cycle.

1. Two parts of the pectoralis are considered as one functional unit. The cranial part arises from the entire shaft of the furcula and has part of its origin on the carina. The caudal part arises from the caudal part of the sternum. The two parts insert on the intramuscular tendon sheet, where the cranial part attaches on the dorsal and the caudal part on the ventral side. The pectoralis inserts on the deltopectoral crest on the anterior part of the humerus at a short distance from the shoulder joint.
2. The origin of the sternocoracoideus is at the cranial part of the sternum and it inserts on the ventral half of the coracoid.
3. The coracobrachialis caudalis originates from the caudal side of the lower part of the coracoid and inserts on the ventral side of the humerus close to the shoulder joint.
4. Humerotriceps connects the proximal dorsal surface of the humerus with the olecranon on the ulna (the olecranon is a process near the joint with the humerus).
5. The biceps brachii originates by a stout tendon from the lateral side of the top of the coracoid and has a second origin on the humerus. It inserts with tendons on proximal sites on radius and ulna.
6. The deltoideus major has a cranial and a caudal part. The origin of the caudal part is on a process on top of the coracoid and adjacent part of the furcula. The muscle contains a sesamoid bone which changes the direction of traction. The origin is on the dorsal anterior part of the humerus. The cranial part of the deltoideus major is attached to the sesamoid bone in the caudal part and to the capsule around the shoulder joint. The insertion is on the anterior part of the humerus.
7. The scapulohumeralis caudalis originates from the caudal part of the scapula; the insertion is by a tendon into the dorsal surface of the proximal humerus.
8. The scapulotriceps runs from the top of the wishbone and the coracoid to the olecranon process on the ulna via a tendon. The tendon contains a sesamoid bone at the elbow.

9. The supracoracoides is a bipinnate muscle; its origin is on the sternum and its tendon runs through the triosseal canal to insert on the dorsal surface of the humerus.

There are many more muscles making the complexity of the structure of the shoulder, overwhelming and for most muscles it is, virtually impossible to determine a possible function from the anatomical details. EMG recordings can provide a very general feeling used to distinguish the muscles involved in the downstroke from those taking part in the upstroke.

Figure 7.6 represents an up and downstroke cycle of the starling lasting on average 72 ms at a wing beat frequency of about 14 Hz. The wind speeds in the tunnel varied between 9 and 20 m s^{-1}. The mean onset, duration, and offset are indicated as thick segments of the circle for the nine muscles. The picture reveals a consistent pattern. Downstroke muscles are the pectoralis, the sternocoracoideus, the coracobrachialis caudalis, and the humerotriceps. The activity starts in each case during the last part of the upstroke and continues into the downstroke. The supracoracoideus and the deltoideus major

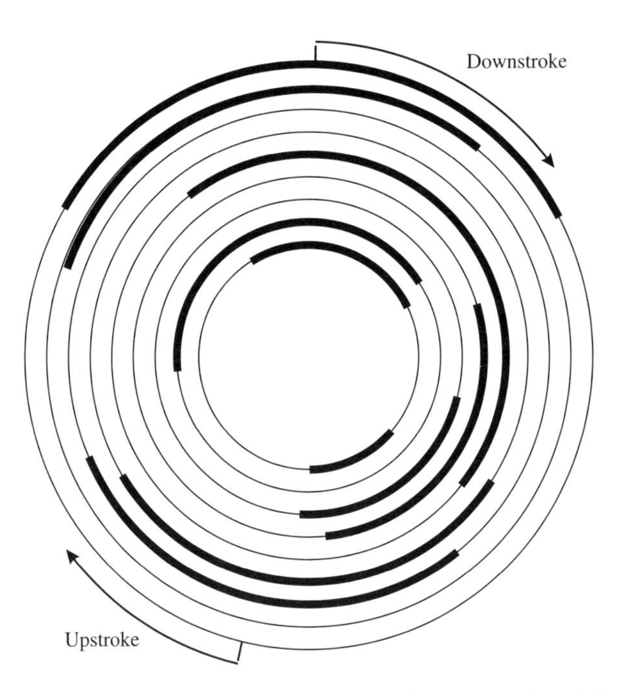

Fig. 7.6 EMG activity of nine muscles from the shoulder girdle of the European starling during one tail beat cycle of 72 ms. The position of the muscles is indicated in Fig. 7.5. The thick lines indicate onset, duration, and offset of the EMG signal of each muscle. The muscles are from the largest circle inwards: pectoralis, sternocoracoideus, supracoracoideus, deltoideus major, coracobrachialis caudalis, scapulohumeralis caudalis, scapulotriceps, humerotriceps, biceps brachii (based on Dial *et al.* 1991, with kind permission of The Company of Biologists).

are typically associated with the upstroke. The activity of the cranial part of the deltoideus major starts and stops earlier than that of the caudal part; the combined active period is indicated here. The active period of supracoracoideus and the deltoideus major starts in the last part of the downstroke and continues into the early upstroke. Electrical activity of the scapulohumeralis caudalis and the scapulotriceps is confined to the second half of the downstroke. The biceps brachii has two active periods in each cycle: one at the beginning and one at the end of the downstroke.

EMGs give an indication of the timing of muscle activity and an approximate period of force production. The function of the separate muscles however is not clarified entirely although some conclusions are probably justified. The humerotriceps is activated before the other muscles involved in the downstroke and might play a role in the extension of the wing in anticipation of the downstroke. The sternocoracoideus follows and the timing is consistent with the idea that it is responsible for the spreading of the furcula as indicated in the previous paragraph. The pectoralis is the dominant downstroke muscle. It not only brings the distal end of the humerus down around the shoulder joint but also pronates the wing by rotating the humerus forward around its longitudinal axis and it pulls the upper arm forward, protracting it prior to the downstroke. The activity starts while the wing is still in the upstroke phase. The muscle is being stretched while it already tries to shorten. The consequences for the force generated will be subject of Section 7.4. The pectoralis might be used during that period to slow the upward movement of the wing down. The activity of the biceps during the beginning of the downstroke (Fig. 7.6) probably helps the pectoralis to protract the humerus. Late during the upstroke three muscles become active: the biceps, the subscapularis, and the coracobrachialis caudalis. All three take part in the lowering of the humerus during the downstroke but that is not the only function. The biceps is activated when the elbow is extending during the end of the upstroke. However, the biceps is expected to flex the elbow and not to extend it. An explanation could be that its function initially is to decelerate the elbow extension. Subsequently it could assist the humerus depression and fix the elbow in its stretched position with a flexor moment counteracting the extensor efforts of the humerotriceps. Near the end of the downstroke three muscles are active. The coracobrachialis caudalis first helps to bring the humerus down (it has with 47% of the cycle the longest contribution of the muscles measured) but changes into a retractor towards the end of the downstroke when the humerus approaches its lowest position. Supination and retraction of the humerus starts during the final stages of the downstroke. The scapulohumeralis caudalis probably rotates and retracts the humerus. It starts to do that before the downstroke is half way and continues to the end of the downstroke. The scapulatriceps is activated next. It is difficult to imagine what it is actually doing because the elbow is already fully stretched. It might control the elbow position working against external forces on the wing. The coinciding activity of its antagonist the biceps probably helps to secure the elbow against torque caused by the aerodynamic forces on the wing. The supracoracoideus and the two parts of the deltoid muscles become active late in the downstroke when they possibly at first decelerate the downward movement of the humerus and subsequently start with the elevation. Wing supination by a rearward rotation of the humerus is also a part of the function of these muscles.

The interpretation of the role of the muscles during the wing beat cycle is complicated by the multifunctional nature of each muscle, by the fact that antagonistic muscles can work simultaneously and by the lack of knowledge about the complex aerodynamic forces on different parts of the moving wing. In conclusion, the cyclic activity patterns combined with the knowledge about the anatomical details do not provide a satisfactory explanation of the mechanics of the wing beat process and we are still far from appreciating the full complexity. First of all we need to know more about the magnitude and the direction of the forces exerted by the flight muscles.

7.4 *Pectoralis force and work: using the deltopectoral crest as a strain gauge*

Force measurements of muscles in an intact bird are technically complicated due to the complex nature of the system and due to the difficulties related to the deployment of force transducers. Force transducers usually do not measure the force directly but translate the strain at the point of attachment to force using strain gauges (see Chapter 3: Section 3.3.2). A strain gauge must be in such a position that the action of the muscle stretches it. This has been only possible so far for the largest flight muscle, the pectoralis. Biewener *et al.* (1992) were the first to record dynamic forces generated by the pectoralis of the starling in flight. The technique allowed the application of the work loop concept developed by Machin and Pringle (1960) and Josephson (1985).

The insertion of the pectoralis on the lower surface of the deltopectoral crest at the anterior side of the humerus (Fig. 7.7) makes it uniquely suitable to apply a strain gauge force transducer. Straight across the insertion point, on the upper side of the humerus there is enough room to attach a strain gauge rigidly to the bone. The deltopectoral crest transfers the force of the pectoralis to the strain gauge on the opposite side of the bone. Calibration of the readings obtained from the strain gauge in that position was done with a separate isometric force transducer firmly attached to the wing of anesthetized birds. Biewener *et al.* (1992) used tetanic contractions of the pectoralis with the humerus in positions covering the normal range during steady level flight, to calibrate the strain gauge readings.

Birds fitted with a strain gauge and an EMG electrode positioned deep inside the pectoralis, were flown in a low turbulence wind tunnel using the starling's preferred speed of 13.7 $\mathrm{m\,s^{-1}}$. High-speed films were made to be able to estimate length changes of the pectoralis and to correlate force changes and muscle activity patterns with the wing beat kinematics.

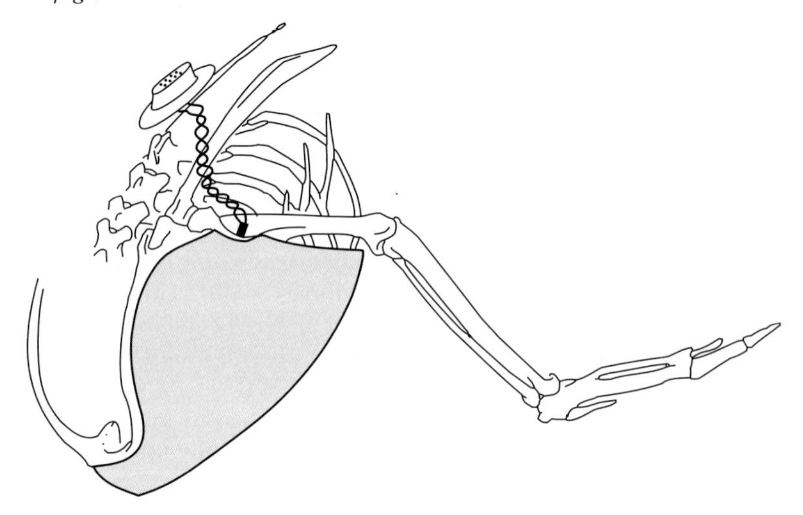

Fig. 7.7 The shoulder and wing skeleton of the European starling and the position of the insertion of the pectoralis (grey) on the ventral side of the deltopectoral crest of the humerus. A strain gauge on the dorsal surface of the crest is indicated as a black rectangle connected subcutaneously by two wires with a connector attached to the animal's back (based on Biewener *et al.* 1992, with kind permission of The Company of Biologists).

The wing beat frequency used at this speed was about 15 Hz corresponding to cycle duration of 67 ms. The results of the measurements during one 67 ms cycle are used as an example in Fig. 7.8. The recording of the pectoralis force during that period is depicted in Fig. 7.8(a). The bird drawings represent from left to right the wing positions half way during the upstroke, at the end of the upstroke, half way during the downstroke and at the end of the downstroke. A thick grey line follows the force trace as long as an EMG is recorded. The force starts to increase during the late upstroke and increases further during the first half of the downstroke reaching a peak value approximately 5 ms after the EMG stopped. There is a rapid decrease during the end of the downstroke, the lowest value is found a few ms after the end. We assume here that there is zero force at that instant when the pectoralis reaches its resting length of 34 mm. Stretching and shortening of the muscle during the cycle are compared with the force development in Fig. 7.8(b). Length change estimates are based on angular displacements of the humerus with respect to shoulder and sternum obtained from the analysis of X-ray films of starlings in flight (see Section 7.1 and Dial *et al.* 1991). The resting length of the muscle was 34 mm and the total excursion 7.31 mm which is 21.5% of that length. Figure 7.8(b) consists of two graphs. In each one the horizontal axis is the strain of the muscle expressed as the percentage of the resting length; the vertical axis is the force measured at the strain gauge on the deltopectoral crest. The data points represent strain and force values at instances obtained by dividing the cycle duration into 13 equal time

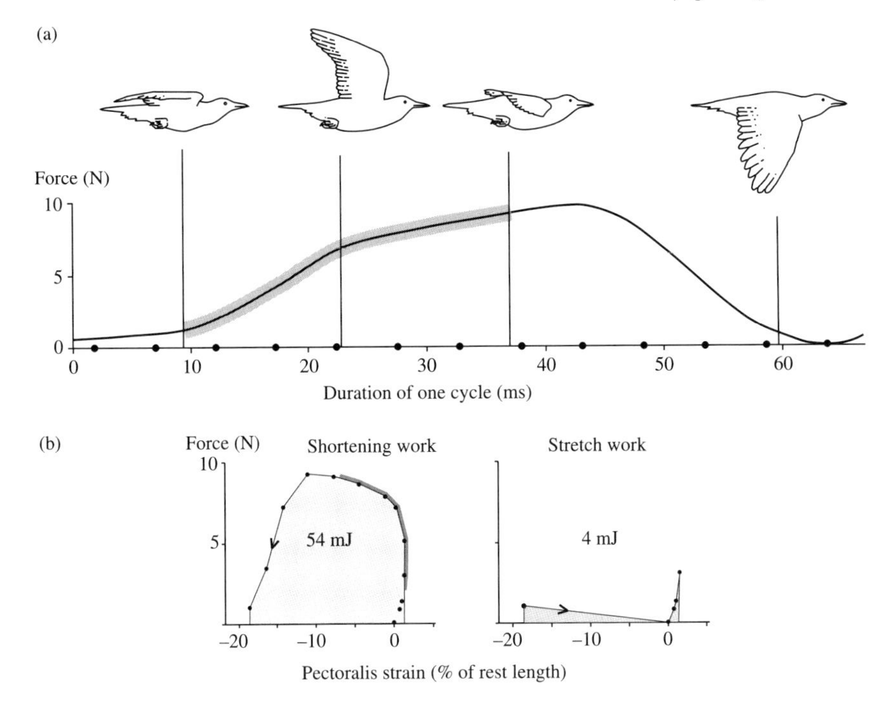

Fig. 7.8 The force exerted by the pectoralis muscle of the starling recorded by a strain gauge on the dorsal side of the deltopectoral crest during one wing beat cycle lasting 67 ms. The bird is flying in a wind tunnel at 13.7 $m\,s^{-1}$ (based on Biewener *et al.* 1992). (a) The force as a function of time. The cycle is divided into 13 equal time intervals by 13 points indicated as black dots on the x-axis. The vertical lines under the bird pictures denote the wing position during the cycle. From left to right: half way during the upstroke, at the end of the upstroke, half way during the downstroke and at the end of the downstroke. EMG activity is indicated by the thick grey line following the force curve. (b) Force as a function of pectoralis strain. The work (force times distance) done to shorten the pectoralis is represented by the surface of the grey area in the left panel, the work done to stretch the pectoralis is indicated on the right. The EMG duration is indicated as a thick outline in the left panel. See text for further explanation.

steps (indicated by black dots on the X-axis in Fig. 7.8(a)). The cycle runs counterclockwise. We assume the force to be zero when the muscle is at its resting length. The total work in N m (J) done by shortening is represented by the surface of the shaded area in the left graph in Fig. 7.8(b). The duration of the EMG is indicated as a dark grey line. The muscle starts shortening when it is about 1.5% longer than the resting length. The overall shortening distance is about 20% of the resting length. The area under the curve represents the total amount of work done by the pectoralis to shorten (54 mJ in this case). A complete cycle requires the muscle to return to the length where it started; this implies that it has to be stretched by an outside force because muscles are unable to stretch themselves. Stretching starts after it is maximally shortened and continues to the resting length and beyond that to the starting point where it reached a length 1.5% longer than the resting length. Stretching requires force to overcome the internal resistance of the muscle. According to the EMG, the muscle starts to generate contracting force towards the end of the upstroke while still lengthening; anticipating the start of shortening by about 4 ms. During that period the pectoralis does negative work actively. The right hand diagram shows the work of about 4 mJ done by an outside force to stretch the muscle. The muscle is mostly passive during the stretch but force required to stretch it increases towards the end of the stretching phase when the pectoralis starts to be active and does negative work. In total the negative work done on the pectoralis is 4 mJ.

A bird has two body halves and sets of flight muscles on the right and the left. The total work required to be done during one wing beat cycle is 0.116 J which is twice the sum of the work done to shorten the pectoralis in one half of the body and the work needed to stretch it back. The supracoracoideus lifts the wing up and stretches the pectoralis. If we neglect the work done by other forelimb muscles we can obtain an estimate of the mechanical flight power. The wing beat frequency of the starling (weighing about 71 g) was 15 Hz and hence the total work per unit time to flap the wing up and down is $0.116 \text{ J} \times 15 \text{ Hz} = 1.7 \text{ W}$.

Similar measurements on pigeons of slightly more than 300 g provided mechanical energy generated by the pectoralis muscles of 3.3 W.

The initial application of this complex experimental method had to deal with uncertainties regarding the measurements of the length changes in the muscle which were determined from wing kinematics based on high-speed cine and X-ray films. In later studies (Biewener *et al.* 1998) length changes and strain cycles are measured directly and more accurately using sonomicrometry, where pairs of piezoelectric crystals are implanted in the muscle. Sound generated by one crystal reaches the other one, travelling at the speed of sound (of about 1540 m s^{-1}) through the muscle. Usually a small anchoring device ensures that the crystals do not move when the muscle goes through cyclic contractions. The distance between the crystals, which is in the order of 10 mm, can be calculated precisely from the time it takes the signal to reach the other crystal. Up to 1000 instantaneous length measurements per second are made. The technique has not been applied to study pectoral muscle performance in starlings yet, but it has been successfully used in other species.

Forces generated by the pectoralis muscle during flight were studied using strain gauges at the deltopectoral crest and length change measurements *in vivo* of European starlings, pigeons, mallards, common magpies, cockatiels, and ringed turtle doves. The last three species were investigated over a range of flight velocities. Table 7.1 summarizes some results and provides the references. The measured power per kilogram bird ranges widely between 9.2 and 53.6 W. Much of this variation is due to different flight speeds. However, it could partly be due to the complexity of the experimental approach. Earlier studies are probably less accurate because the techniques were not

Table 7.1 Mechanical power generated by the pectoralis muscle measured in flight.

Species	Mass (g)	Velocity ($m\,s^{-1}$)	Frequency (Hz)	Mechanical power W	Mechanical power W/kg bird	Techniques	Reference
Starling	70–73	13.7	15	1.7	24.2	Tunnel, film, EMG, strain gauge	Biewener *et al.* 1992
Pigeon	301–314	6–9	8.6	3.3	10.6	Free, film, EMG, strain gauge	Dial and Biewener 1993
Pigeon	649	5–6	8.7	12.6	19.4	Free, Sono, EMG, strain gauge	Biewener, Corning and Tobalske 1998
Mallard	995	3	8.4	20.7	20.8	Free, Sono, EMG, strain gauge	Williamson, Dial and Biewener 2001
Magpie	174	0	8.2	3.6	20.7	Tunnel, film, EMG, strain gauge	Dial *et al.* 1997
	174	4–12	7–8	1.6	9.2	Tunnel, film, EMG, strain gauge	Dial *et al.* 1997
	174	14	8	2.1	12.1	Tunnel, film, EMG, strain gauge	Dial *et al.* 1997
Cockatiel	78.5	5	8.4	1.3	16.6	Tunnel, Sono, EMG, strain gauge	Hedrick *et al.* 2003; Tobalske *et al.* 2003
	78.5	14	8.2	3.7	47.1	Tunnel, Sono, EMG, strain gauge	Hedrick *et al.* 2003; Tobalske *et al.* 2003
Ringed turtle-dove	139.8	7		4.3	30.8	Tunnel, Sono, EMG, strain gauge	Tobalske *et al.* 2003
	139.8	17		7.5	53.6	Tunnel, Sono, EMG, strain gauge	Tobalske *et al.* 2003

The pectoralis forces are measured by strain gauges at the deltopectoral crest on the humerus. The data selected are for level flight only.

yet fully developed. The effect of flight speed on pectoralis power output has been investigated in three species so far. Values tend to be high during both slow and high velocity flights and moderate at intermediate speeds following a more or less U-shaped curve. The differences could be caused in principle by different wing beat frequencies, muscle shortening distances or force production. The wing beat frequencies turn out to vary surprisingly little over the ranges of speeds measured. Variation in muscle stretch caused by variations in amplitude is also limited. The only factor remaining is hence the force produced. Larger force could be provided by an increase of the number of pectoralis muscle fibres involved in the contraction but in fact we do not know.

The possibility to attach a strain gauge on the upper side of the deltopectoral crest makes it possible to measure the pectoralis force in flight. Other flight muscles do not have that advantage and forces generated by these muscles have not been measured in flight.

7.5 *The main upstroke muscle*

The supracoracoideus is generally considered to be the main muscle involved in the upstroke. This muscle is situated on the sternum underneath the pectoralis. It inserts with a long tendon via the triosseal canal in the shoulder joint dorsally on the humerus close to the shoulder joint. It is a strong bipinnate muscle. Poore *et al.* (1997*a,b*) studied its function in the European starling and in pigeons in detail by stimulating either the muscle or the nerve in anaesthetized birds. The movements of the humerus in lateral and frontal view were measured. Forces were measured by removing the bone with the tendon attached and connecting it to a force transducer.

The supracoracoideus of both species elevates and simultaneously rotates the humerus backwards. At the end of the downstroke the humerus is retracted towards the body, starts to elevate and rotates. The elbow flexes and the wing supinates (Fig. 7.3(c)). Rotations about the longitudinal axis up to 80° and elevations up to 60° above the horizontal were measured. The maximum forces measured were 6.5 and 39.4 N for starling and pigeon respectively. These values correspond to 10 times the body weight in both birds.

It has not yet been technically possible to study the function of the upstroke muscle in flight, but the measurements made here allow reconstruction of the force record in time in combination with the measured activity of the pectoralis. The EMG timing and durations are indicated. EMG's do not overlap (see Fig. 7.6) but the proposed supracoracoideus force starts to build up before the pectoralis force has disappeared. This would indicate that towards the end of the downstroke pectoralis and supracoracoideus simultaneously generate force, obviously to control and stiffen the movements of the shoulder joint.

7.6 *Tail steering*

Electrical activity has been measured during walking, and during different flight stages in the tail muscles of the pigeon. The position of the muscles is shown in Chapter 2 in Fig. 2.9 (Gatesy and Dial 1993). A brief anatomical description is included in Section 2.6 and tail movements have been described in Section 4.6. EMG patterns may show if a muscle is involved in a certain activity. The relationship between EMG and force production is different for each muscle. Muscles can act in complex combinations and may be involved in positive and negative work. These considerations imply that exact functions are difficult to assess but it is possible to detect different patterns of activity during various forms of locomotion.

Circular diagrams in Fig. 7.9 represent such patterns for the tail muscles shown in Fig. 2.9 together with the electric activity of the most important flight muscle, the pectoralis. Each circle represents a full cycle of the activity; thick segments indicate the onset, the offset, and duration of the EMG. Circular arrows on the outside of each diagram indicate the direction of movement.

During walking the pectoralis and most tail muscles are inactive. The longissimus dorsi is continuously firing but the EMG amplitude increases twice during each cycle. The iliotrochantericus caudalis starts to show activity in the swing phase of the walk and continues through much of the propulsion phase. The caudofemoralis is active half way down the step; both levator caudae muscles start to be active during the propulsion phase of the walking cycle and continue well into the swing phase.

More muscles are active during all three flight modes. The pectoralis starts activity towards the end of the upstroke and the EMG continues well into the downstroke. We saw that maximum force generation in this muscle occurs after the EMG stopped. The iliotrochantericus caudalis is usually not active during flight. The other hind leg muscle, the caudofemoralis, is active during the downstroke phase in all three flight stages, but the signal is less strong than during walking. The bulbi recticium are continuously active during take-off, flapping flight, and landing.

During take-off most other muscles are active twice during the cycle; three (longissimus dorsi, levator caudae pars vertebralis, and lateralis caudae) towards the end of both upstroke and downstroke and three (depressor caudae, pubocaudalis externus and internus) around or just after the transitions between upstroke and downstroke. The levator caudae pars recticalis is not biphasic and fires only towards the end of the upstroke.

Landing shows a completely different pattern. All muscles, except the inactive iliotrochantericus caudalis and the continuously active bulbi rectricium, have one burst of activity per cycle. The caudofemoralis brings the femur forward during the downstroke and is not active during upstroke. The other ones are either active during the end of the downstroke or near the end of the upstroke.

Two muscles remain silent during slow level flight: the iliotrochantericus caudalis and the levator caudae pars rectricalis. Two show continuous

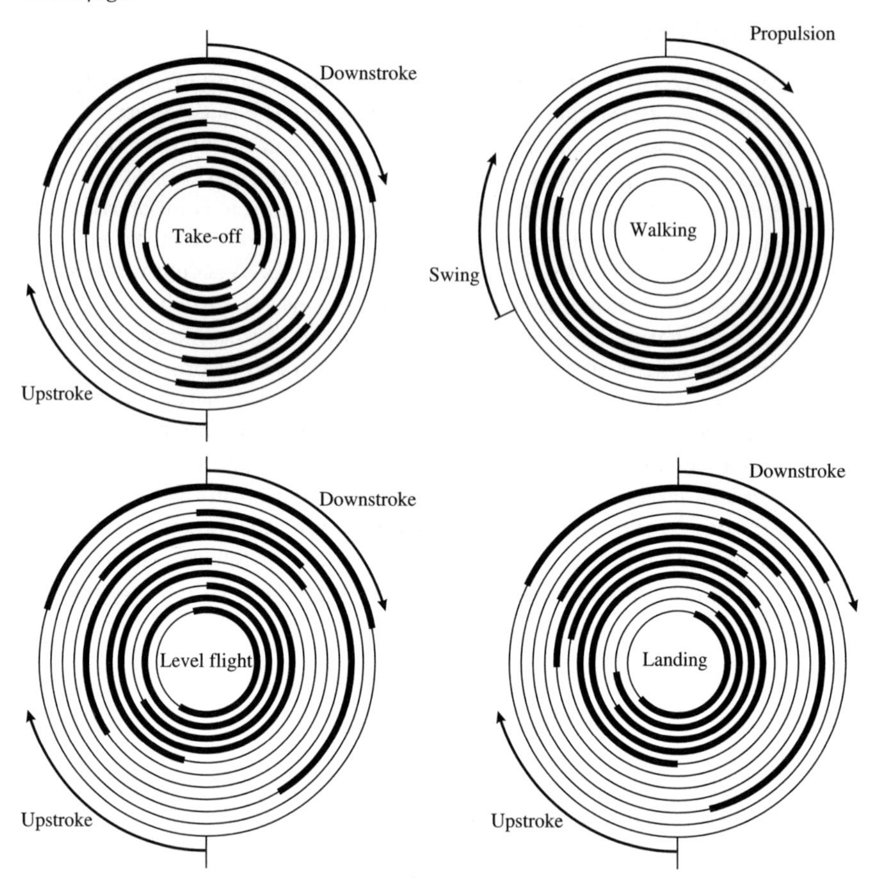

Fig. 7.9 EMG activity of tail muscles during walking, take-off, slow level flight, and landing. Every circle represents a full cycle for one muscle, the thick part indicates EMG activity. The muscles are from the outer circle inwards: pectoralis, iliotrochantericus caudalis, caudofemoralis, longissimus dorsi, levator caudae pars vertebralis, levator caudae pars rectricalis, lateralis caudae, bulbi rectricium, depressor caudae, pubocaudalis externus, and pubocaudalis internus. See Fig. 2.9 for the position of these muscles in the tail (based on Gatesy and Dial 1993, with kind permission of The Company of Biologists).

activity: typically the bulbi rectricium and also the pubocaudalis externus. The other muscles have single bursts during different parts of the cycle. The pubocaudalis internus and the depressor caudae fire more or less in synchrony during the downstroke.

The biomechanics of the tail function are still far from established. We know the origins and insertions of the muscles but have hardly any idea about the actions of the complex skeletal elements they may steer. The patterns of EMG activities must be regarded as a first step towards understanding of form and function in bird tails.

7.7 *Wing beat cycles and respiration*

The movements of the wishbone and sternum in common magpies during flight were investigated using the X-ray filming method described in Section 7.2 (Boggs *et al.* 1997*a*,*b*). The results were similar to those obtained for the starling: the furcula bends outward laterally during the downstroke and back during the upstroke; the sternum tilts up and down during the down and upstroke respectively. There are usually more wing beat cycles during each respiratory cycle. Faster wing beats coincide with faster respiration with prolonged breathing times during glides. The air sac pressure, measured through inserted cannulae, is positive during expiration and negative during inspiration. When inspiration coincides with the downstroke the negative air sac pressure is further reduced. During expiration positive pressures are reduced whenever there are upstrokes. Downstrokes are related to increase and upstrokes to decrease of pressures in the interclavicular and the posterior thoracic air sacs.

Ratios between wing beat and respiratory cycles are commonly in the order of 3 : 1 but can vary from 1 : 1 to 5 : 1. Figure 7.10 gives a general scheme of three coordination patterns among air sac pressure changes, wing

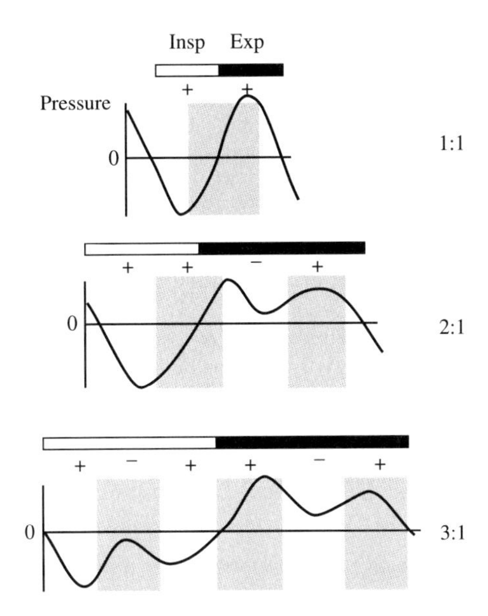

Fig. 7.10 The complex relationships between respiration cycles and the effect of movements of the furcula and sternum during up and downstroke cycles on the pressures in the air sacs. Typical pressure recordings are given for three cases. In the upper panel there is one wing beat cycle per breath (inspiration plus expiration), the middle panel shows the effect of two wing beat cycles per breath and the bottom graph represents what happens when a bird makes three wing beats during one respiration cycle. See text for further explanation (from Boggs *et al.* 1997*b*, with kind permission of The Company of Biologists).

beat, and respiration cycles. The bar on top of each of the three graphs gives the duration of inspiration (not filled) and expiration (black). The grey blocks indicate the duration of the downstrokes. Blanc spaces in between are the upstrokes. The numbers of wing beat cycles per respiratory cycle are: 1 : 1, 2 : 1, and 3 : 1. The graphs show typical traces of pressure recordings in the air sacs. When the downstroke coincides with the start of expiration or during expiration the increased air pressure helps the respiration (indicated by a +). During the first downstroke in the 3 : 1 case the tendency to increase the pressure coincides with inspiration, which makes it less effective. The minus above that downstroke indicates the negative effect of the downstroke. A similar story can be told about the pressure reducing effect of the upstroke. When the upstroke coincides with inspiration the pressure reduction helps, but when it occurs during expiration it diminishes the net output of the expiration at that instant. The three cases show that the positive effects of the wing beat cycle prevail and that the net effect of the interference is positive.

7.8 *Summary and conclusions*

Movements of the skeleton of European starlings and common magpies have been visualized and quantified during flight in a variable speed wind tunnel using high-speed X-ray film techniques. The up and down movements and rotations of the wing bones during wing beat cycles can be determined from these films. The X-ray technique also showed that the wishbone acts as a spring by being bent outward during the downstroke and recoiling inward during the subsequent upstroke. The sternum also appeared to make cyclic up and down excursions in synchrony with the wing beat cycle.

The electrical activity of the larger flight muscles has been measured in flight and could be related to the different phases of the wing beat cycle. The origin and insertions of these muscles is also well known in a number of species. However, the precise functions during flight remains enigmatic in most cases mainly because of a lack of insight in the force generation.

The pectoralis is the largest flight muscle and responsible for the downstroke of the wing and for forward rotation of the humerus at the start of the downstroke. The insertion of this muscle on the deltopectoral crest offers the unique opportunity to measure forces directly. Data on shortening and lengthening during each wing stroke are combined with the timing of force generation to determine the amount of work produced per wing beat. This approach offers the opportunity to estimate the largest part of the mechanical power involved in flight at different speeds.

Experiments with the supracoracoideus of pigeons and European starlings revealed that this strong bi-pinnate flight muscle is not only involved in powering the upstroke of the wings but also plays an important role in the backward rotation of the wings at the beginning of the upstroke.

The timing of EMG activity in tail muscles during walking and flight has been investigated in the pigeon. Interpretation of the patterns found is not straightforward because the relationship with force production and the complex kinematics remains unclear.

X-ray films of starlings and common magpies in flight show movements of the wishbone and sternum in harmony with the wing beat cycles. Measurements of pressure changes in the anterior air sacs reveal a complex relationship with respiration.

We are still far from understanding the flight engine of birds. In fact we have only reached the stage where we have a vague idea of form and function of the most important parts.

8 *Energy required for flight*

8.1 *Introduction*

During flight, the beating wings are motorized by muscle activities. In Chapter 7 we saw how muscles shorten and stretch, do work, and power the wing beat cycles. The splitting of adenosine triphosphate (ATP) into adenosine diphosphate (ADP) provides the energy for the dynamic connections between the thick and thin filaments involved in shortening (see for more detail Box 7.1). Different metabolic pathways are responsible for the continuous supply of these energy rich phosphates to the muscles. Complex pathways essentially free energy from foodstuff by burning it with the use of oxygen. Water and carbon dioxide are byproducts of this process. The food burnt may consist of lipids, carbohydrates, or proteins, single or in various combinations. In principle, flight costs can be deduced directly from the amount of fuel combusted or from the heat produced. Indirectly, energy expenditure can also be derived from the exchange of oxygen and carbon dioxide or even from the amount of water produced. This sounds easier than it is. In most cases, we do not know precisely which substrates the animals are burning, let alone how much energy they gain from the combustion processes.

This chapter gives an overview of attempts made to estimate and measure the energetic costs of flight. The variety of approaches will serve as a guide. It is good to be aware of the fact that all the figures are to be treated as estimates. Laboratory measurements will introduce biases that are fundamentally different from the uncertainties connected with field approaches. The methods will be critically assessed to find out, which figures can be more or less trusted. These figures are used in the next chapter to search for general trends among birds and to find empirical equations with some predictive value. Now a feeling is generated for the order of magnitude of the cost of forward and hovering flapping flight and for the main factors determining the level of energy expenditure. Manoeuvring and the use of intermittent flight techniques will of course affect the flight costs but data indicating how much that is are lacking.

The SI ('Système Internationale') (see Box 1.1) unit for energy or work is the Joule (J); it is defined as the amount of work done or energy needed to lift a weight of 1 Newton (N) over 1 metre (m). Calories are used in the older literature and might be more familiar to many readers. A calorie is the amount of heat required to raise the temperature of 1 g water from 14.5°C to 15.5°C. The international rules will be obeyed and the SI system is used throughout; but it is good to keep in mind that 1 calorie equals about 4.2 J (Pennycuick's (1974) 'Handy matrices of unit conversion factors for biology and mechanics' are frequently used). In the SI system, the rate at which energy is consumed during flight is expressed in Watt (W), a derived unit equivalent to one $J\,s^{-1}$.

The idea that mass loss can be used to find out how much flight costs is obvious and there have been many attempts to measure it. The next paragraph summarizes these, discusses the reliability, and selects the most confident measurements.

Oxygen consumption and carbon dioxide production as indicators of energy turnover found a widespread use although inventing a method to measure these gases appeared to be a great challenge especially under field conditions. During measurements in the field, the conditions are natural but vary tremendously. There the birds will be using various energy saving techniques and deal with numerous abiotic conditions. Field measurements usually include some resting, as well as take-off and landing manoeuvres and other aerobatics. For the moment the bias caused by the variation is unavoidable and unfortunately there is no single definition for flight costs. Therefore, the relevant publications will be carefully studied to obtain some feeling for the implications and complications connected with each of the flight cost figures found.

Heartbeat rates also indicate the level of exercise and can be used as an indicator of flight costs if a relationship with oxygen consumption and carbon dioxide production can be established.

In wind tunnels, birds fly at constant average speeds although some birds use intermittent flight. The unnatural conditions might affect the measurements but the direct access to the birds is highly advantageous. A variety of masks has been used to collect breaths from birds in flight; an overview of the various shapes will be given.

Heavy stable isotopes of hydrogen, oxygen, and carbon have been used in cleverly designed experiments, including some on oceanic birds, to measure energy turnover during exercise. The methods are explained and an overview is given of the most promising results.

The majority of studies into the energetics of flight concentrated on forward flight but hovering flight of hummingbirds (and one sun bird) was also the subject of investigation. The nature of hovering behaviour where a bird remains in one position with respect to the ground seems to facilitate respiratory measurements. The scientific efforts to obtain trustworthy figures will be reviewed.

Estimates of metabolic costs which are considered more or less trustworthy are collected in Table 8.1.

8.2 *Mass loss estimates*

The first attempts to estimate the metabolic costs of flight date back from the early 1950s. The use of mass loss figures measured during migration

Table 8.1 Accumulated data on the cost of forward flapping flight.

Species	Body mass (g)	Flight costs (W)	Technique	Flight speed ($m\,s^{-1}$)	Flight duration	Gliding (%)	Source
Green violet-ear	5.5	1.82	R, W	11	1–8 min		Berger (1985)
Palestine sunbird	6.2	1.64	C, I		2 min		Hambly *et al.* (2004)
Sparkling violet-ear	8.5	2.46	R, W	11	1–8 min		Berger (1985)
Pine siskin	12.5	3.03	M, F	15	56 min	0	Dolnik and Gavrilov (1973)
Sand martin	13.7	1.60	D, F		12.7 h	21	Westerterp and Bryant (1984)
Zebra finch	14.5	2.24	C, I		2 min		Hambly *et al.* (2002)
Barn swallow	17.3	1.34	M, F		>2 h		Lyuleeva (1970, 1973)
Northern house martin	17.8	1.01	D, F			54	Hails (1979)
Barn swallow	19.0	1.30	D, F				Hails(1979)
Barn swallow	19.0	1.62	D, F	11			Turner (1982*a*,*b*)
Northern house martin	19.7	1.08	M, F		>2 h	54	Lyuleeva (1970, 1973)
Chaffinch	22.3	4.51	M, F	15	56 min	0	Dolnik and Gavrilov (1973)
Brambling	23.2	4.60	M, F	16	52 min	0	Dolnik and Gavrilov (1973)
Thrush nightingale	24.7	1.75	M, W	7.9	15 min	0	Kvist *et al.* (1998)
Thrush nightingale	26	1.91	M, W	10	12 h	0	Klaassen *et al.* (2000)
Eurasian bullfinch	29.5	*5.60*	M, F	14	60 min		Dolnik and Gavrilov (1973)
Swainson's and hermit thrushes	30	4.3	D, F	13	7.7 h		Wikelski *et al.* (2003)
Budgerigar	35.0	4.12	R, W	12	0.5–2 h		Tucker (1968, 1972)
Common swift	38.9	1.80	M, F		>2 h	70–80	Lyuleeva (1970)
Wilson's storm petrel	42.2	1.82	D, F		2–4 d	0	Obst *et al.* (1987)
Purple martin	50	*4.1*	D, F	8	4–8.5 h		Utter and LeFebvre (1970)
Rosy starling	71.6	8.05	D, W	11.1	>6 h		Sophia Engel, personal communication
European starling	73	9.0	R*, W	16		17	Torre-Bueno and LaRochelle (1978)
European starling	77	10.5	D, F	14	3.5 h		Westerterp and Drent (1985)
European starling	89	12	R, W	9.9	12 min		Ward *et al.* (2001)
Red knot	128	13.5	D, W	15	6–10 h		Kvist *et al.* (2001)
Common kestrel	180	13.8	B, I	9	49 d	30	Masman and Klaassen (1987)
Sooty tern	187	4.8	D, F	10	8–23 h	5–25	Flint and Nagy (1984)

Common kestrel	213	14.6	D, F	8	2–5 h		Masman and Klaassen (1987)
Common teal	237	13.2	M, W	11.5	15 min	0	Kvist *et al.* (1998)
Fish crow	275	24.2	R, W	11	15–20 min		Bernstein *et al.* (1973)
Laughing gull	277	18.3	R, W	12	20–30 min		Tucker (1972)
Bar-tailed godwit (m)	282	17.8	M, F	16	>24 h		Piersma and Jukema (1990); Lindström and Piersma (1993)
Laughing gull	322	26.3	R, W	13	20–30 min		Tucker (1972)
Bar-tailed godwit (f)	341	24.2	M, F	16	>24 h		Piersma and Jukema (1990); Lindström and Piersma (1993)
Pigeon	394	31.9	D, F	17	7–8 h		LeFebvre (1964)
Pigeon	394	33.1	M, F	17	7–8 h		LeFebvre (1964)
Pigeon	425	34.1	R, F	19	3 h		Polus (1985)
Pigeon	442	26.8	R, W	10	10 min		Butler *et al.* (1977)
Chiuahuan raven	480	32.8	R, W	11	30 min		Hudson and Bernstein (1983)
Red-footed booby	1001	24.0	D, F		5–28 h		Ballance (1995)
Barnacle goose	2100	102	H, F	14–20			Ward *et al.* (2002)
Cape gannet	2580	*81*	D, F		1 d		Adams *et al.* (1991)
Bar-headed goose	2600	135	H, F	16-21			Ward *et al.* (2002)
Laysan albatross	3064	24	D, F		3 d		Pettit *et al.* (1988)
Northern gannet	3210	97	D, F		4–11 h	0	Birt-Friesen *et al.* (1989)
Black-browed albatross	3580	22	H, F	8	9 d	91	Bevan *et al.* (1995)
Grey-headed albatross	3707	28	D, F		3–4 d	97	Costa and Prince (1987)
Southern giant petrel	3885	68	D, F			76	Obst and Nagy (1992)
Wandering albatross (f)	7300	31	D, F		3–9 d	97	Adams *et al.* (1986)
Wandering albatross (m)	9310	45	D, F		2–8 d	97	Adams *et al.* (1986)
Wandering albatross (f)	9360	43.8	D, F	5	4–7 d	97	Arnould *et al.* (1996)
Wandering albatross (m)	10740	38.1	D, F	5	4-9 d	97	Arnould *et al.* (1996)

Only the most reliable figures are included; see text for the justification. Italic flight costs figures are considered the least reliable. Other italic figures have been derived from sources other than the papers indicated. Flight duration indicates the time span of the measurements. (R = respirometry (R* without mask), W = wind-tunnel measurements, M = mass loss figures, F = measurements in free unrestrained flight, D = doubly labelled water (DLW) technique, I = indoor flight in a corridor, H = heart rate telemetry, B = material balance, and C = heavy carbon method, f = female, m = male.)

seemed a promising approach then. Nisbet (1963) summarized the knowledge collected before 1963. Accurate measurements of the mass of migrants at stopover sites after a presumed non-stop long-distance flight were collected. For some species, there are hundreds of measurements, in other cases, data for only a few birds were available. The lack of information on the mass before departure and on the conditions during the flight was the main problems in all these studies. Passerines crossing large stretches of water are usually, rightfully, not supposed to have landed and so the non-stop condition will be fulfilled. The exact flight distance and duration was usually by no means clear. Radar observations provided additional information in some studies. Wind conditions and flight altitude could usually not be taken into account. In a few cases, some birds out of the migrating population could be measured before departure, but the same birds were never caught at arrival. Estimates of flight costs expressed as mass loss per unit flying time seriously suffered from large standard deviations. In some cases, only a few specimens were taken into account. Bird species involved in these studies were goldcrest, European robin, blackpoll warbler, Northern wheatear, and song sparrow.

8.2.1 Early attempts

One category of mass loss studies used the sad fact that illuminated towers, lighthouses, and other high buildings pose serious threats to nocturnal migrants. For example, Graber and Graber (1962) collected victims of a television tower in Illinois, Hussel (1969) and Hussel and Lambert (1980) used those of a lighthouse near Lake Erie. They collected dead or traumatized birds throughout the night and found that there appeared to be a decrease of the average mass for each species in the course of the night. The data sets involved are on various American passerines. The assumption was that all birds departed at the same instant from an unknown distant location and that the animals that collided at 4 o'clock in the morning flew 3 h more than those that struck the tower at 1 a.m. The costs of flight are supposedly the calculated rate of mass loss. In all these cases, the picture is far from clear. For example, Hussel's (1969) 80 body mass values of veeries were collected during the night of 6–7 May 1965 at 23.00, and every hour from 00.30 to 03.30. The series of average mass values found was 32.3, 32.3, 31.0, 30.6, and 30.8 g. The heaviest bird of 37 g was found at 23.00 and the leanest of 25.7 g was among the 13 animals collected at 01.30 h. The overlap between values for each sample is very large and there is no proper statistical proof that there is actually a trend. Other reasons for mistrusting flight cost estimates from these data sets are obvious. There is no proof for the assumption that the animals departed simultaneously nor do we know that birds dying at different times during the night were equally heavy at departure. There is also no indication that all the birds met the same flight conditions on the way from the unknown location. The time of bird collection was not always the same as the time of arrival at the obstacle. The information available does not offer the possibility to proof that the probability that a certain bird could be collected was not dependent of its body mass. The use of these data in comparisons of flight costs is therefore not recommended.

Not all the early attempts to estimate flight costs from mass loss are useless. The Neringa Spit is a narrow strip of sand dunes, about 100 km long and 0.7–3.5 km wide, covered with trees and shrubs, separating the Gourlandic Haff from the Baltic Sea (Fig. 8.1). It is a route of very intense daytime migration. At two sites, birds were trapped, measured, and released to establish the energy expenditure during flight over a stretch of 50 km where conditions could be monitored. Small passerines required about an hour to cover that distance non-stop during calm weather conditions. The average flight speed for each species could be calculated under windless conditions. The mass and fatness of large numbers of birds were measured before and after the flight. Wind speeds and directions were measured midway between the traps and were assumed constant along the spit and similar to that at the monitoring point. Wind data are used to calculate an equivalent flight distance in still air. Table 8.2 summarizes the data sets obtained.

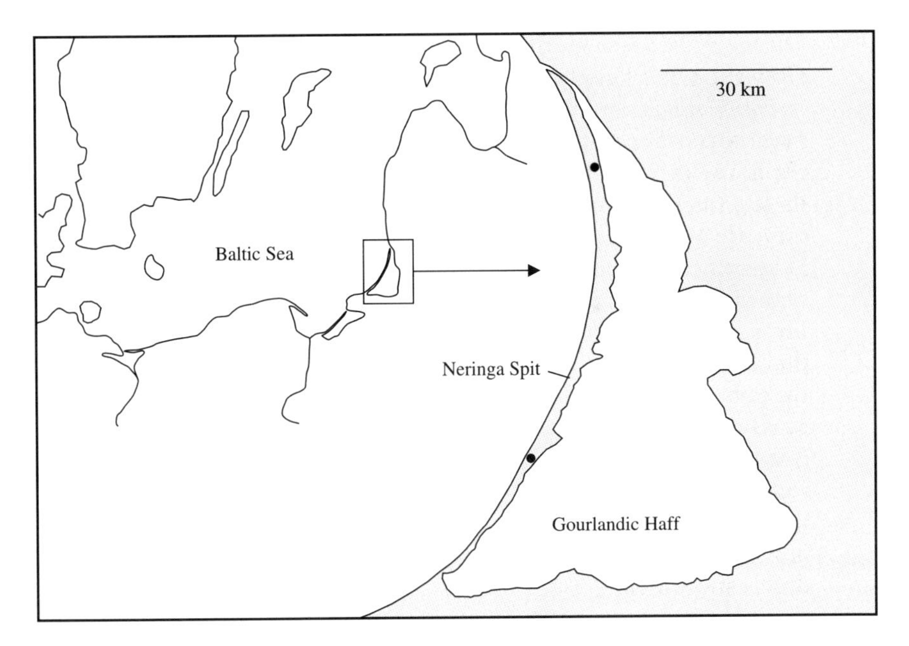

Fig. 8.1 The Gourlandic Haff, based on satellite pictures taken in August 1987 (Beekman *et al.* 1994). The trapping sites are indicated as black dots (Dolnik and Blyumental 1967) The Neringa Spit, a hairline of beaches, dunes and forest, separates the Haff from the Baltic Sea. The legend goes that long time ago a girl-giant Neringa helped the fishermen in their fights against the storms. She created the sandy spit by emptying an apron full of sand, leaving a small gap in the North. Migrating birds use it as a shortcut or are following the coastline during North–South migrations.

Table 8.2 Average mass loss data and flight cost estimates of small passerines during flights over a distance of 50 km (from: Dolnik and Gavrilov 1973).

Species	Flight speed ($km\ h^{-1}$)	Start sample (n)	Average start mass (g)	End sample (n)	Average end mass (g)	Average mass lost (g)	Average costs (W)
Chaffinch	52.2	1623	22.7	3452	21.9	0.8	4.5
Brambling	57.6	89	23.5	325	22.8	0.7	4.6
Pine siskin	54.0	284	12.7	233	12.3	0.4	3.0
Eurasian bullfinch	50.4	8	29.9	6	29.1	0.9	5.6

The energetic equivalent for mass loss was assumed 25.3 $kJ\ g^{-1}$. This figure was based on the fraction of fat in the mass loss during 50 km flights of 1288 chaffinches. Pure fat combustion would provide 39.8 $kJ\ g^{-1}$ of energy. On average slightly more than 63% of the mass lost was fat in chaffinches, which provides the figure of 25.3 $kJ\ g^{-1}$ assuming that no other substances were burnt. It is obvious that the energy equivalent of mass loss cannot be as simple as that and requires the special attention given in Box 8.1 on 'Fuel for flight'. The sample sizes in Table 8.2 for the chaffinches, the brambling and the pine siskin are large enough to accept the data with some confidence. The high cost value for the Eurasian bullfinch should be used with some care given the small sample size.

The Neringa Spit was also the location where Lyuleeva (1973) did her experiments during the early 1960s to determine the costs of flight of Northern house martins, barn swallows, and common swifts. Birds belonging to these groups pose an extra problem to the investigator of flight costs because they are aerial feeders. They probably forage and drink while travelling. Lyuleeva caught birds near the nests and released them 40 or 70 km south of the nesting area. To make sure that the birds would not obtain energy en route, she secured their bills with a thread through the nostrils and around the bill. The victims were deprived of food for 3 h before release to empty the guts: a measure was taken to prevent defecation during the experiment. Mass losses could be calculated for the re-trapped birds; the others either freed their bills or died of starvation. The time span between release and recapture varied considerably from 2 h to up to 18 h, and flight speed could not be estimated. One Northern house martin was found 33 h after release in a state of torpidity. Its mass was 14.2 g and it had lost 5 g. Lyuleeva's observation was:

> It reacted to shaking and pushing with a feeble fluttering of its wings. Despite the fact that it did not appear excessively emaciated, and its weight had not decreased to a critical point, it perished after some time. This episode suggested that considerable losses of weight, characteristic of swallows deprived of food for a long period of time, induce torpidity and subsequent death.

The figure obtained for this animal was, of course, not included in the average rates of mass loss given in Table 8.3.

Box 8.1 ***Fuel for flight***

The oxidation of proteins, lipids, and carbohydrates produces the energy to make high-energy compounds such as ATP, which is the fuel used by muscles to generate force and heat during flight (see Box 7.1). Per unit mass, each of these substrates provides different amounts of energy, produces different amounts of water and carbon dioxide (CO_2) and uses different amounts of oxygen. The respiratory quotient (RQ), the ratio of CO_2 produced over O_2 uptake, reaches specific values in each case. The figures given in the table can only be approximations, because they depend on the chemical composition of the substrates and on the precise pathways of complex combustion processes. In principle, flight cost estimates could be based on measurements of mass changes, oxygen use, CO_2 production, or even heat production. If the reduction of body mass is used, it is important to realize that the production and losses of water can play an important role. Note that the energetic equivalents for oxygen consumed are close together for the three substrates. CO_2 production as a predictor of energy consumption requires a more accurate assessment of the katabolized metabolic substrate.

Accurate RQ measurements of house sparrows and verdins (Walsberg and Wolf 1995) revealed three important aspects:

1. The RQ may change drastically over periods of many hours after the feeding event regardless of the food source in caged birds.
2. The RQ values do not simply reflect the expected values based on the food sources given (either millet or meal worms for the house sparrows and meal worms for the verdins).
3. RQ values below 0.71 may occur probably reflecting non-pulmonary losses of CO_2.

Substrate	Energy	CO_2		O_2		RQ
	($kJ\,g^{-1}$)	($l\,g^{-1}$)	($kJ\,l^{-1}$)	($l\,g^{-1}$)	($kJ\,l^{-1}$)	
Lipids (fat)	39.7	1.43	27.8	2.01	19.8	0.71
Proteins	17.8	0.70	25.4	0.95	18.7	0.74
Carbohydrates	16.7	0.80	20.9	0.80	20.9	1.00
Commonly used figures for unknown mixed substrate					20	0.79

Lyuleeva used an energetic equivalent of 20.1 $kJ\,g^{-1}$ for the barn swallows and 20.7 $kJ\,g^{-1}$ for Northern house martins. These figures are based on results of oxygen consumption measurements by Kespaik (1968). The energetic mass loss equivalent of the common swift is assumed to be the

Table 8.3 Flight cost estimates by Lyuleeva (1970) of three species of aerial planktivorous birds.

Species	Sample size (n)	Average mass (g)	Average mass lost ($g\,h^{-1}$)	Flight costs (W)
Northern house martin	8	19.7	0.19	1.1
Barn swallow	8	17.3	0.24	1.3
Common swift	4	38.9	0.32	1.8

average of the swallow and martin value to find an estimate of the swift flight costs using Lyuleeva's (1973) data.

Lyuleeva closed the beak of her experimental birds to avoid disturbance due to food intake of the mass-based estimates of flight costs. Pigeons are not supposed to feed in the air, but are notorious bombers. Pearson (1964) therefore sealed the cloacae of tippler pigeons to find an accurate figure for mass loss during flights of more than 3 and up to 6 h. Tippler pigeons are inclined to fly many hours over the home loft, circling without soaring and without aerobatics. The average mass at the start was 254 g ranging from 221 to 293 g. Five birds were shot straight after landing (the weight of the shot in the carcass was taken into account). Mass losses varied from 1.96 to 3.95 $g\,h^{-1}$. Pearson assumed that all mass loss was due to fat combustion at 39.6 $kJ\,g^{-1}$. If that were the case, the flight costs would vary between 21.5 and 43 W with an average value of 33 W at the average mass loss rate of 3 $g\,h^{-1}$. There is one more interesting data point in this chapter. A 419 g homing pigeon flew from Redding to Martinez in California in 5 h at an average ground speed of 56 $km\,h^{-1}$ (15.6 $m\,s^{-1}$). The cloaca of this bird was not sealed and it lost 39.1 g, which in case of fat combustion would indicate flight costs in the order of 86 W (twice as high as the least economic tippler). These early pigeon points should be treated with great caution and are not used in Table 8.1.

8.2.2 Fat as fuel during long distance travel

Bar-tailed godwits fly during spring migration from the Banc d'Arguin in Mauretania to staging sites in the Dutch Wadden Sea. Piersma and Jukema (1990) collected enough data over a number of years on the average mass decrease of males and females during that flight to enable estimates of the flight costs. Mass increases prior to departure were 2.8 $g\,day^{-1}$ for the males that leave on average weighing 350 g on 25 April, and 3.2 $g\,day^{-1}$ for the females on average weighing 430 g when they leave 2 days later. Unfortunate casualties of the trapping procedure were used to measure fat content (both absolute and relative to the body mass). This has to be an average figure for males and females in the population because it is impossible to kill the same bird twice. Lindström and Piersma (1993) related the mass and fat data to the time before departure and found that 64% of the mass increase is due to the deposition of fat in males and 67% in females. Captures from the same population straight after arrival in the Netherlands show that males lost on average 136 g and females 178 g. Assuming that the ratio of fat over non-fat tissue used is the same as that accumulated before departure, flight cost calculations can be made although a fair number of additional assumptions are needed. Average wind conditions en route indicate that the best strategy for the birds would be to fly at different altitudes up to 5.5 km where they could meet average tail winds of 18 km h^{-1}. The estimated airspeed of bar-tailed godwits is 57 km h^{-1}, resulting in a speed over the ground of 75 km h^{-1}. It would be the best strategy to fly along the shortest (great circle) route, which is 4300 km long, and there is some evidence that these birds actually do that. At 75 km h^{-1} the journey would last 57.3 h. Furthermore, Piersma and Jukema assumed that fat combustion yields 39.4 kJ g^{-1} and fat-free body tissue with a water content of 75%, 5.1 kJ g^{-1}. If all the suppositions were realistic, the flight costs of the bar-tailed godwits would be 17.8 W for males and 24.2 W for females.

8.2.3 Mass loss measurements in a wind tunnel

A large wind tunnel has been set up at Lund university designed to study long migratory flights under controlled conditions. Its design and construction was supervised by two outstanding bird flight experts: Thomas Alerstam whose main research interest is directed towards migration and Colin Pennycuick, ultimately interested in both physical and biological aspects of the flight. Klaassen *et al.* (2000) estimated flight costs of a thrush nightingale named 'Blue' during eight experimental flights; seven of these lasted 12 h nonstop, one flight took 16 h. The flight speed was kept constant at 10 m s^{-1}. Fuel combustion was used as a measure for the flight costs. Blue's diet consisted of mealworms containing 44% fat and 56% proteins. Its mass loss during the 12 h flights was 3.82 g on average from a mean starting value of 27.82 g. It took Blue 3 days to recover completely from each 12 h flight bout and to regain its starting weight. The energy consumed minus the energy lost during these 3 days was considered equivalent to the costs of the experimental flight. The metabolic equivalent for each gram burnt turned out to be 21.6 kJ. The average flight power generated by the 26 g bird during the 12 h experiment turned out to be 1.91 W.

Blue was also involved in another study in Lund where the rate of mass loss was measured during flights at different speeds (Kvist *et al.* 1998). Blue was one of two thrush nightingales weighing on average 25 g. A common teal of 237 g was treated the same way. The cloaca of the teal was covered with hydrophobic cotton and tape during flights to avoid mass loss by defecation. Flight periods during which thrush nightingale defecated were excluded from the results. Two sophisticated methods were used to convert mass loss to energy used (Box 8.1 summarizes the basic components of the conversion). The thrush nightingales were assumed to have moderate heat loss which they could regulate without evaporating water. The water content of the birds was assumed to remain constant. The minimum power calculated for the thrush nightingales was 1.7 W at 7.9 m s^{-1}. Different conditions were assumed for the teal. The heat loads of the ten times bigger animal were considered so big that it had to evaporate water to keep the body temperature about constant. A different model used the mass loss data at speeds varying from 10 to 15 m s^{-1} to find a minimum cost estimate of 13.2 W at 11.5 m s^{-1}.

Much earlier Marcel Klaassen, one of the authors of the previous papers, was involved in our laboratory in food balance experiments with common kestrels in a 'wind-tunnel' without wind. Chapter 6 describes how the 142 m long corridor in our institute was used to fly trained kestrels up and down between the gloves of two falconers. The corridor was windless and the kestrels flew at the speed of their choice, which was electronically recorded (see Chapter 6). Masman and Klaassen (1987) studied the food balance during periods with daily training sessions of three kestrels, one male and two females, flying up to 20 km day^{-1}. The daily energy intake, the energy lost through the excretion of faeces and pellets and the fluctuations in body mass were taken into account. The kestrels stayed in a respirometer when they were not flying in the corridor. The average flight speed of 8.7 m s^{-1} included starting and landing (Videler *et al.* 1988). The mean mass of the three birds was 180 g ($\pm$14 g). The energy used during the flight sessions was determined at 13.8 W ($\pm$3.1 W). This value could be directly compared with flight costs estimates based on field experiments with kestrels where the turnover of stable isotopes of oxygen and hydrogen were used to measure oxygen consumption and carbon dioxide production (see Section 8.5.1).

8.3 *Respirometric results from cunning experiments*

The rate of oxygen consumption and/or carbon dioxide production is also a potential tool to measure flight costs although there are again many pitfalls. We meet most of these while discussing the various approaches used. One very extensive and widely used data set is that of Teal (1969). It is important to examine it more closely because it offers flight cost data of 13 species, a number that has been statistically dominating many allometric studies on bird flight costs. Measurements of respiratory gases suffer from sampling problems. Teal's birds were flying in polyethylene tubes, 11 and 17 m long with diameters of 0.6 and 1 m respectively. The birds were induced to fly between perches at the ends of the tubes by alternating the illumination of the perches. Flying speeds were calculated from hand-clocked flight durations between the perches. Carbon dioxide production was measured from the increase in concentration in the surrounding air during flight and corrected for resting rate levels. During the most economic flights production varied between 40 and 77 ml g^{-1} h^{-1}. Teal, assuming a ratio of carbon dioxide produced over oxygen consumed (respiratory quotient, RQ) of 0.8, figured that the production of 1 ml carbon dioxide would provide 24.8 J of energy. The average flight cost figure found in this way was at 0.34 W g^{-1} extremely high compared to other data. The problem of this set-up was that the birds could not fly properly. The confined space and short distance induced abnormal behaviour. The average flight speed for example was only about half the speed considered typical for each of the species involved. The impact on the flight budget of starting, slow flight, landing, and fluttering manoeuvres must have been tremendously large. We do not use these data in our search for the costs of natural bird flight.

It is not easy to measure respiratory gas exchange directly of a free flying bird that is not confined to a small container or to the measuring section of a wind tunnel. Berger *et al.* (1970) flew the American black duck, ring-billed gull, and evening grossbeak with latex rubber masks over the beak to measure the oxygen content of the expired air (Fig. 8.2(a) and (b)). The mask of the duck had an inlet and an outlet valve and was fitted with an oxygen electrode in the expired airflow. The masks of the smaller birds contained only one opening from which air was drawn through a long polyethylene tube over an oxygen electrode on the ground. It is not clearly described but flights were probably restrained by the wires from the electrode in the duck and by the length of the tube in case of the other birds. Anyway, the tests were very short, lasting 7–5 s only. The 1026 g duck was panting after a flight of 8 s, burning away at a rate of 78 W. The results for the other birds were 11.1 W for the 59.3 g evening grossbeak and 21.6 W for the 427 g ring-billed gull. The figures obtained with this experimental set-up, resulting in extremely short flights of frightened birds trying to escape, probably do not represent the flight cost figures we are looking for.

There is one attempt published where the expired air is collected on board of a free flying bird. Polus (1985) flew 17 pigeons of 400–450 g during 4000 tests over a total distance of up to 200 km. The pigeons expired via a valve in a mask and a short tube into a thin polyethylene air-collecting bag dangling under the belly just before the tail. The bag could be closed and ejected after a predetermined time interval. The mass of the device was 4 g and the aerodynamic drag was estimated to be less than the effect of 0.28 $m\,s^{-1}$ (1 $km\,h^{-1}$) headwind, perhaps representing 2% of the measured costs. The RQ was 0.91 during take off and reached a level of 0.86 after about 10 min. The birds were flown in a flock over home territory during calm weather conditions with wind speeds less than 0.2 $m\,s^{-1}$. Observers, who had to collect the bags, estimated the flight speed (from the times used to cover known distances between landmarks) at about 19 $m\,s^{-1}$. The estimate for flight costs from these experiments was 34.8 W for a 425 g pigeon.

8.3.1 The use of heart rate measurements

Heart rates can reflect oxygen consumption but the relationship has to be established in each case. This technique has been used in studies of black-browed albatrosses, barnacle geese, and bar-headed geese. Barnacle geese travel in September from Hornsund in southern Spitsbergen to Scotland over a distance of about 2400 km. Two male geese were fitted with ECG recorders.

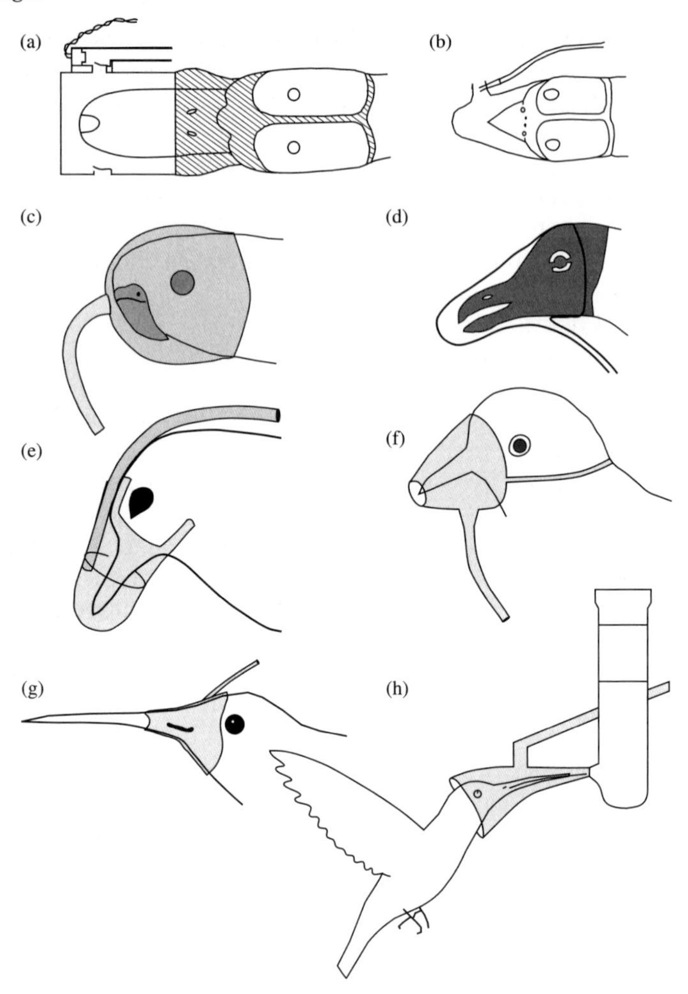

Fig. 8.2 Respiratory masks used in wind-tunnel experiments. (a) American black duck; (b) evening grossbeak (Berger *et al.* 1970); (c) budgerigar and (d) laughing gull (Tucker 1968, 1972); (e) pigeon (Butler *et al.* 1977); (f) pigeon (Rothe *et al.* 1987); (g) sparkling and green violet-ear (Berger 1985); (h) glittering-throated emerald (Berger and Hart 1972). See text for more information.

At the start of the journey the mean heart rate was 317 beats min^{-1}, dropping to 226 beats min^{-1} towards the end (Butler *et al.* 1998).

In a wind tunnel study Ward *et al.* (2002) found linear relationships between oxygen consumption, measured through a mask, and heart rates of barnacle and bar-headed geese flying at velocities between 14 and 20 $m\,s^{-1}$ and 16 and 21 $m\,s^{-1}$ respectively. Heart rates and oxygen consumption did not vary with flight speed. These relationships made it possible to calculate oxygen consumption of birds flying without a mask using measured heart rates of 423 and 434 beats min^{-1} respectively. Metabolic power estimates based on the calculated oxygen consumptions were on average 102 W for a 2.1 kg barnacle goose and 135 W for a 2.6 kg bar-headed goose. These figures are used in Table 8.1. The heart rates measured during migration are considered unrealistically low and are not supposed to reflect metabolic power accurately.

Albatrosses are extreme travellers, renowned for their ability to soar in high wind conditions. Flight cost data are available for three Antarctic and one tropical species. Bevan *et al.* (1995) determined the energy expenditure of free-ranging black-browed albatrosses using heart rate telemetry. Reliable relationships between heart rates and energy expenditure could be established from birds walking on a treadmill (Bevan *et al.* 1994). The field experiments were performed on 25 birds with an average mass of 3.58 kg at colonies on Bird Island, South Georgia during incubation, brooding and chick rearing. Data loggers, recording ECG's and temperature, were implanted in the abdominal cavity. They were removed after recapture on average 23 days after release. Foraging costs were determined of five birds carrying data loggers and salt water switches to determine the resting time at sea. Energy expenditure at sea increased from 4.63 to 5.80 $W\,kg^{-1}$ depending on the phase of the reproductive cycle. The energy expenditure during flight (69% of the time at sea) was 6.2 $W\,kg^{-1}$. This figure was statistically indistinguishable from the 5.8 $W\,kg^{-1}$ found for the costs of floating on the water. Both values represent about twice the calculated basal metabolic rate of 3.1 $W\,kg^{-1}$.

8.4 *Gas exchange measurements in wind tunnels*

Wind tunnels offer the opportunity to fly birds at different speeds at one spot. However, it is not always easy to train birds to fly in a wind tunnel. Space is often confined, the airflow can be turbulent and the engines of the blowers usually make a tremendous noise. Electrified perches or grids are often used to keep the birds from landing. Metabolic measurements are possible by analysing the air directly in closed tunnels or by using breathing masks and tubes in systems that are more open. The air temperature can be under control of the experimenter. Flight costs may vary with speed. We will concentrate on flight costs at maximum range speeds in this chapter.

The flight metabolism of the budgerigar and that of the laughing gull were published by Tucker in 1968 and 1972 respectively. The budgies were purchased from a local pet shop and kept in small (22 × 26 × 40 cm) cages. The wind tunnel was rather small too, with a 30 cm long working section measuring 30 cm in diameter. A transparent mask (Fig. 8.2(c)) fitted over the head was kept in place with a rubber band around the back of the head. A flexible tube entered the front of the mask. The whole contraption weighed 1.48 g. To collect the expired respiratory gases air was sucked through the tube. The air from the room entered the mask from behind. Flight was initiated by withdrawing the perch in a working wind tunnel. It took the budgies a few hours to learn to fly regularly. The flights ended when the perch was put back in position. After 6 weeks of daily training, the birds would fly for 20 min or more. Flight speed was varied between 19 and 48 km h^{-1} (5.3 and 13.3 m s^{-1}). During flight tests at 23°C, the budgerigars, weighing on average 35 g, used oxygen at a rate of 32.5 and 34.2 ml g^{-1} h^{-1} at the lowest and highest speeds respectively. The oxygen consumption reached a minimum value of 21.9 ml g^{-1} h^{-1} at 9.7 m s^{-1} (35 km h^{-1}) which is about 13 times the estimated resting rates. The RQ measured was 0.78 indicating an energetic equivalent of 20.1 J ml^{-1} oxygen used. In the original paper (Tucker 1968) no correction was made for the extra drag of mask and tube. This was rectified in Tucker (1972) where estimates for the best performance are 3.67 W at 9.7 m s^{-1} and 4.12 W at the maximum range speed of 11.7 m s^{-1}.

Tucker kept two tame hand-reared laughing gulls, in a 4 m high outdoor cage of 6 by 6 m. These birds with wingspans of about 0.78 m were trained to fly in a 1.4 m wide measuring section of the wind tunnel. During the experiments, they did that for 0.5 h or more. The oxygen consumption and carbon dioxide production were measured from expired air collected with the use of a tightly fitting mask and tube weighing 4.4 and 6 g respectively (Fig. 8.2(d)). The whole wind tunnel could be tilted around a transverse axis. This property was used to compensate for the extra drag from mask and tube by tilting the wind tunnel 1.5° downward. Flight with a mask resulted in higher than normal wing beat frequencies of about 3.8 Hz. The energetic equivalent, assumed to be 20 kJ l^{-1} oxygen consumed, was based on an in flight RQ of 0.74 (it was 0.70 during rest). Tucker did two types of experiments. First, the effect of body mass on flight costs was measured at one speed of 10.8 m s^{-1} at temperatures ranging from 25°C to 35°C. The body mass varied between 328 and 420 g. Flight costs were found to increase with mass to the power of 0.325. This figure was used to correct for small differences in the mass of birds flown in tests at a range of speeds. The exercise resulted in two data sets: one for a 277 and one for a 322 g bird. Both values include the mass of mask and tube. The lowest flight costs were 18.9 and 15.0 W at minimum power speeds of 8.62 and 8.64 m s^{-1} for the 322 and 277 g gulls respectively. The maximum range speeds were about 12.5 m s^{-1} for both weight classes but the rate of energy used was 19.0 and 23.2 W for the lightest and the heaviest birds respectively.

The same wind tunnel and mask were later used by Bernstein *et al.* (1973) to study the fish crow and by Hudson and Bernstein (1983) to measure the gas exchange and the energy cost of flight in the chihuahuan raven. Two out of five adult fish crows with an average mass of 275 g were willing (induced by mild electroshocks) to fly steadily for 15–20 min at speeds varying between 7 and 11 m s^{-1}, in flights descending 2°, 4°, and 6° relative to the horizontal. For some reason the crows refused to fly level long enough to make reliable measurements possible. Level flight figures therefore had to be extrapolated from the performance at various descent rates. Flight patterns varied from steadily flapping to a combination of flapping and gliding. The rate of oxygen consumption did not vary much with speed but decreased with increasing angle of descent as one would expect. The RQ was assumed to be at 0.8. This figure was not critical since an RQ of 0.7 would give only 2% and an

RQ of 1, 4% error in the flight cost estimates. The cost of level flight as a function of speed did not show a U-shaped curve but decreased linearly with increasing speed to reach a minimum value of 24.2 W at 11 $\mathrm{m\,s^{-1}}$ (this is the figure corrected for the drag of mask and tube according to Tucker (1972)).

Seven chihuahuan ravens (named white-necked raven in the article) were caught as fledglings in New Mexico. They grew up in captivity for 2–3 years in an outdoor 4 m wide, 8 m long, and 3 m high aviary, to an average adult weight of 480 g. No electric shocks were needed to train these birds to fly in the wind tunnel. The tests lasted 10–30 min, usually two or more times per day. The RQ was 0.77 (on average, $n = 73$). The flight angle was called zero after correction for mask and tube using a small downward angle. At speeds varying from 8 to 11 $\mathrm{m\,s^{-1}}$, oxygen consumption increased again linearly with speed and so did the power figures reaching values of 32.8 W.

Butler *et al.* (1977) used a much larger wind tunnel with a 24 m long 2.5 m wide and 2 m high test-section. Pigeons, with a wingspan of 72 cm, were flown in a 1.8 $\mathrm{m^3}$ wire-mesh cage placed inside the tunnel. The mask used to sample the expired air was connected to the respirometer by a tube running over the head and back of the birds (Fig. 8.2(e)). Mask and tube weighed 18 g and were estimated to require 12% of the oxygen used. The pigeons were not primarily used for flight cost studies. EMG's of flight muscles, heart beat, and respiration frequencies and temperatures were measured as well. The tests were only at one speed of 10 $\mathrm{m\,s^{-1}}$. The RQ rose from 0.85 at rest to 0.99, 30 s after take off, and reached a constant value of 0.92 after 7 min. Oxygen consumption was high during the first 6 min and stabilized at a lower value after that initial period. The average flight costs after 6 min were 30.5 W for a 442 g pigeon. A correction for mask and tube of 12% (based on Tucker (1972)) yields flight costs of 26.8 W.

The working section of the Saarbrücken wind tunnel was 1×1 m and 1.4 m long. Rothe *et al.* (1987) flew a special breed of racing pigeons, gripplers, with a wingspan of only 60 cm weighing between 300 and 350 g, in it. (For the pigeon racing experts, grippler pigeons are hybrids between grivuni and tippler pigeons.) The masks used were made of polyethylene centrifuge tubes and weighed only 0.7 g (Fig. 8.2(f)). The interior of the mask was large enough to permit opening of the beak. The inflow was at the front. The flexible silicon rubber sampling tube (weight 15 $\mathrm{g\,m^{-1}}$) was attached to the lower part of the mask and led out through the front of the working section of the tunnel. This tube arrangement caused problems because it fluttered quite a bit. Rothe *et al.* indicated that it probably added 15–30% to the metabolic rates during flight. The tests lasted more than 1 h. The RQ values were shown to depend on the carbohydrate/fat contents of food, the time of year, and the time since last feeding. Pigeons start to burn carbohydrates with RQ's of about 1 and gradually change to pure fat combustion during prolonged flights where RQ's are close to 0.72. Flight cost data are based on five pigeons in 41 flights. The cost varied with speed but not in a clear-cut U-shaped or straight fashion. The maximum range speeds were somewhere between 11 and 13 $\mathrm{m\,s^{-1}}$. The best estimate for an average 330 g grippler during steady maximum range free flight seems to be 25.4 W after a correction for mask and tube on the measured values of about 33 W representing 130% of the actual flight costs. The uncertainty about the correction factor affects the reliability of this point and it is therefore not included in Table 8.1.

Level flight costs of hummingbirds were measured in a wind tunnel by Berger (1985). He used a silicon mask with a hole in the front fitting over the bill leaving the nostrils inside. Air was sucked through the tube, emerging from the top of the mask. Surrounding air could enter the mask from the back (Fig. 8.2(g)). The working space in the tunnel was 65 cm long and 33 cm in diameter. Six birds belonging to two species (the sparkling violet-ear, average mass 8.5 g, and the green violet-ear, average mass 5.5 g) were trained. They started to fly when the wind tunnel was switched on and landed on perches in the rear when the blower stopped. Wind velocities could reach a value of 11.2 $m\,s^{-1}$. The oxygen consumption of both species was in the order of 45 $ml\,g^{-1}\,h^{-1}$ at velocities from zero (hovering) up to 8 $m\,s^{-1}$. The lowest flight cost values are 2.07 W for the green and 2.79 W for the sparkling violet-ear, both flying at 10.8 $m\,s^{-1}$ (assuming an equivalent of 20 $J\,ml^{-1}$ oxygen). No correction was made for the mask and tube system. If we assume (following Tucker 1972) that 12% of the total costs is caused by energy needed to overcome drag forces on mask and tube, the flight cost values are 1.82 and 2.46 W.

The extra drag and the discomfort for the birds caused by wearing masks and tubes may undermine the value of wind tunnel respirometry. Torre-Bueno and LaRochelle (1978) found a way to escape from this disadvantage. They trained wild European starlings to fly in the (71 cm wide, 40 cm high, and 92 cm long) working section of a closed circuit wind tunnel. The birds were tossed upstream in the tunnel with their feet taped so that they could not land easily. Only 5 out of 100 birds managed and learned to fly for more than 90 min without being chased. Carbon dioxide production and oxygen consumption were measured for at least 90 min from air drawn from the wind tunnel every 15 min. The air was pumped back into the respirometer after the samples were analysed. Oxygen levels never dropped below values that would affect respiration. Average RQ after 30 min of flight was about 0.7, indicating the burning of fat. Three birds with a mean mass of 72.8 g were tested during 72 flights at speeds between 8 and 18 $m\,s^{-1}$. The metabolic rate was on average 8.9 ± 1 W and not changing with speed. The wing beat frequency was constant at 12 ± 0.5 Hz. Wing amplitude showed a U-shaped curve as a function of speed. The angle swept declined from 130° at 6 $m\,s^{-1}$ to a minimum of 95° at 14–16 $m\,s^{-1}$ increasing to about 125° at 18 $m\,s^{-1}$. The tilting angle between the body and the horizontal plane decreased linearly with increasing speed from about 30° at 6 $m\,s^{-1}$ (only one bird would fly briefly at that speed where the observation was made that the feathers at the back of the neck were lifted) to about 8° at velocities between 14 and 18 $m\,s^{-1}$. Earlier experiments showed that 13.5 $m\,s^{-1}$ was the preferred flight speed of starlings. The costs of 9 W at the maximum range speed of 16 $m\,s^{-1}$ are included in Table 8.1.

8.5 *Measurements based on the turnover of stable isotopes*

Two methods using stable isotopes have been used to estimate energy turnover in birds: the doubly labelled water (DLW) technique and a heavy carbon method (HC). The DLW technique, determines the rate of change of concentrations of heavy isotopes of oxygen and hydrogen in the blood. The technicalities of this technique are explained in Box 8.2. The HC method involves injection of $NaHCO_2$ with the stable carbon isotope ^{13}C instead of the lighter ^{12}C. The ratio of the two isotopes could be measured in the expired carbon dioxide (Hambly *et al.* 2002).

Box 8.2 ***The doubly labelled water (DLW) technique to measure energy expenditure***

Animals use oxygen to burn their food and produce carbon dioxide (CO_2) and water (H_2O) in the process. The amount of energy metabolized can be estimated from the production rate of these molecules if the substance burnt is known (see Box 8.1).

For an individual animal, the energy expenditure can be measured by labelling oxygen and hydrogen in the body with stable (i.e. non-radioactive) isotopes. Labelling is commonly done by injecting $^2H_2{}^{18}O$ or $^3H_2{}^{18}O$. Deuterium (^2H) or tritium (^3H) and the oxygen isotope ^{18}O are heavier than hydrogen (^1H) and normal oxygen (^{16}O). ^2H and ^{18}O naturally occur in all types of water including water in the body at levels of roughly 0.015% and 0.2% respectively. See for example Lifson and McClintock (1966); Nagy (1980); Nagy and Costa (1980); and Visser *et al.* (2000) for more detailed information than given in this simplified representation.

A short time after the injection, a blood sample is taken to establish the figures for the initial amounts of stable isotopes in the body. The animal is subsequently released and its activities are monitored until it is recaptured usually one or more days later. The final blood sample then taken contains less isotope labels than the initial one. ^2H has left the body as expired H_2O, whereas ^{18}O was removed in both CO_2 and H_2O. The total H_2O flux can be calculated from the ratio of ^2H concentrations in the final and the first sample, times the total amount of H_2O in the body. Knowing the total amount of H_2O lost provides the possibility to calculate the amount of oxygen lost through the expiration of H_2O. ^{18}O isotopes left the body as CO_2 and H_2O. The concentration in expired H_2O is assumed the same as that in the body H_2O, so the amount of ^{18}O isotopes that left with the H_2O is known. Twice as many ^{18}O isotopes will have left as CO_2. The rate of CO_2 production can now easily be estimated from the difference between the total loss of ^{18}O isotopes and the isotopes lost with the expired H_2O. Finally, the energy expenditure can be calculated from the CO_2 production when the foodstuff that was burnt is known (see Box 8.1).

The original method required a number of assumptions to be met:

a. The water content and body composition of the animal must remain constant during the experiment.
b. All rates of intake and output must be constant. In any case the outcome is only an average figure over the experimental period.
c. All the body water is assumed to be uniformly labelled.
d. The H and O_2 in the body water behaves the same as in the removed substrates.
e. No CO_2 enters the body through the skin or through respiration.

Nowadays the calculations have been adapted to circumvent these conditions.

Blood samples in birds are usually taken from some vein. In kestrels for example the posterior tibial vein was used. Injections with mixtures of $H_2{}^{18}O$ and $^{2}H_2O$ (2 : 1) are commonly given subcutaneously in the abdomen. The amount injected depends on the duration of the experiment and on the body mass of the animal. Blood samples sealed in glass capillaries may conveniently be stored at 5°C. The treatment in the laboratory includes water extraction by vacuum distillation and mass spectroscopy. Body water volume can be directly measured of desiccated carcasses or calculated from the dilution after the enrichment with a known number of isotopes in the initial sample after injection. Simultaneous use of different methods to obtain figures for CO_2 production in captive kestrels including the DLW method showed deviations in the order of 2.2%. Energy expenditure figures obtained with direct respirometry and DLW measurements of Albatrosses exercised on a treadmill did not differ significantly (Bevan *et al.* 1994).

So far the method has been successfully used in mammals (including man), birds, reptiles, and even insects.

The two methods provide figures for total energy turnover during prolonged periods. Estimates of flight costs have to be distilled from these figures using detailed knowledge of the behaviour of the animal preferably in terms of time budgets during the periods measured. Flight is usually the most expensive item on the budgets. Flight costs are commonly determined from the increase in total energy expenditure per period taking the increased proportion of time flown into account. An example is given for the kestrel below. An advantage of the methods is that one gets an estimate of the cost in real life; a disadvantage is the obvious lack of information about the details regarding for example speed, amount of gliding, wind conditions, and use of up draughts. DLW and HC measurements provide only a single value over the monitoring period, whereas heart rate telemetry offers continuous recordings. DLW and heart rate measurements should be compared with other flight cost estimates with this background information in mind. Table 8.1 contains results of 19 studies based on DLW measurements. The HC method is less well established with only four papers with results by the group who invented the method; two of these are included in Table 8.1.

8.5.1 DLW results

The first flight cost data points based on the DLW technique are from pigeons measured by LeFebvre (1964). His story sounds like the nursery rhyme on 10 little Indians. Thirty-one racing pigeons were the experimental animals. Nine pigeons were randomly chosen and sacrificed just prior to the flight to provide an estimate of the pre-flight water, fat, and protein composition. The experimental group was deprived of food for 24 h or more before racing 22 of them over 483 km from Allerton (Iowa) to St Paul (Minnesota). Probably due to thunderstorms late in the day of the experiment, only eight pigeons returned. Of these, four pigeons did not show signs of stopovers (mud on the feet, food in the crop) and presumably flew the whole distance non-stop. Estimates based on the reduction of fat content (assuming that only fat was burnt at an RQ of 0.71) were compared with flight costs calculated with the DLW method. LeFebvre used the time interval between blood samples to represent flight time.

We try to be more precise and take the real flight time assuming resting metabolism during the short periods when they were not flying, after and before the blood samples were taken. The resulting data, shown in Table 8.4, are highly variable. The DLW and fat measurements of the first bird (3208) were not in agreement. Pigeon 4012 took more than 12 h to cover the distance, about 3 h more than the rest of the ones that made it. We end up with two trustworthy data points only. The average values for flight costs at 17 m s^{-1} were 31.9 and 33.1 W, based on DLW measurements and fat combustion respectively.

Utter and LeFebvre (1970) used DLW to measure carbon dioxide production in four purple martins released 100 miles from their nests and shot at return. The birds foraged and drank during the return flight. The duration of

Table 8.4 Data from four pigeons based on non-stop 483 km flights (LeFebvre 1964).

Pigeon (*nr*)	Average mass (g)	Average velocity (m s^{-1})	Flight costs (DLW) (W)	Flight costs (fat) (W)
3208	*345*	*18.8*	*18.0*	*36.3*
4012	*361*	*11.2*	*18.5*	*21.0*
4051	390	17.9	34.9	36.2
1285	398	16.1	28.8	30.0
Mean values of 3 and 4				
	394	17.0	31.9	33.1

Figures in italics should not be trusted as explained in the text.

the return flights varied between 4 h and 15 min for the fastest and slightly more than 6 h for the slowest bird. Assuming an RQ value of 0.78, the average flight costs were 4.1 W for the birds of slightly more than 50 g.

Swallows and martins were the subject of DLW studies at the University of Stirling (Scotland). Hails (1979) measured the average daily metabolic rate of 27 adult house martins and of three barn swallows during the nestling period. Birds away from the nest were supposedly foraging and flying continuously. Aerial hunting for insects involves twisting, turning, short glides, and active pursuit. Uni-directional flight may only have been a small portion of the total flight time. Male birds rearing the largest broods had the highest energy expenditure due to high flight costs. It is not surprising that figures for flight costs obtained in this way are rather variable, most probably reflecting real facts of Hirundine life. The house martins used between 0.83 and 1.18 W and the three swallows needed an average of 1.3 W.

Turner (1982*a*) found that swallows fly fast (11 $m\,s^{-1}$) to catch large prey and slowly at 5 $m\,s^{-1}$ to collect small insects. DLW analyses showed that the fast foraging flights cost 1.62 W and the slow flights 0.68 W. The energy gained exceeds the costs about tenfold in both foraging modes. However, a swallow uses more than twice as much time to catch one unit of energy while hunting for small prey at low speed. Westerterp and Bryant (1984) added a data point on collared sand martins to the list by measuring the energy expenditure in relation to the percentage of 24 h day spent flying. In fact, the estimate was based on a comparison between incubating birds, flying only 25–30% of the time, and nestling rearers who flew twice as much.

The same authors also published DLW data on two non-hirundine species. Westerterp and Drent (1985) established the metabolic rate of flight in the starling. The best flight figures were obtained from four animals displaced between 10 and 30 km from their nest box. Average return speed of these birds was 10 $m\,s^{-1}$. Flight costs were 10.5 W for a starling of an average body mass of 77.5 g. Tatner and Bryant (1986) caught six European robins from the wild and kept these in outdoor aviaries. The activities in the cages were recorded using an electronic event recorder. Flights in the confinement of the cages were short, lasting on average 0.78 s. The time spent flying per day varied from 72 s to 1.62 h. The flight costs estimates, based on regression of daily energy expenditure on the time in flight, ended up to be extremely high at 7.1 W for birds weighing on average 18.6 g. This figure is probably too high because it includes a substantial amount of starting, manoeuvring, and landing, probably reflecting foraging costs of robins in dense bushes.

To continue our story on the kestrels, Masman and Klaassen (1987) injected 10 birds, two females just before egg laying and four males and three females (one individual twice) during the nestling phase, with DLW. The kestrels were breeding in the last reclaimed polder of the Netherlands. The Lauwersmeer polder offered open flat countryside where the behaviour of the injected birds could be monitored precisely well over 90% of the daylight time. The daily energy use was calculated from the mass-specific carbon dioxide production using an energy equivalent of 23.6 $kJ\,l^{-1}$ carbon

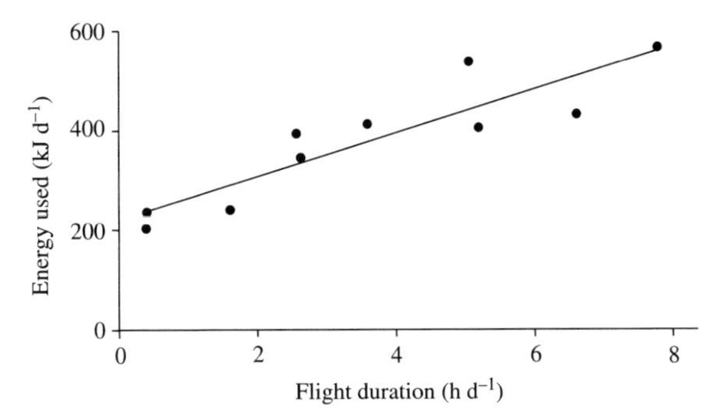

Fig. 8.3 The average daily energy expenditure of kestrels as a function of the daily time spent in flapping flight (windhovering and directional flight). The metabolic rate without flight is 2.6 W and the flight costs are 14.6 W. Data from Masman and Klaassen (1987).

dioxide. Figure 8.3 shows the collected data for the 10 experimental birds with a mean mass of 213 g. The slope of the regression line, representing the average daily energy used, as a function of the number of seconds spent in flapping flight per day, provides an estimate of the extra costs for flapping flight of 12.0 W. The *Y*-intercept indicates that a kestrel of 213 g uses 2.6 W if it does not fly. This adds up to the total costs of flapping flight including windhovering of 14.6 W (±2.1 W). It is very close to the value for flight costs found during the corridor experiments.

Migration cost of swainson's and hermit thrushes were measured by injecting 38 birds of about 30 g with DLW at a spring migration stopover site. Six birds flew 600 km in 7.7 h during the night following the injection and were captured the next morning. The isotope concentrations could be compared with birds which did not fly that night and stayed at the stopover site. The airspeed of the birds was estimated at 13 $m\,s^{-1}$. The total flying energy expenditure was 4.3 W which is slightly more than five times resting metabolism.

8.5.2 Flight costs of oceanic birds

The labelled water technique has offered opportunities to measure the energy involved in extended foraging journeys of oceanic birds. Estimates of the cost of flight of seabirds require answers to at least two major questions: how much time do birds rest at sea and to what extent is gliding involved?

Flint and Nagy (1984) injected 18 sooty terns with DLW. The birds were incubating on Tern Island (French Frigate Shoals in the Hawaiian Archipelago). Males and females breed in turns, the incubation spans lasting 2–3 days. The time that the foraging birds actually spent in flight was estimated by periodically surveying the area for marked birds. They were considered to be at sea when they were not seen sitting in the nesting area.

The time at sea can safely be regarded as flight time because these terns hardly ever rest on the water. At low to moderate wind speeds 94.3% of the flight time is flapping flight. When the wind blows faster than 5 $m\,s^{-1}$ flapping is used during 75.4% of the time. Wind speeds were low during the experiments. The outcome of the measurements can therefore be regarded as an approximation of the cost of flapping flight. Linear regression of the metabolic rates measured against percentage of time spent flying gave a significant result showing that on average 187 g sooty terns use 4.8 W for foraging flight and 1.6 W during incubation.

Brown and Adams (1984) used oxygen consumption measurements to determine the BMR of wandering albatrosses at Marion Island in the Antarctic region. They found an average value of 20.3 W for males and females with an average body mass of 8130 g. Adams *et al.* (1986) used this result to calculate the ratio of foraging metabolic rates over basal metabolic rates of wandering albatrosses. Five males and four females feeding chicks were injected with DLW before the start of a foraging trip. The duration of these trips varied from about 2 to almost 9 days with an average value of 5.11 days. The energy expenditure during the foraging trips was 45 W for males and 31 W for females. The same species was the subject of experiments by Arnould *et al.* (1996) using DLW, satellite telemetry, and leg-mounted activity recorders. The use of energy during foraging trips varying in duration from 1.5 to almost 8 days was determined. The proportion of time spent flying did not change the total rate of energy expenditure, confirming Bevan *et al.*'s (1995) finding that the cost of flight is the same as the cost of sitting on cold water. The wandering albatrosses of Arnould *et al.* were slightly heavier than the ones of Adams *et al.* but the flight costs are in the same order of magnitude. Costa and Prince (1987) found similar figures for the grey-headed albatrosses at Bird Island, South Georgia, using the DLW technique combined with activity budgets at sea. Activity recorders, attached to the legs of this species, measured that 35% of the time was spent on the sea surface (Prince and Morgan 1987). The total costs of foraging for grey-headed albatrosses, weighing on average 3707 g, was 27.8 W.

Antarctic albatrosses breed and forage under extreme conditions. The weather is fiercely cold and rough, but food is abundant. Around the Hawaiian Islands air and sea temperatures of the sub-tropical and tropical oceans are considerably higher and the wind conditions are less extreme, but food resources are meagre. Pettit *et al.* (1988) studied the incubation and foraging energetics of the laysan albatross, also breeding on Tern Island. Four foraging birds weighing on average 3064 g (SD 413 g) made sea trips of about 3 days producing 84 $l\ CO_2\ d^{-1}$. An estimated equivalent of 24.7 $kJ\,l^{-1}$ carbon dioxide was based on the composition of the diet of the laysan albatrosses. The energy expenditure during foraging was 24 W. Due to the higher water temperatures the costs of sitting on the water could well be lower than for the Antarctic albatrosses. In that case the figure for cost of flight would be somewhat higher than 24 W.

We move now from the largest to the smallest oceanic birds. Obst *et al.* (1987) studied the energetics of Wilson's storm petrels weighing about 40 g using respirometric and DLW techniques. These birds live up to their scientific name and are purely oceanic. While away from the nest they are virtually continuously on the wing, flying in a flapping and gliding mode sometimes windhovering briefly (using the webbed feet as sea anchor) to pick up food from the sea surface (Pennycuick 1982; Withers 1979). Recorded behaviour of Wilson' storm petrel at sea showed that the species rarely rests on the water and then only for a few seconds while feeding. Consequently, metabolic rates of foraging Wilson's storm petrels breeding near Palmer station, Antarctica, represent the costs of free flight. Resulting figures from 12 birds are very close together. The average mass was 42.2 g and the metabolic rate 1.82 W, with standard deviations of 0.9 g and 0.1 W respectively. The frequent hovering bouts (every 40 s) lasting 1–20 s do not seem to be very costly. Withers (1979) suggests that Wilson's storm petrels have an energetically inexpensive foraging strategy based on the use of shear forces in wind gradients, ground effects, and a special wing-flip mechanism.

Birt-Friesen *et al.* (1989) measured the field metabolic rates and the activity of 20 free-living Northern gannets (mean mass = 3.21 kg) rearing chicks at Funk Island, Newfoundland, using respirometers, DLW, and activity timers. The activity timers were digital watches attached to the legs of the birds and modified to switch off in contact with water. The durations of the trips to sea varied from <1 h to several days. Metabolism during foraging trips averaged 6000 kJ d^{-1}. An estimate of 97 ± 30 W for flight costs was found from the regression of field metabolic rates of nine gannets against their flying times.

The energy expenditure during foraging of the closely related Cape gannets was determined using DLW by Adams *et al.* (1991). A 2.58 kg Cape gannet is expected to produce 181 l carbon dioxide per day during field trips which is equivalent to 4670 kJ day^{-1} in this species. No activity measurements are available but we might assume that the increment of the field metabolic rate due to flight is the same as that of the Northern gannet. If that assumption is true the flight costs of Cape gannets amount to 81 W.

Johnston Atoll consists of four tiny islands in the Central Pacific. Ballance (1995) reports on a study of the energetics of a colony of the white morph of red-footed boobies on East Island. The 0.1 km^2 Island was created by dredging in 1964, it consists of coral rubble covered by grass and some sparse bushes. It offers a breeding site for seven species of seabirds. Red-footed boobies are extremely pelagic ranging out 250 km or more from the colony. Daily travel must exceed 500 km because adults return to feed their chicks on average once each day. Flight costs of 10 red-footed boobies were measured using the DLW technique. The time spent sitting at sea was measured with adapted digital watches attached to the feet of the birds. Resting metabolic rates based on oxygen consumption were measured separately in a dark metabolic chamber. The recorders of 6 out of 10 birds worked properly. Data of two of these six birds are suspect because the total field metabolic rates are very close to the resting values. The data shown in Table 8.5 of the

Table 8.5 Foraging and flight energetics of red-footed booby (Ballance 1995).

Red-footed body (individual)	Body mass (g)	Field metabolic rate (W)	Flight time (%)	Resting metabolic rate (W)	Average flight costs (W)
Mohawk	1050	14.9	55.2	9.8	19.0
Hiawatha	1080	15.9	66.8	10.3	18.6
Fern	1080	18	42.3	10.3	28.4
Alfonso	795	12	26.1	5.7	29.8
Average	1001				24.0

four remaining birds are used to obtain an average figure of 24 W for the power required for flight of a 1 kg bird.

Southern giant petrels are large seabirds (mean body mass 3.89 kg), confined to the waters of the Southern Hemisphere. Their large size, shape, and dynamic soaring habits suggest a convergent evolution with the albatrosses (Obst and Nagy 1992). These authors did fieldwork on birds from a nesting colony on Humble Island (Antarctica). Field metabolic rates were obtained using DLW techniques of eight individually colour marked birds during 13 brooding–foraging cycles. The foraging bouts lasted between 9 and 37 h. There was a significant positive correlation between foraging duration and metabolic rates measured, yielding an average value of 68.5 W ($\pm$18.0 W) for the cost of foraging of 3.89 kg ($\pm$0.5 kg) birds. No indications are given of how much of the foraging time the birds were actually airborne. There is some evidence indicating that the giant petrels travelled as far as 250 km away from the colony on a single foraging trip. Giant petrels use flapping flight during substantial proportions of the time at wind speeds up to 8 m s^{-1}. Figures of 3.5–4 times values found for albatrosses are quoted. That probably can partly explain why foraging costs in this species is so much higher than the figure found for flight costs of albatrosses (Table 8.1). The convergence of the evolution of giant petrels and albatrosses might not be so striking after all if energy budgets are included in the comparison.

8.5.3 The use of DLW in wind tunnels

The use of DLW to measure energy turnover during flight requires enough time to allow the levels of heavy isotopes to drop to levels that are sufficiently different from the starting concentrations. The method can be applied in wind tunnels if the tested birds are willing to fly for long periods. As usual, some individuals are more cooperative than others.

In the Lund wind tunnel, four red knots flew 28 sessions of 6–10 h with DLW added to their blood (Kvist *et al.* 2001). The average metabolic rate of the birds was 13.5 W while flying at 15 m s^{-1}.

Rosy starlings were flying 27 bouts of about 6 h in the wind tunnel of the Max Planck Institute for Ornithology in Andechs (Germany). The average weight of 8 birds was 71.6 g, the flight speed 11.1 m s^{-1}, and the energy expenditure 112.3 $kJ\,h^{-1}$ or 8.05 W (Sophia Engel, personal communication). The Centre for Isotope Research (Groningen University) was, in this and many other studies mentioned above, responsible for the analysis of the blood samples.

8.5.4 The heavy carbon method

The zebra finch is the first species where the metabolic cost of flight was determined by following the decay of the concentration of the stable isotope ^{13}C after an injection with a solution of $NaH^{13}CO_3$ (Hambly *et al.* 2002). The expired air of an injected bird can be used to determine the ^{13}C concentration. Breath samples were taken every minute during 14 min after the injection. The bird was subsequently induced to fly up and down a 20 m long corridor between two perches. The time spent on the perches (varying from 18% to 49%) was subtracted from the flight time. Flight speeds were not measured. During the 14 min before the flight the concentrations of ^{13}C gradually diminished linearly. After the 2 min flight period the heavy carbon concentration in the expired air showed a considerable drop. The difference in concentration corrected for the decline that would have occurred without activity, could be used to calculate the energy turnover during the flight period. The outcome for the zebra finch with an average mass of 14.5 g was 2.24 W. Hambly *et al.* (2004) used the same technique to analyse the flight costs of a starling and a sunbird. The energy expended by 8 Palestine sunbirds (mean mass 6.17 g) flying during periods of 2 min between perches at 6 m distance was 1.64 W on average. The data obtained for the starlings must be treated with greater care. The distance flown by the nine birds (average weight 70.11 g) was only 5 m which means that starting and landing took most of the time and probably of the energy. Furthermore, the results of the measurements of the isotope elimination after flight sessions showed a peculiar trend. During the first 3 or 4 min after the measurements started the concentration of ^{13}C increased before a decline commenced. This phenomenon makes it difficult to decide what the level actually was after the flight period stopped. The flight cost estimated after extrapolation of the isotope concentrations during the first post-flight minutes was at about 20.6 W extremely high compared to other birds of similar body mass. This figure is not included in Table 8.1 because of these disturbing factors.

In principle the method is very promising but it obviously needs careful treatment of the results.

8.6 *The cost of hovering*

It might seem easy to measure the energy expenditure of birds that can fly on one spot for prolonged periods. There are however only about a dozen studies providing insight in the metabolic cost of flight at zero forward speed in still air. Three respirometric techniques have been used, each with its own way to measure oxygen and carbon dioxide concentrations in the expired air. Some studies use closed containers as respiratory chambers. Others collect air through tightly fitting breathing masks. A third technique uses loose masks disguised as bird feeders. Results are compiled in Table 8.6 and will be briefly discussed below.

Pearson (1950); Lasiewski (1963); Hainsworth and Wolf (1969); Schuchmann (1979*a*,*b*); and Epting (1980) used containers as respiratory chambers. The advantage is that there is no need for a respiratory mask and the bird flies unattached. However, the jet of air generated by the birds will disturb the air in the chamber drastically. There will be ground effects if the chamber is not very high and all kinds of first and higher order wall effects in a confined space. Data points obtained in containers smaller than 5 l are therefore considered unreliable. Berger (1985) used a tightly fitting mask (Fig. 8.2(g)) to fly hummingbirds in a windtunnel. An original solution was introduced by Berger and Hart (1972) who trained captive hummingbirds to feed on a respirometry mask disguised as a hummingbird feeder (see Fig. 8.2(h)). Epting (1980) compared data obtained in a 10.1 l container with those measured with a feeder mask using the same birds and showed that the figures found are similar (Table 8.6). Bartholomew and Lighton (1986) induced wild free ranging Anna's hummingbirds to feed and breathe from a similar feeder mask placed outdoors. The mass could be measured whenever the birds landed close to the feeder on a trapeze perch suspended from a force transducer. A possible disadvantage here is that wind conditions could have influenced the outcome of the measurements.

Schuchmann (1979*a*,*b*) has shown convincingly that the costs of hovering and the costs of resting in hummingbirds decrease linearly with increasing ambient temperatures between 10°C and 40°C. But his oxygen consumption values during hovering are extremely low compared to the figures found by others.

Hummingbirds are capable of hovering at high altitudes. Chai and Dudley (1995) cleverly investigated the capability of hummingbirds to hover in thin air. Ruby-throated hummingbirds were trained to hover in front of a feeding mask in a respirometric chamber where the sea-level density of the air (1.2 $kg\,m^{-3}$) was decreased by replacing air with heliox (a mixture of 79% helium and 21% oxygen, density 0.4 $kg\,m^{-3}$). The minimum density at which hovering was still performed without failure was 0.6 $kg\,m^{-3}$. That is half the density at sea level and equivalent to an altitude of 6000 m. With densities decreasing from 1.2 $kg\,m^{-3}$ (normal air at sea level) to 0.54 $kg\,m^{-3}$ (6000 m) the duration of hovering bouts decreased from about 30 s to less than 5 s. Wing beat frequency went up slightly from about 49 to 52 Hz. The stroke amplitude showed considerable increase from 145° to almost 180°. Oxygen consumption went up from 48.5 to 61.5 $ml\,g^{-1}\,h^{-1}$.

The oxygen consumption of broad-tailed and rufous hummingbirds was actually measured while hovering at 2195 m above sea level (Wells 1993). At

Table 8.6 The metabolic costs of hovering of hummingbirds (Trochlidae) and one species of sunbird (Nectariniidae).

English name	Body mass (g)	Metabolic costs (W)	Source	Temperature (°C)	Method
Hummingbirds					
Racquet-tailed puffleg	2.7	*0.51*	Schuchmann (1979*a*)	25	2.8 l container
Costa's hummingbird	3.0	*0.71*	Lasiewski (1963)	24	3.8 l container
Ruby-throated hummingbird	3.0	0.89	Chai and Dudley (1995)	25	Feeder mask
Allen's hummingbird	3.4	0.93	Epting (1980)	20	16.65 l container
Allen's hummingbird	3.5	*1.65*	Pearson (1950)	24	4.6 l container
Ruby-throated hummingbird	3.6	0.98	Chai and Dudley (1995)	25	Feeder mask
Ruby-throated hummingbird	3.6	1.05	Chai and Dudley (1995)	25	Feeder mask
Black-chinned hummingbird	3.6	1.05	Epting (1980)	20	Feeder mask
Ruby-throated hummingbird	3.9	1.07	Chai and Dudley (1995)	25	Feeder mask
Broad-tailed hummingbird	4.0	1.01	Wells (1993)	22	Feeder mask
Anna's hummingbird	4.1	*1.55*	Pearson (1950)	24	4.6 l container
Rufous hummingbird	4.3	1.08	Wells (1993)	22	Feeder mask
Black-chinned hummingbird	4.3	1.24	Epting (1980)	20	16.65 l container
Rufous-tailed hummingbird	4.4	*0.50*	Schuchmann (1979*b*)	25	2.8 l container
Anna's hummingbird	4.6	1.06	Bartholomew and Lighton (1986)	20–25	Feeder mask
Anna's hummingbird	4.6	1.27	Epting (1980)	20	16.65 l container
Indigo-capped hummingbird	4.8	*0.63*	Schuchmann (1979*b*)	25	2.8 l container
Anna's hummingbird	5.0	1.36	Epting (1980)	20	Feeder mask
Glittering-throated emerald	5.7	1.36	Berger and Hart (1972)	>20	Feeder mask
Purple-throated carib	8.3	2.00	Hainsworth and Wolf (1969)	20	10.1 l container
Sparkling violet-ear	8.5	1.90	Berger (1985)	18–24	mask in tunnel
Sunbirds					
Bronzy sunbird	15	4.15	Wolf *et al.* (1975)	20	10.1 l container

Italic metabolic cost figures are considered unreliable because these are from measurements in containers smaller than 5 l.

that altitude the hovering costs were 1.17 and 1.25 W for 4.0 g broad-tailed and 4.3 g rufous hummingbirds respectively. Chai and Dudley's (1995) data show that the oxygen consumption at 2000 m reaches 116% of the values measured at sea level. The figures for the cost of hovering of broad-tailed and rufous hummingbirds in Table 8.6 have been recalculated using this percentage and represent values at sea level.

Sunbirds are considered the old world equivalents of the hummingbirds of the American continent. They share the small size, beautiful plumage, and food preferences with the hummingbirds, but their foraging strategy is rather different. Sunbirds only scarcely hover in front of flowers; they land instead, sometimes piercing the flowers to reach the nectar. They can hover briefly like most other small passeriformes but lack the special anatomy that the hummingbirds have to turn the hand wings upside down during the upstroke. The bronzy sunbird measured by Wolf *et al.* (1975) was kept in a 10.1 l chamber where oxygen consumption was continuously recorded. The short hovering bouts were timed with a stopwatch. The percentage of time spend hovering varied from 0 to up to 14%. The steady-state oxygen consumption increased with increasing percentage of hovering time. Extrapolation to 100% of the time hovering gave a value of 47 $\text{ml g}^{-1}\,\text{h}^{-1}$ (4.15 W for the 15 g sunbird).

Hovering costs of larger birds are difficult to measure because flight at zero airspeed is obviously so strenuous (figures approaching infinity according to the aerodynamic models) that birds either avoid it or do it during very short periods only. Hovering episodes are so brief that a steady-state measurement of oxygen consumption is virtually impossible. The use of oxygen consumption as tool to measure metabolic costs under these conditions is also precluded by the fact that muscles most probably will be used anaerobically.

8.7 *Summary and conclusions*

The metabolic costs of forward flapping flight have been more or less accurately measured in 37 species. The data set is not a uniform one and should be used with caution. It contains measurements under laboratory conditions in wind tunnels or in still air as well as estimates derived from data obtained in the field under natural but uncontrolled circumstances. A variety of techniques has been designed to measure flight costs as accurately as possible. Fuel consumption rates potentially should provide a straight-forward figure for the energy turnover required to fly over a certain distance. There are however serious pitfalls attached to this approach. It is necessary to determine what kind of fuel the birds have been using. Fat and proteins are the most likely candidates but in some cases the use of carbohydrates cannot be ruled out entirely. The loss of body mass after long distance non-stop flights and food balance experiments under controlled flight conditions have probably provided the most reliable figures.

Flight costs have also been derived from measurements of the exchange of respiratory gases, oxygen and carbon dioxide. A variety of masks has been designed to collect these gasses allowing continuous measurements of concentrations in the expired and the inhaled air.

Changes in the concentrations of stable heavy isotopes of oxygen, hydrogen and carbon injected in the blood offer the possibility to calculate the amount of oxygen used over prolonged periods. The main advantage of this method is that it can be used in free-ranging birds in the field. Hcart rate telemetry is another indirect way to detect oxygen consumption and hence to measure the costs of exercise. For each species a relationship between heartbeat frequencies and oxygen consumption must be established before the method can be applied in the field using telemetric techniques. The use of both direct and indirect respirometry to determine flight costs demands knowledge of the substance burnt; a condition which is difficult to meet fully.

Hummingbirds and sunbirds are able to hover for more prolonged periods. There are 11 reliable metabolic costs studies of respirometry of hovering hummingbirds and one of a sunbird.

In this chapter we established a set of figures for the metabolic costs of flight that we can more or less trust. The struggle to determine the reliability of measured values gave us a feeling for the order of magnitude of the cost of forward and hovering flapping flight. In the next chapter the figures collected in Tables 8.1 and 8.6 will be used to search for general trends among birds and to find empirical equations with some predictive value.

9 *Comparing the metabolic costs of flight*

9.1 *Introduction*

The data set accumulated in Chapter 8 represents the tremendous amount of effort of many investigators to measure flight costs directly. This deserves further exploitation. Comparison of the best measurements available over a wide range of species should reveal general trends in bird flight costs if there are any. General trends can be expected since flight is dominated by severe aerodynamic and physiological constraints common to all flyers. On the other hand, the substantial morphological variety hatches the expectation of aerodynamically distinct designs. Each design will be closely related to the dominant flight style of a species.

Comparisons to reveal both the common and the distinct features can be made in a variety of ways. Fair and meaningful ones have to be selected. First, metabolic costs are related to body mass for all measured values in an attempt to discover trends and limits that deepen our insight into the causes of the variation or the lack of it. Fair comparisons can further be made by considering the costs of flight per unit weight lifted into the air or by looking at the amount of work per unit distance flown. It is probably even more fair to compare the energy required to transport one unit of weight over one unit distance in a dimensionless way by dividing the flight costs (Watts, W) by the product of speed ($\mathrm{m\,s^{-1}}$) and weight (Newtons, N). Dimensionless figures can also be used to compare birds with insects, bats, and aircraft.

Flight costs expressed as multiples of BMR provide a feeling of how strenuous the exercise is for each species and of the impact of flight costs on the daily energy budget.

Flight energetics can also be approached from a different perspective. A large amount of metabolic energy is required to deliver the relatively small amount of mechanical power for the interaction between the bird and the air. Methods were developed to measure the mechanical power by taking the amount of work per wing beat cycle into account. This was done by measuring the force exerted by the dominant downstroke muscle and multiplying that with the shortening distance of the muscle (Chapter 7). The forces required to move the wings are a large portion of the mechanical power required for flight.

Aerodynamic theory can also be used to estimate the mechanical costs to overcome drag and to generate lift during flight. It could predict the metabolic costs if the efficiency of converting metabolic energy into mechanical energy would be known. The conversion efficiency is defined as the ratio of mechanical costs required over the metabolic energy birds have to produce. We are interested in that fraction. There are only a few measurements of both costs for a same species. Instead, aerodynamic model predictions of the mechanical power required and empirically determined metabolic data can be used to find conversion efficiencies.

In the last paragraph of this chapter, the energy required by hovering hummingbirds is related to body mass and is compared to hovering costs of sunbirds and bats.

9.2 *How to make fair comparisons?*

Would it be fair to compare the flight costs of a siskin with those of an albatross, or to compare a swift with a magpie? Birds fly at different speeds, vary in body mass and shape and use distinct flight styles. A comparison can only be fair if we take various constrains into account.

Theoretical considerations (Chapter 2) and measurements in variable speed wind tunnels have made it clear that the mechanical and metabolic costs of flight may vary with air speed and body weight. Costs will be high for hovering or for flight at very low speeds. They are expected to reach minimum values at intermediate velocities and to increase again when a bird approaches its maximum speed. Heavier birds have to generate more lift and must use more mechanical and metabolic energy to stay airborne. The data for flapping forward flight accumulated in Table 8.1 have been selected assuming best performance. Those based on variable speed wind-tunnel studies are chosen from a range of figures, mostly collected over short periods. The maximum range speed, where the amount of work per unit distance covered was minimal, was considered the best performance in these cases. Birds measured flying over prolonged periods in the wild will give an average figure for the total costs during the measurement including all activities. An underlying assumption usually is that birds use their best performance during long distance flights. We have to live with the fact that the data in Table 8.1 are biased by the different approaches and most probably by the inclusion of other unknown artefacts. On the other hand, this empirical data set is what we have at this moment in time.

Interpretation of the accumulated data is not an easy task. The variation is caused by several factors for which the impact on the flight cost figure is usually not known. Body mass is an obvious one. The impact of speed on the outcome of the comparisons is probably partly eliminated by the best performance criterion. The different methods used to measure flight costs are not equally accurate which introduces unknown bias. It would be most important for our understanding of the biology of flight to know the impact of design features and flight styles. Strong effects of body weight will appear in a graph where flight costs in W are plotted against body mass. General trends found give an indication of the magnitude of the impact of body mass.

Effects of body mass can be partly eliminated by plotting the costs per unit weight against body mass. This trick will not diminish the scatter due to the other effects. Another option is to compare the amount of work per unit distance flown. This comparison can be made by dividing the metabolic flight costs in W by the maximum range speed in $m\,s^{-1}$. The unit of the amount of work per unit distance covered ($J\,m^{-1}$) is in fact force (*N*) since one J is one Nm by definition (see Chapter 1). The flight costs in W ($J\,s^{-1}$) are almost two orders of magnitude higher for large birds than for the smallest ones. In an ultimate attempt to eliminate all the size effects, dimensionless cost of transport can be calculated by dividing the flight costs in W ($J\,s^{-1} = Nm\,s^{-1}$) by the weight (N) and the speed ($m\,s^{-1}$) of a bird.

In 'fish swimming' dimensionless costs of locomotion were compared (Videler 1993). It is now time to do it again for the costs of flight using additional empirical data of insects, birds, bats, and aircraft.

9.3 *Flight costs related to body mass*

The results of all reliable measurements compiled in Fig. 9.1 show that the rate at which energy is used increases with body mass of the birds. However, this broad trend includes a lot of variation. The data points seem to fit inside a wide ribbon with a rather sharp edge on the high and a ragged shape on the lower edge. The sharp upper border closely follows the line drawn in Fig. 9.1. Note that this is not a regression line but a hand drawn straight line as close as possible to the data points. This line ($y = 60x^{2/3}$) seems to represent an upper limit which seems to follow the *two-thirds-power* law. This law suggests that the maximum flight costs depend on some limiting surface area. (When, for different sized birds, the linear scale varies as l, the mass m should vary as l^3 and surface areas as l^2 (assuming isometry). Therefore, if size l is proportional to $m^{1/3}$, surfaces will be proportional to $(m^{1/3})^2 = m^{2/3}$.) The surface area of the lungs is a likely candidate. Duncker and Güntert (1985) investigated the volume and area of the lungs of a large variety of birds and found that the lung surface area of galliform birds increased as mass to the power of 0.69 which is close to two-third. The exponent of all other birds however was 0.96. Both lung surface areas and volumes of non-galliform birds increased directly proportional to mass. In conclusion, isometric scaling rules seem to make sense for galliforme birds but not for the others. This implies that the slope of the line is still enigmatic.

Hummingbirds, finches, starlings, corvids, seagulls, bar-tailed godwits, and pigeons seem to have been measured flying close to the maximum capacity predicted by the *two-thirds-power* law line. The other species were not performing near that maximum. In some cases, it seems obvious how they did that. The cluster with the aerial feeders including swallows, martins, and swifts managed to fly at between one-half and a quarter of the maximum capacity probably because they use many short gliding bouts. The gliding times estimated for house martins vary between 21% and 54% and swifts are

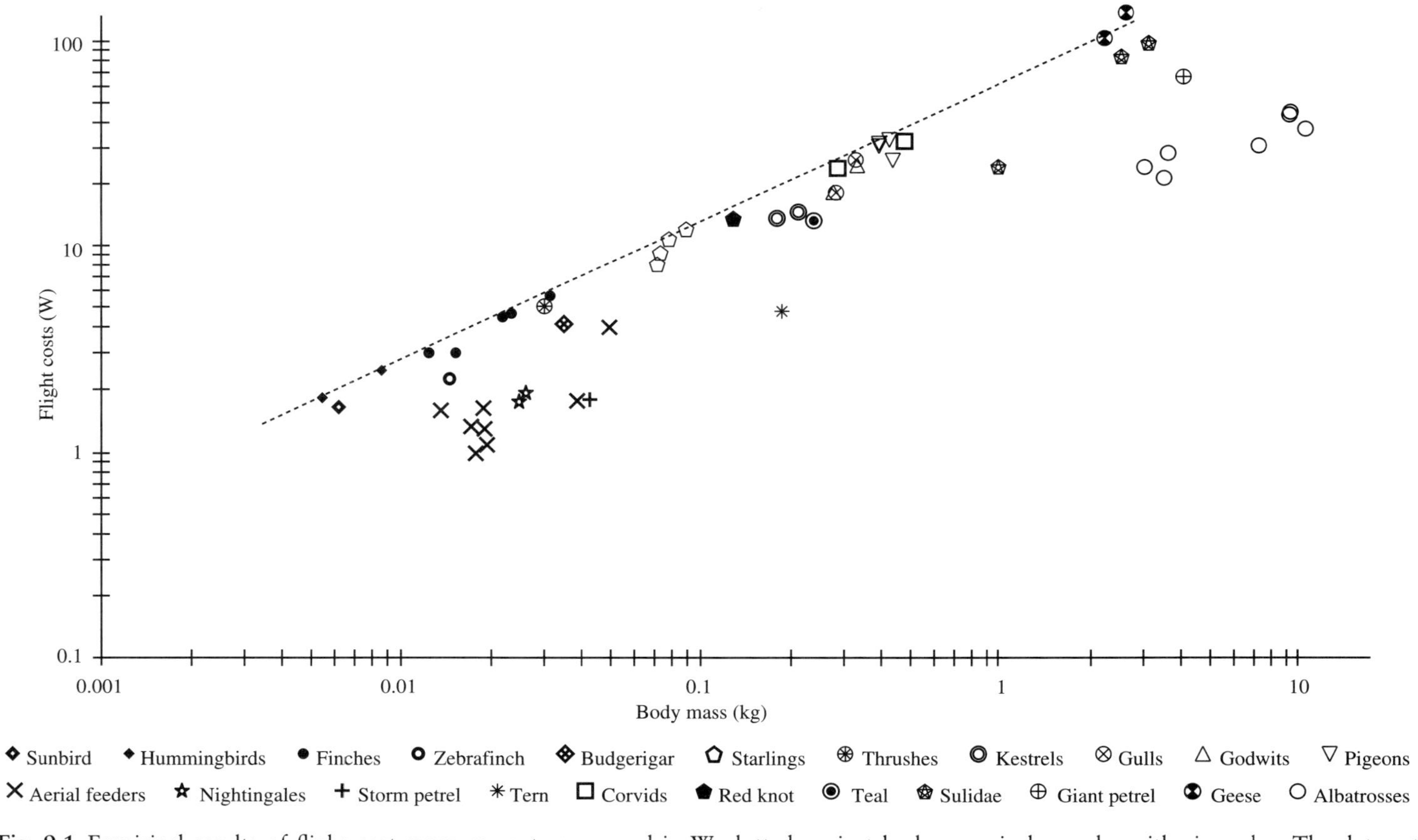

Fig. 9.1 Empirical results of flight cost measurements expressed in W plotted against body mass in kg on logarithmic scales. The data set includes 38 species and it is accumulated in Table 8.1. Detailed information on each data point is given in Chapter 8.

found to glide during 70–80% of the flight time. Seabirds manage to underscore the upper limit by their well-known capacity to glide a lot. Albatrosses seem to fall even in a separate category. Their foraging flights consist for 97% of gliding including the various forms of soaring (Chapter 6). The rate of energy expenditure is between $\frac{1}{8}$ and $\frac{1}{5}$ times the upper limit predicted by the *two-thirds-power* law. Both gannets seem to fly at 75% of their upper limit, the red-footed booby at only 40%. Energy saving techniques allow the sooty tern and Wilson's storm petrel to travel across the oceans at about one-quarter of the flight costs required by continuously flapping bird species of the same body mass. Flying and windhovering kestrels use gliding bouts to save energy (Videler *et al.* 1983). This strategy could be resulting in a flight cost reduction of about one-third of the maximum rate of expenditure (Chapter 6).

There are a few low flight cost results, which are not easily explained by the hypothesis that the birds were using energy saving tricks such as bounding, gliding, or soaring. The thrush nightingales flew long hours non-stop in two studies (see Chapter 8) in the Lund wind tunnel and did not show any obvious energy saving trick. It could be that the circumstances were optimal: a large wind tunnel with constant airflow conditions at a very favourable speed. Klaassen *et al.* (2000) suggest that migratory birds use less energy than non-migratory birds. However, the question remains as to how the thrush nightingales managed to use only about one-third of the energy required by finches of approximately the same body mass. The method used was based on mass loss measurements and was used to estimate flight costs of the teal as well. Results for that bird were also low, even lower than the figures found for the kestrels. One other enigmatically low point is that of the budgerigar flapping continuously wearing a mask in Tucker's wind tunnel. An easy way out could be to suggest that the negative effect of the mask was negligible. However, that would not solve the problem completely because the total power input at the maximum range speed of 11.7 $\mathrm{m\,s^{-1}}$ was only 5.01 W. Only the 6.67 W used by the budgerigar flying at 13.3 $\mathrm{m\,s^{-1}}$ was close to the predicted maximum. The zebra finch and the Palestine sunbird points are based on the new ^{13}C stable isotope technique. It could be that this method underestimates flight costs systematically but it is not obvious why that would be the case.

It is clearly impossible to explain all the variation in Fig. 9.1. What can be obtained from it is insight in the order of magnitude of the flight costs of birds over a wide range of body masses. It offers an equation predicting some kind of upper limit:

$$\text{Maximum flight costs} \approx 60m^{0.667}$$

where the costs are expressed in W when the mass (m) is in kg. This formula provides a useful and simple tool to obtain a first estimate of metabolic flight costs but one has to keep in mind that birds have all kinds of tricks to fly cheaper than at the maximum price. We have also to be aware of the speed-related variation mentioned earlier. Aerobic capacity limits prolonged flights only and not short bursts.

The maximum flight costs increase with mass at a rate of two-third, that is much lower than one. The cost per unit weight (the cost in W divided by the weight in N) will therefore decrease with body mass as illustrated in Fig. 9.2(a). The upper limit line has of course a slope of minus one-third. A weight difference of a factor 10 means an approximate difference in flight costs per unit weight of slightly more than two. In other words, 10 times lighter birds use more than twice as much energy per unit weight.

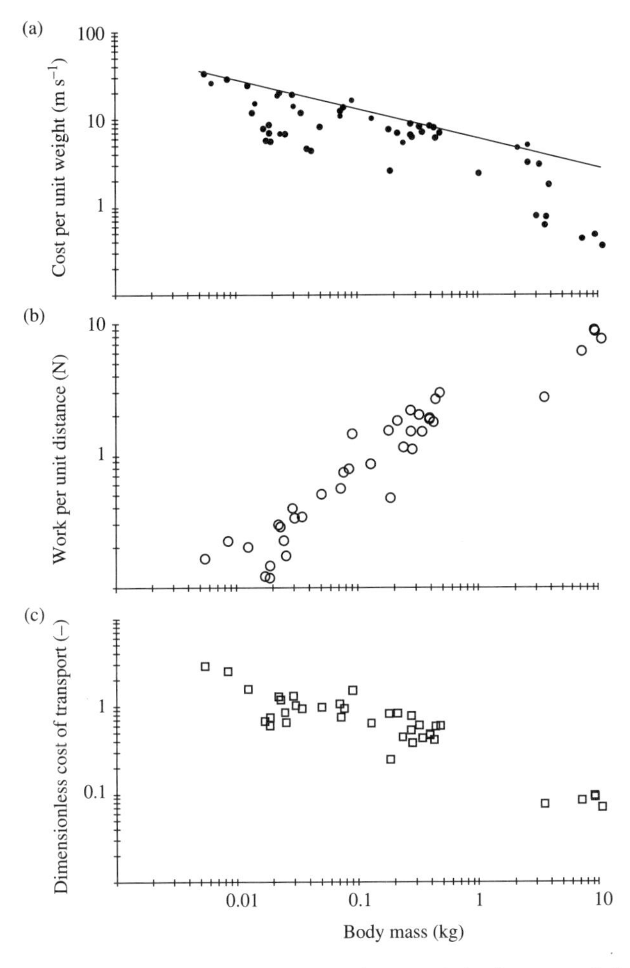

Fig. 9.2 Three ways to compare the relationships with body mass of the empirical flight cost estimates of Table 8.1 and Fig. 9.1: (a) Flight cost per unit weight transported ($W : N = Nm\,s^{-1} : N = m\,s^{-1}$). (b) The amount of work per unit distance travelled ($J : m = Nm : m = N = W : m\,s^{-1}$). (c) Dimensionless cost of transport or the amount of work per unit weight and distance covered ($J : Nm = W : Nm\,s^{-1} =$ *n.d.*).

In those cases where the flight speed was measured or estimated it is possible to calculate the amount of work done per unit distance covered ($\text{J m}^{-1} = \text{N}$) by dividing the cost of flight ($\text{W} = \text{J s}^{-1}$) by the speed. Figure 9.2(b) shows the increasing trend. The energy required to cover a unit distance in J m^{-1} is larger for the heavier species. A 10 times heavier bird does about 4 times as much work per unit distance.

A dimensionless comparison can be made by dividing the cost of flight ($\text{W} = \text{J s}^{-1} = \text{N m s}^{-1}$) by the product of body weight (N) and velocity (m s^{-1}). The results of this exercise are in Table 9.1 and Fig. 9.2(c). The amount of energy used per unit weight and distance covered decreases with increased body mass. The rate of decrease of these costs of transport (COT) is somewhere in the order of minus one-half to one-third. Both the trend and the amount of variation are unexpected because we are comparing dimensionless figures and we should expect flight costs per unit distance and weight to be approximately the same. What it does show is that birds are not exact scale models of each other with body mass being the only variable. The design of the flight apparatus as well as the flight kinematics and the use of energy saving strategies all contribute to the variation in dimensionless flight costs. If we exclude the obvious gliders there is still a decrease with body mass hinting at fundamental design changes with increasing mass.

This dimensionless approach can also be used to compare birds with other flyers such as insects, bats, and aircraft.

9.4 *Birds compared with other flyers*

In this comparison, (Table 9.1 and Fig. 9.3) albatrosses stick out as the world's cheapest flyers. They need about half the amount of energy used by a Boeing 747 to transport one N over one m. This is surprising because we see an overall decreasing trend with increasing mass and the aircraft weighs 25,000 times more than the albatrosses. The albatrosses were measured flying over large distances at a low average velocity (Arnould *et al.* 1996). They save energy by using dynamic soaring, sweeping flight and gust soaring techniques (Chapter 6). The comparison with commercial aircraft is of course somewhat unfair because the limiting conditions are incomparably different.

There is a lot of scatter among the flying vertebrates; bats show a trend close to the observed upper limit line for birds. The flight costs in W of small bats tend to be lower than the costs of birds of similar mass (Winter and von Helversen 1998), but the flight speed is also estimated to be much lower (Winter 1999) which brings the cost of transport above or at the level of that of small birds. The energy expenditure of big fruit bats is of the order of magnitude of birds of similar size but flight speed is lower. Surprisingly, the insect data points do not decrease with body mass. Although the scatter among insects is large,the lack of a decreasing trend indicates that fruit flies

Table 9.1 Comparison of dimensionless flight costs (COT in J $(Nm)^{-1}$) of birds with those of insects, bats, and aeroplanes.

	Mass (kg)	Speed ($m\,s^{-1}$)	COT	Source
Insects				
Fruitfly	7.2×10^{-7}	1	8.5	Hocking (1953)
Blackfly	2.3×10^{-6}	1	7.6	Hocking (1953)
Mosquito	2.5×10^{-6}	1	8.7	Hocking (1953)
Horse fly	1.7×10^{-4}	2	4.2	Hocking (1953)
Honey bee	8.0×10^{-5}	4	12.2	Nachtigall *et al.* (1989)
Honey bee	8.0×10^{-5}	4	14.0	Wolf *et al.* (1989)
Honey bee	8.4×10^{-5}	2	10.0	Hocking (1953)
Honey bee	1.4×10^{-4}	4	8.0	Wolf *et al.* (1989)
Bumble bee	5.0×10^{-4}	4	19.3	Ellington *et al.* (1990)
Locust	2.0×10^{-3}	3	3.4	Weis-Fogh (1952)
Birds				
Green violetear	0.0055	11	3.1	Berger (1985)
Sparkling violetear	0.0085	11	2.7	Berger (1985)
Pine siskin	0.0125	15	1.6	Dolnik and Gavrilov (1973)
Barn swallow	0.019	11	0.8	Turner (1982*a,b*)
Chaffinch	0.0223	15	1.4	Dolnik and Gavrilov (1973)
Brambling	0.0232	16	1.3	Dolnik and Gavrilov (1973)
Thrush nightingale	0.0247	7.9	0.9	Kvist *et al.* (1998)
Thrush nightingale	0.0259	10	0.7	Klaassen *et al.* (2000)
Eurasian bullfinch	0.0295	14	1.4	Dolnik and Gavrilov (1973)
Thrushes	0.03	13	1.1	Wikelski *et al.* (2003)
Budgerigar	0.035	12	1.0	Tucker (1968, 1972)
Purple martin	0.05	8	1.0	Utter and LeFebvre (1970)
Rosy starling	0.0716	11	1.0	Sophia Engel, personal communication
European starling	0.073	17	0.7	Torre Bueno and LaRochelle (1978)
European starling	0.077	14	1.0	Westerterp and Drent (1985)
European starling	0.089	9.9	1.4	Ward *et al.* (2001)
Red knot	0.128	15	0.7	Kvist *et al.* (2001)
Common kestrel	0.18	9	0.9	Masman and Klaassen (1987)
Sooty tern	0.187	10	0.3	Flint and Nagy (1984)
Common kestrel	0.213	8	0.9	Masman and Klaassen (1987)
Common teal	0.2373	11.5	0.5	Kvist *et al.* (1998)
Fish crow	0.275	11	0.8	Bernstein *et al.* (1973)
Laughing gull	0.277	12	0.6	Tucker (1972)
Bar-tailed godwit	0.282	16	0.4	Piersma and Jukema (1990), Lindström and Piersma (1993)
Laughing gull	0.322	13	0.6	Tucker (1972)
Bar-tailed godwit	0.341	16	0.5	Piersma and Jukema (1990), Lindström and Piersma (1993)
Pigeon	0.394	17	0.5	LeFebvre (1964)
Pigeon	0.394	17	0.5	LeFebvre (1964)
Pigeon	0.425	19	0.4	Polus (1985)
Pigeon	0.442	10	0.6	Butler *et al.* (1977)
Chiuahuan raven	0.48	11	0.6	Hudson and Bernstein (1983)
Black-browed albatross	3.58	8	0.1	Bevan *et al.* (1995)
Wandering albatross	9.36	5	0.1	Arnould *et al.* (1996)
Wandering albatross	10.74	5	0.1	Arnould *et al.* (1996)

Table 9.1 *Continued.*

	Mass (kg)	Speed ($m\,s^{-1}$)	COT	Source
Bats				
Pipistrellus pipistrellus	0.0067	4	4.1	Speakman and Racey (1991), speed Norberg (1990)
Plecotus auritus	0.0079	4	3.6	Speakman and Racey (1991), speed Norberg (1990)
Glossophaga commissarisi	0.0086	7	2.2	Winter and von Helversen (1998), speed Winter (1999)
Hylonycteris underwoodi	0.0087	5	2.6	Winter and von Helversen (1998), speed Winter (1999)
Glossophaga soricina	0.0116	7	2.0	Winter and von Helversen (1998), speed Winter (1999)
Phyllostomus hastatus	0.0930	8	1.2	Thomas (1975)
Hypsignathus monstrosus	0.2580	8	1.1	Carpenter (1986)
Eidolon helvum	0.3150	8	0.9	Carpenter (1986)
Pteropus poliocephalus	0.5740	9	0.8	Carpenter (1975, 1985, 1986)
Pteropus gouldii	0.8200	9	0.7	Thomas (1975, 1981)
Aeroplanes				
Cherokee	978	56	0.5	Stanfield (1967)
Cessna 340A	2717	99	0.3	Pilot manual
Grand Commander	3860	103	0.4	Stanfield (1967)
Cessna 550 Citation II	6033	198	0.4	Jane's all the world's aircraft (1989)
Cessna 650 Citation III	9979	243	0.3	Jane's all the world's aircraft (1989)
Fokker 50	19,000	145	0.3	Fokker, aircraft brochures
Fokker 100	40,000	165	0.3	Fokker, aircraft brochures
DC9-10 (transport)	41,300	244	0.3	Stanfield (1967)
BAe 200QT	43,381	186	0.3	Jane's all the world's aircraft (1989)
Boeing 737-400	50,000	220	0.3	KLM, flight technical department
Airbus 310	100,000	235	0.2	KLM, flight technical department
DC8 (transport)	107,000	268	0.2	Taylor (1968)
Boeing 747-400	300,000	274	0.1	Jane's all the world's aircraft (1989)

The figures given for insects and bats are averages for each species. (If a species was measured by different authors or by the same author under different body mass conditions it is mentioned more than once.)

and honey bees are better scale models of each other than hummingbirds and pigeons or Cessna's and Boeing 747's. The commercial aircraft are more or less on one descending line. The slope of the aircraft tendency is approximately minus one-sixth, which is half as steep as the tendency among the vertebrate flyers. These rough trends are obvious despite the fact that the doubly logarithmic scales (the x-axis is spreading over more than 12 orders of magnitude) have the tendency to conceal the variation. This difference must therefore reveal some fundamental contrasts in design. This has been insufficiently appreciated both by aircraft engineers and by biologists looking for the fundamental principles governing animal flight.

Fixed wing aircraft are in fact motorized gliders, generating lift and drag forces with virtually unchangeable fixed wings; engines generate thrust. Aircraft designers stick to a number of strict and well-established rules imposed

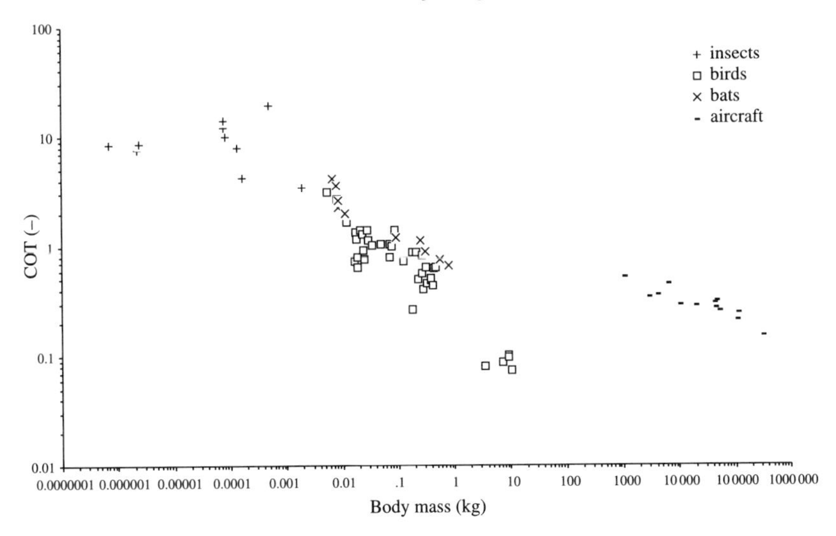

Fig. 9.3 Dimensionless cost of transport in flight at the maximum range speed of insects, birds, bats, and aircraft as function of body mass. (Both scales are logarithmic).

by a desire for safety and stability. The basic material used to shape aircraft is bended plate, which seriously limits freedom of design.

Flying vertebrates use their wings to generate both lift and thrust. The wings are not fixed in shape but extremely versatile. They move during every wing beat through a cycle of extreme changes in shape and aerodynamic properties. The interactions between wings and air are fundamentally different from those around fixed wings (Chapter 4). Aircraft wings are designed for a stable steady state lift/drag relation at a small range of speeds. The wings of birds and bats operate intrinsically unsteady. Only extreme gliding birds use the design principles of aircraft but even under gliding conditions the performance of animals can be much more versatile due to the far greater capacity to change the shape and attitude of the wings and the use of leading edge vortex lift and drag on the hand wing.

9.5 *Metabolic rates as units of energy expenditure*

The official SI currency is not always satisfactory. Costs of locomotion are often expressed as multiples of basal or resting metabolic rates (BMR, RMR). These provide an easy measure of how high the exertion for a particular animal must be. For example, heavy exercise (100 or 200 m running) of an athlete requires almost 15 times the resting values of energy expenditure. How are BMR and RMR exactly defined in birds? Aschoff and Pohl (1970) studied the energy used by birds sitting motionless in the dark with empty guts at temperatures not exceeding their thermo neutral zone. (That is the zone of ambient temperatures where the metabolic energy required for thermoregulation is constant at a minimum level.) The birds were allowed to drink water but did not spend energy on the digestion or absorption of food. The oxygen uptake under these conditions changes rhythmically over daily cycles. During the time when the birds would have been active, it is higher than during the time the birds would have been resting anyway. BMR is the minimum rate of energy expenditure under the described conditions during the subjective resting period, whereas the RMR reflects the energy used during what would have been the active period. Gavrilov and Dolnik (1985) regressed the BMR and body mass values of 263 birds and found the allometric relation:

$$\mathrm{BMR} = 4.2m^{0.677}$$

where BMR is in Watts and mass (m) in kilograms. The exponent of about $\frac{2}{3}$ indicates that surface to volume ratio's dominate this relationship which could make it biologically meaningful. There is of course scatter around that regression line because measurements of BMR vary considerably. Part of the variation is caused by different experimental designs. Biological variation among species, but also between and even within individuals is substantial. Daan *et al.* (1990) discovered that during the breeding season, when parental care demands the highest levels of energy expenditure, there is a positive correlation between the relative dry mass of heart and liver and deviations from the average BMR values. Furthermore, the daily energy expenditure (DEE) of the species with the highest mass of highly active tissues is, around 4 BMR, about twice as high as that of species with relatively more inactive tissue (bone, feathers) in their bodies. In other words, some species have intrinsically higher activity levels than other species. Most BMR values found in the literature should be treated as approximate minimum levels of energy expenditure.

Many other metabolic rates were defined. The doubly labelled water technique (Box 8.2) yields average energy expenditure figures over prolonged periods of, for example, one day or any other time span conveniently related to the behaviour of a particular species. Sustained metabolic rates (SusMR) have been defined as the time-averaged energy budget that an animal maintains over periods during which the average body mass is kept constant because intake and expenditure are balanced. Among vertebrates sustained metabolic rates never seem to exceed seven times the resting rates (Hammond and Diamond 1997). Tour de France cyclists (peculiar birds) reach sustained metabolic rates of 5.6 times resting values (Westerterp *et al.* 1986).

Table 9.2 shows BMR values of all birds for which we have empirical data on flight costs. Most of the values are measured directly according to Aschoff and Pohl's (1970) definition. In some indicated cases daytime resting rates were measured only. There are no empirical data available for sixteen species. In these cases the equation of Gavrilov and Dolnik (1985) was used.

Table 9.2 Flight costs expressed as multiples of BMR.

Common name	Body mass (kg)	Flight costs (W)	BMR		Flight costs		BMR source
			Measured (W)	Calculated (W)	Measured (xBMR)	Calculated (xBMR)	
Green violet-ear	0.0055	1.82		0.12		14.7	Calculated
Palestine sunbird	0.0062	1.64		0.13		12.2	Calculated
Sparkling violet-ear	0.0085	2.46		0.17		14.8	Calculated
Pine siskin	0.0125	3.03	0.26		11.7		Calculated
Sand martin	0.0137	1.6	0.23		6.8		Calculated
Zebra finch	1.0145	2.24		0.24		9.4	Calculated
Barn swallow	0.0173	1.34	0.29		4.6		Calculated
Northern house martin	0.0178	1.01	0.31		3.2		Calculated
Barn swallow	0.019	1.3	0.32		4.1		Calculated
Barn swallow	0.019	1.62	0.32		5.1		Calculated
Northern house martin	0.0197	1.08	0.35		3.1		Bryant and Westerterp (1980)
Chaffinch	0.0223	4.51	0.36		12.4		Aschoff and Pohl (1970)
Brambling	0.0232	4.6	0.42		11.0		Aschoff and Pohl (1970)
Thrush nightingale	0.0247	1.75		0.34		5.1	Calculated
Thrush nightingale	0.0295	1.91		0.35		5.4	Calculated
Eurasian bullfinch	0.0295	5.6	0.54		10.5		Calculated
Budgerigar	0.035	4.12	0.42		9.9		Calculated
Common swift	0.0389	1.8	0.38		4.8		Calculated
Wilson's storm petrel	0.0422	1.82	0.50		3.6		Obst *et al.* (1987)
Purple martin	0.05	4.1		0.55		7.4	Calculated
Rosy starling	0.0716	8.05		0.70		11.5	Calculated
European starling	0.073	9	0.87		10.3		Calculated
European starling	0.077	10.5	0.92		11.4		Calculated
European starling	0.089	12		0.81		14.7	Calculated

Table 9.2 *Continued*

Common name	Body mass (kg)	Flight costs (W)	BMR		Flight costs		BMR source
			Measured (W)	Calculated (W)	Measured (xBMR)	Calculated (xBMR)	
Red knot	0.128	13.5		1.04		13	Calculated
Common kestrel	0.18	13.8	0.86		16.0		Daan *et al.* (1990)
Sooty tern	0.187	4.8	1.01		4.7		Gavrilov and Dolnik (1985*)
Common kestrel	0.213	14.6	1.02		14.3		Daan *et al.* (1990)
Common teal	0.237	13.2		158		8.4	Calculated
Fish crow	0.275	24.2	3.58		6.8		Bernstein *et al.* (1973*)
Laughing gull	0.277	18.3	2.82		6.5		Tucker (1972*)
Bar-tailed godwit	0.282	17.8	1.79		10.0		Daan *et al.* (1990)
Laughing gull	0.322	26.3	3.28		8.0		Tucker (1972*)
Bar-tailed godwit	0.341	24.2	2.16		11.2		Daan *et al.* (1990)
Pigeon (Rock dove)	0.394	31.9	1.77		18.0		Calculated
Pigeon (Rock dove)	0.394	33.1	1.77		18.7		Calculated
Pigeon (Rock dove)	0.425	34.1	1.91		17.8		Calculated
Pigeon (Rock dove)	0.442	26.8	1.99		13.5		Calculated
Chihuahuan raven	0.48	32.8		2.55		12.9	Calculated
Red-footed booby	1.001	24	8.93		2.7		Ballance (1995*)
Barnacle goose	2.1	102		6.92		14.7	Calculated
Cape gannet	2.58	81	8.30		9.8		Adams *et al.* (1991)
Bar-headed goose	2.6	135		8.00		16.9	Calculated
Laysan albatross	3.064	24	9.08		2.6		Grant and Whittow (1983)
Northern gannet	3.21	97	8.58		11.3		Birt-Friesen *et al.* (1989)
Black-browed albatross	3.58	22		11.08		2.0	Ellis (1984)
Grey-headed albatross	3.707	28		11.37		2.5	Ellis (1984)
Southern giant petrel	3.885	68	10.86		6.3		Adams and Brown (1984)
Wandering albatross	7.3	31	18.25		1.7		Brown and Adams (1984)
Wandering albatross	9.31	45	23.28		1.9		Brown and Adams (1984)
Wandering albatross	9.36	43.8	23.40		1.9		Brown and Adams (1984)
Wandering albatross	10.74	38.1	26.85		1.4		Brown and Adams (1984)

For calculated BMRs, Gavrilov and Dolnik's 1985 equation was used: $BMR = 4.2m^{0.677}$, with mass m, in kg.
* = daytime resting rate.

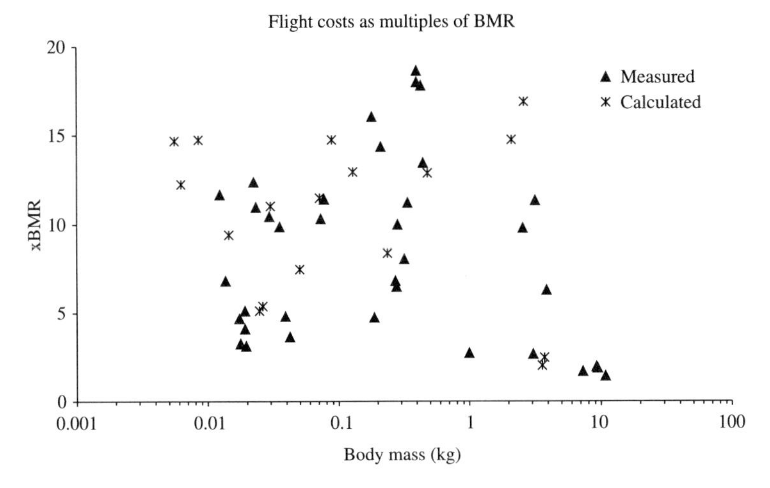

Fig. 9.4 Flight costs expressed as multiples of BMR plotted against body mass (On a logarithmic scale). The data points are taken from Table 9.2.

Flight costs range from 1.4 BMR for a male wandering albatross up to 18.7 BMR measured for a pigeon. Figure 9.4 displays the scatter between these two values but note that the *y*-axis is linear. Aerial feeders and gliders score consistently low. Comparison with the daily energy expenditure during periods of intensive parental care provides a feeling of the relative value of the cost of flight (Daan *et al.* 1990). During that period kestrels are living at a rate of about 5.6 times BMR. The contribution of flight costs of on average 15 BMR must be substantial limiting the period of the day spent in flight to about one-third. The average value of daily energy expenditure of starlings during the period when they feed their offspring is at about four times BMR slightly less than one-third of the flight costs of about 10 or 11 BMR. Frequent flyers such as sand martins, house martins, swallows, storm petrels, and albatrosses fly during most of the day and hence the daily rate of energy expenditure approximately equals the flight costs.

9.6 *Predictions from aerodynamic models*

Aerodynamic models, developed to calculate mechanical cost of flight of birds, use basic principles explained in Chapter 4, Section 4.2 and conventional fixed wing theory used in aircraft design. The models provide rough estimates of the mechanical power output required for flight, which is the rate at which mechanical work is done. These estimates consider the amount of work involved in the interactions between the bird and the air. Muscles output the forces required (Chapter 7). The work done by these forces is only a fraction of the total metabolic energy needed by the bird to fly. The ratio of mechanical over metabolic power is the measure of efficiency of the bird as a flying machine. The processes determining the efficiency of the conversion of mechanical power into metabolic power are very complex. Precise figures for the two powers are hard to get.

9.6.1 Properties of bird flight models

Models used to calculate the total mechanical power required for bird flight (P_{tot}) are based on the sum of the estimates of:

1. The induced power (P_i) needed to generate lift.
2. The profile power (P_{pro}) needed to overcome the drag on the wings.
3. The parasite power (P_{par}) required to overcome the drag on the body and tail.

The sum of the profile power and the parasite power is the total drag power P_d indicated in Fig. 4.3 (Chapter 4).

Several marginally different models have been based on this addition of components of mechanical costs. Greenewalt's (1975) model does not differentiate between drag and lift components but treats the power required for the wing action as a whole. This is a reasonable approach because animals unlike aircraft generate thrust and lift as one entity by wing activity. The parameters required by Greenewalt's model are the weight of the bird, the wing area and span, the air density, and the velocity. Tucker (1974) takes also the viscosity of the air into account. Rayner's (1979 *a*,*b* and *c*) model derives the power estimates from the vorticity in the wake and includes drag and lift coefficients of body and wings. Pennycuick (1989) published his 1975 model in a practical manual complete with a floppy disk. It became widely used by ecologists interested to know the cost of flight either during migration or as part of the energy budget. Pennycuick modified his model twice since the manual appeared. The first modification was induced by the fact that in the original version wing shape, expressed as the aspect ratio of the wings (AR = the span squared over the wing area) did not influence the outcome of the profile power calculations. The model used a fixed value of seven, which is the AR of the pigeon. Adaptation of the model (Pennycuick 1995) changed the results obtained for birds with extremely low and high aspect ratio wings. For example, the profile power of the Southern giant petrel, with an AR of 12 instead of seven, is 20% higher after the modification. The second change (Pennycuick *et al.* 1996) affected the parasite power calculations. The body drag coefficient, $C_{d\,body}$, is the ratio of the drag of a streamlined body over the drag of a flat plate with the same frontal area (S_b) as the streamlined body (Chapter 1). The frontal area of a streamlined body is less effective in stopping the flow than a flat plate of the same size put square in the flow because the air is diverted smoothly past the body by the streamlined shape. The value for $C_{d\,body}$ used in the calculation manual varied between 0.25 for large birds such as geese and swans to 0.4 for small and intermediate size birds. These figures indicate that the larger birds are better streamlined than the smaller ones. However, results of wind tunnel measurements relating wing beat frequencies with speed of the common teal and the thrush nightingale would only be predicted by the model if $C_{d\,body}$ would be about 0.05. This indicates that the streamlining of large and small birds is much more effective than previously thought. The new drag coefficient estimate divides the previous P_{par} calculations by a factor 8. Frontal areas of birds in flight can be measured from head on photographs or calculated using allometric equations offered in the literature. Hedenström and Rosén (2003) compared the allometric relationships with body mass of frontal areas of passerines with those of non-passerines and found significantly different relationships. The equation for passerines was $S_b = 0.0129 m_b^{0.614}$ (m^2) and for non-passerines $S_b = 0.00813 m_b^{0.666}$ (m^2). In these equations m_b is the body mass in g. For a 10 g bird the first equation predicts a frontal area twice as large as the second equation. For a 100 g bird the passerine equation predicts a S_b of 31 cm^2 and the non-passerine equation gives an estimate of 17.5 cm^2. The variation among species of equal body mass can be large. The 17 g dunnock had a S_b of 7.7 cm^2 and the reed bunting of the same weight had a frontal surface area of 17.8 cm^2, a factor 2.3 larger. All this means that model predictions of mechanical power have to be treated with as much care as the results of empirical measurements of metabolic power.

Aerodynamic models calculate the energy needed to generate lift and thrust at various speeds. The induced power to generate lift is high at very low speeds and decreases with increasing velocities. The power to overcome drag increases with speed. This behaviour predicts a U-shaped cost-speed curve explained in Chapter 4.

9.6.2 Conversion between mechanical (output) and metabolic (input) power

Aerodynamic models provide estimates of the power output (P_o), being the rate at which mechanical work is done. Measurements of metabolism during flight produce figures indicating the level of power input (P_i). The ratio of P_o over P_i is the conversion efficiency (η). It depends on some poorly known parameters: that is, the muscle efficiency to transfer fuel into work done by flapping wings and the amount of energy lost by the wings moving the air in directions that do not contribute to lift or thrust. Muscle efficiency is not expected to vary much among birds because there are no obvious differences at the level of the muscle fibres. There is also a great amount of similarity in the design of the connection between pectoral muscles and the wings. It is reasonable to expect that the interactions between wings and air during steady forward flight generate losses of the same order of magnitude among birds of approximately the same size. It seems therefore reasonable to assume that the conversion efficiencies should be of the same order of magnitude for similar sized birds. The studies on mechanical power output of the pectoralis (see Chapter 7, Table 7.1) yield values close to P_o. For two species, starling and pigeon, empirical P_o and P_i data are available although originating from completely independent studies. Biewener *et al.* (1992) estimated the P_o of 73–77 g starlings at 1.7 W. Table 8.1 (Chapter 8) has a P_i values for 73 and 77 g starlings 9 and 10.5 W respectively. The combined data predict a conversion efficiency of 17.4%. Dial and Biewener (1993) measured an average P_o of 3.3 W for 301–314 g pigeons. The four pigeon data points in Table 8.1 have an average P_i of 31.5 W. The pigeons of that data set were about 100 g heavier. The efficiency for pigeons based on these studies is 10.5%.

Wingspan, wingarea, and body mass are needed to calculate the power output using Pennycuick's computer programme. Table 9.3 shows the subset of the data collected in Table 8.1 where wingspan and wingarea are available to calculate P_o using Pennycuick's (1989) equations including the modifications made by Pennycuick (1995) and Pennycuick *et al.* (1996). The empirical values under 'flight costs' in Table 8.1 are taken as P_i. The efficiency increases with body mass but the scatter is large. Figure 9.5 visualizes the data in a graph with a double logarithmic scale. Both scatter and trend are obvious. The increasing trend with mass seems to be equally valid for predominantly flapping flight and for birds that glide more than 50% of the time.

Direct measurements were made for the budgerigar and the laughing gull by Tucker (1972), for the fish crow by Bernstein *et al.* (1973) and the chihuahuan raven by Hudson and Bernstein (1983) using a clever trick involving a tilting wind tunnel. At small downward sloping angles the gravitational force contributes part of the thrust force needed to balance the drag at a given speed. The difference between mechanical power output during level flight and downward flight can now be calculated. At the same time oxygen consumption measurements provided a figure for the difference in metabolic costs for level and sloping flight. The (partial) conversion efficiency is the ratio of the difference in mechanical power over the difference in metabolic power. Values for the budgerigar were found to vary with speed from 19% to 28% at flight speeds between 5.3 and 13.3 m s^{-1}. Data obtained at maximum range speeds are given in Table 9.3 and Fig. 9.5.

9.7 *Hovering flight*

The hovering flight cost data collected in Table 8.6 are plotted along linear scales against body mass in Fig. 9.6. Three data points on nectar-feeding bats hovering under laboratory conditions at a feeder mask are added for comparison. *Hylonycteris underwoodi*, *Glossophaga*, and *Choeronycteris mexicana* weigh 7.0, 11.9, and 16.5 g and hover at 1.1, 1.9, and 2.6 W respectively (Voigt and Winter 1999; Winter 1999). Oxygen consumption in small bats increases towards a steady-state value after about seven seconds

Table 9.3 Comparison between measured power input and calculated power output data.

Species	Mass (kg)	Wingspan (m)	Wing area (m^2)	P_i Measured (W)	P_o Calculated (W)	η Mainly flapping (%)	η >50% gliding (%)	Source
P_o from Pennycuick								
Pine siskin	0.0125	0.214	0.0068	3.03	0.10	3		Dolnik and Gavrilov (1973)
Northern house martin	0.0178	0.292	0.0092	*1.01*	0.09		*9*	Hails (1979)
Barn swallow	0.0190	0.330	0.0135	*1.30*	0.09	7		Hails (1979)
Barn swallow	0.0190	0.327	0.0135	*1.62*	0.10	6		Turner (1982*a*,*b*)
Barn swallow	0.0173	0.330	0.0135	*1.34*	0.08	6		Lyuleeva (1970, 1973)
Northern house martin	0.0197	0.292	0.0092	*1.08*	0.11		*10*	Lyuleeva (1970, 1973)
Chaffinch	0.0223	0.285	0.0102	4.51	0.16	3		Dolnik and Gavrilov (1973)
Brambling	0.0232	0.281	0.0123	4.60	0.19	4		Dolnik and Gavrilov (1973)
Thrush nightingale	0.0259	0.263	0.0130	1.75	0.29	16		M. Klaassen personal communication
Common swift	0.0389	0.420	0.0165	*1.78*	0.19		*10*	Lyuleeva (1970)
Wilson's storm petrel	0.0422	0.376	0.0192	1.82	0.31	17		Obst *et al.* (1987)
European starling	0.0728	0.384	0.0192	9.01	0.72	8		Torre-Bueno and LaRochelle (1978)
European starling	0.0775	0.395	0.0192	10.50	0.74	7		Westerterp and Drent (1985)
Common kestrel	0.180	0.74	0.0708	13.80	1.21	9		Masman and Klaassen (1987)
Sooty tern	0.187	0.84	0.0626	4.79	0.88	18		Flint and Nagy (1984)
Common kestrel	0.213	0.74	0.0708	14.60	1.62	11		Masman and Klaassen (1987)
Bar-tailed godwit	0.282	0.66	0.0465	17.80	2.74	15		Piersma and Jukema (1990), Lindström and Piersma (1993)
Laughing gull	0.277	1.00	0.1207	18.30	1.53	8		Tucker (1972)
Laughing gull	0.322	1.00	0.1207	26.30	1.97	8		Tucker (1972)
Bar-tailed godwit	0.341	0.71	0.0555	24.20	3.42	14		Piersma and Jukema (1990), Lindström and Piersma (1993)
Pigeon	0.394	0.66	0.0630	32.50	5.44	17		LeFebvre (1964)
Pigeon	0.442	0.72	0.0698	26.80	5.79	22		Butler *et al.* (1977)
Red-footed booby	1.001	1.52	0.1992	24.00	5.84	24		Ballance (1995)
Northern gannet	3.210	1.94	0.2900	96.90	26.80	28		Birt-Friesen *et al.* (1989)
Southern giant petrel	3.885	1.95	0.2748	*68.50*	47.70		*70*	Obst and Nagy (1992)
Tilting wind tunnel measurements								
Budgerigar	0.035					19		Tucker (1968)
Fish crow	0.277					29		Bernstein *et al.* (1973)
Laughing gull	0.322					30		Tucker (1972)
Chiuahuan raven	0.480					33		Hudson and Bernstein (1983)

The conversion efficiency η is the ratio of P_o over P_i times 100%. P_o calculations are based on Pennycuick (1989, 1995) and Pennycuick *et al.* (1996). The bottom of the table has figures of conversion efficiencies based on direct measurements using a tilted wind tunnel. See text for further explanation.

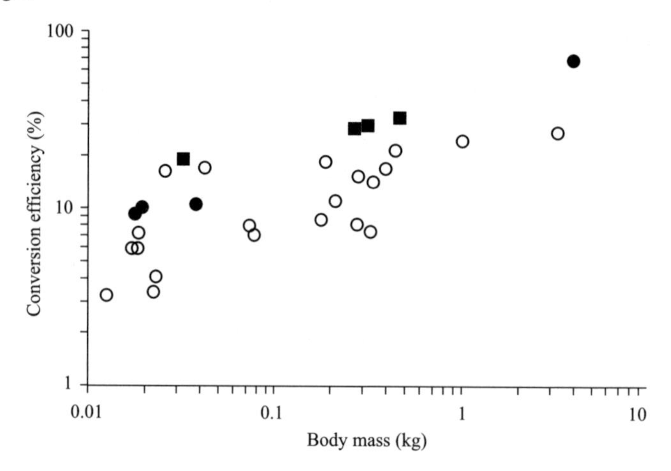

Fig. 9.5 Conversion efficiencies in relation to body mass (both on logarithmic scales). The circles are based on measured power input and on calculated power output using Pennycuick's aerodynamic model. Closed circles indicate birds that glide more than 50% of the time and open circles represent the predominantly flapping birds. The black squares represent the figures for the partial conversion efficiencies comparing flight energetic measurements during level and descending flight in a wind tunnel. See text for further explanation.

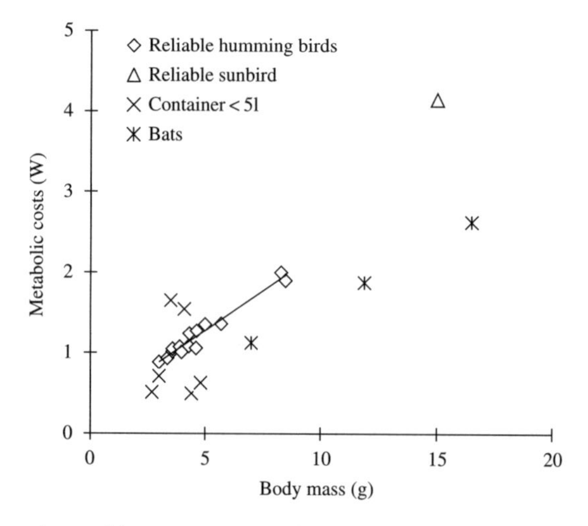

Fig. 9.6 Comparison of hovering costs of small birds and bats against body mass. Note the linear scales. The straight line is drawn through the reliable hummingbird points. See text and Table 8.6 for further information.

of flight. Hovering cost values indicated and shown in Fig. 9.6 are steady-state values.

The six unreliable hummingbird data points are either much higher or lower than those of the birds tested in a more adequate experimental set-up. The latter ones are grouped along a straight line best described by the equation:

$$\text{Metabolic costs (W)} = 0.2 \text{ Body mass (g)} + 0.3 \quad (n = 14,\ r^2 = 0.96)$$

The costs of hovering of bats is much lower than that of birds, on average at about 70% of the costs for hummingbirds predicted by the equation. This is surprising since hummingbirds seem to be better adapted for hovering with wings that can turn upside down during the upstroke. Both groups are obviously specialized in feeding by hovering. Microbats are probably in some way physiologically adapted to decrease the cost of foraging. Sunbirds do usually not hover for food and are apparently not adapted to this kind of exercise at all. They use about twice as much energy to hover as a bat of approximately the same weight.

9.8 *Summary and conclusions*

Comparisons are made using the empirically determined flight costs in birds. Costs as a function of body mass shows that there is a lot of scatter among species, some of it can be explained by differences in flight strategies: Frequent gliders fly cheaper because they use energy saving techniques. Some species are intrinsically more energetic than others are. However it is not possible to explain all the variation. Flight costs in general and the amount of work per unit distance increase with body mass but cost per unit weight and the energy required to transport 1 N over 1 m (the dimensionless COT) decrease.

A simple allometric equation predicts an upper limit for flight costs in all birds. Several groups are capable to fly at much cheaper rates.

A comparison of the dimensionless cost of transport among insects, bats, birds, and aircraft show some marked differences. Albatrosses are the world's cheapest flyers in terms of the energy used to transport one unit of weight over one unit of distance. Dimensionless COT decrease drastically in birds and bats with increasing body mass, showing that small and large flying vertebrates are not scale models of each other. Larger vertebrates are designed for lower costs of transport. This trend also occurs among aircraft but less explicit. Insects varying in mass over about four orders of magnitude seem to have dimensionless costs of transport that are in the same order of magnitude.

Albatrosses and other frequent flyers raise their basal metabolic rates only a few times no matter whether they are breeding, caring for offspring, resting at sea or flying. Flying can be as cheap as that because energy is saved by gliding. On the other side of the scale birds like pigeons have to increase their metabolic rates to up to almost 19 times BMR to fly. This high expenditure limits the amount of time these species can fly per day. Aerodynamic models are based on fixed wing aircraft theory and designed to predict the mechanical flight costs of birds. The predictions based on these models were used to estimate how efficient birds are. The ratio of mechanical power over empirically determined metabolic power roughly increases with body mass. For flapping flight the efficiency increases from a few per cent in small birds to about 30% in large birds.

The cost of hovering of hummingbirds increases linearly with body mass over a range of weights between 3 and 9 g. Surprisingly, nectar-feeding bats hover at about 70% of the costs of hummingbirds. One measurement on a sunbird suggests that this group is not well adapted to hovering flight at low costs.

Appendix 1. Common names of birds

African white-backed vulture	*Gyps africanus*
Allen's hummingbird	*Selasphorus sasin*
Alpine swift	*Tachymarptis melba*
American black duck	*Anas rubripes*
Amazilia hummingbird	*Amazilia amazilia*
Anna's hummingbird	*Calypte anna*
Audouin's gull	*Larus audouinii*
Bar-headed goose	*Anser indicus*
Barn owl	*Tyto alba*
Barn swallow	*Hirundo rustica*
Barnacle goose	*Branta leucopsis*
Bar-tailed godwit	*Limosa lapponica*
Black Jacobin	*Florisuga fusca*
Black-browed albatross	*Thalassarche melanophrys*
Black-chinned hummingbird	*Archilochus alexandri*
Blackpoll warbler	*Dendroica striata*
Black-tailed godwit	*Limosa limosa*
Black-winged kite	*Elanus caeruleus*
Blue grouse	*Dendragapus obscurus*
Blue rock thrush	*Monticola solitarius*
Blue-throated hummingbird	*Lampornis clemenciae*
Brambling	*Fringilla montifringilla*
Brandt's cormorant	*Phalacrocorax penicillatus*
Broad-tailed hummingbird	*Selasphorus platycercus*
Bronzy sunbird	*Nectarinia kilimensis*
Brown-backed needletail	*Hirundapus giganteus*
Budgerigar	*Melopsittacus undulatus*
Calliope hummingbird	*Stellula calliope*
Canada goose	*Branta canadensis*
Cape gannet	*Morus capensis*
Chaffinch	*Fringilla coelebs*
Chicken (red junglefowl)	*Gallus gallus*
Chihuahuan raven	*Corvus cryptoleucus*
Cockatiel	*Nymphicus hollandicus*
Collared sand martin	*Riparia riparia*
Common kestrel	*Falco tinnunculus*
Common magpie	*Pica pica*
Common pheasant	*Phasianus colchicus*
Common quail	*Coturnix coturnix*
Common raven	*Corvus corax*
Common swift	*Apus apus*

Continued

Common teal	*Anas crecca*
Costa's hummingbird	*Calypte costae*
Crag martin	*Ptyonoprogne rupestris*
Downy woodpecker	*Picoides pubescens*
Dunnock	*Prunella modularis*
Eleonora's falcon	*Falco eleonorae*
Eurasian black grouse	*Lyrurus tetrix*
Eurasian blackbird	*Turdus merula*
Eurasian bullfinch	*Pyrrhula pyrrhula*
Eurasian collared dove	*Streptopelia decaocto*
Eurasian jackdaw	*Corvus monedula*
Eurasian siskin	*Carduelis spinus*
Eurasian sparrow hawk	*Accipiter nisus*
Eurasian woodcock	*Scolopax rusticola*
Eurasion griffon	*Gyps fulvus*
European goldfinch	*Carduelis carduelis*
European robin	*Erithacus rubecula*
European shag	*Phalacrocorax aristotelis*
European starling	*Sturnus vulgaris*
European storm petrel	*Hydrobates pelagicus*
European white stork	*Ciconia ciconia*
Evening grossbeak	*Hesperiphona vespertina*
Fish crow	*Corvus ossifragus*
Glittering-bellied emerald	*Chlorostilbon aureoventris*
Glittering-throated emerald	*Amazilia fimbriata*
Goldcrest	*Regulus regulus*
Goliath heron	*Ardea goliath*
Goosander	*Mergus merganser*
Great white pelicans	*Pelecanus onocrotalus*
Greater kestrel	*Falco rupicoloides*
Green sandpiper	*Tringa ochropus*
Green violet-ear	*Colibri thalassinus*
Grey heron	*Ardea cinerea*
Grey partridge	*Perdix perdix*
Grey-headed albatross	*Thalassarche chrysostoma*
Hairy woodpecker	*Picoides villosus*
Hermit thrush	*Catharus guttatus*
Hooded crow	*Corvus corone cornix*
House sparrow	*Passer domesticus*
Indigo-capped hummingbird	*Amazilia cyanifrons*
Laughing gull	*Larus atricilla*
Laysan albatross	*Diomedea immutabilis*
Lewis' woodpecker	*Melanerpes lewis*
Light-mantled albatross	*Phoebetria palpebrata*
Long-tailed hawk	*Urotriorchis macrourus*

Continued

Long-tailed jaeger	*Stercorarius longicaudus*
Long-tailed tyrant	*Colonia colonus*
Magnificent frigatebirds	*Fregata magnificens*
Magnificent hummingbird	*Eugenes fulgens*
Mallard	*Anas platyrhynchos*
Manx shearwater	*Puffinus puffinus*
Marmora's warbler	*Sylvia sarda*
Mistle thrush	*Turdus viscivorus*
Mute swan	*Cygnus olor*
Neotropical cormorant	*Phalacrocorax brasilianus*
Northern flicker	*Colaptes auratus*
Northern fulmar	*Fulmarus glacialis*
Northern gannet	*Morus bassanus*
Northern giant petrel	*Macronectes halli*
Northern goshawk	*Accipiter gentiles*
Northern house martin	*Delichon urbicum*
Northern wheatear	*Oenanthe oenanthe*
Osprey	*Pandion haliaetus*
Ostrich	*Struthio camelus*
Palestine sunbird	*Cinnyris osea*
Pallid swift	*Apus pallidus*
Peregrine falcon	*Falco peregrinus*
Pied kingfisher	*Ceryle rudis*
Pigeon (rock dove)	*Columba livia*
Pileated woodpecker	*Dryocopus pileatus*
Pine siskin	*Carduelis pinus*
Pink-footed geese	*Anser brachyrhynchus*
Purple martin	*Progne subis*
Purple-throated carib	*Eulampis jugularis*
Racquet-tailed puffleg	*Ocreatus underwoodii*
Red-footed booby	*Sula sula*
Red-legged partridge	*Alectoris rufa*
Red-naped sapsucker	*Sphyrapicus nuchalis*
Reed bunting	*Emberiza schoeniclus*
Ring-billed gull	*Larus delawarensis*
Ringed turtle dove (Barbary dove?)	*Streptopelia risoria*
Rock dove (pigeon)	*Columba livia*
Rook	*Corvus frugilegus*
Rosy starling	*Sturnus roseus*
Rough-legged buzzard	*Buteo lagopus*
Ruby-throated hummingbird	*Archilochus colubris*
Rufous hummingbird	*Selasphorus rufus*
Rufous-tailed hummingbird	*Amazilia tzacatl*
Scissor-tailed kite	*Chelictinia riocourii*
Secretary bird	*Sagittarius serpentarius*

Continued

Song sparrow	*Melospiza melodia*
Sooty tern	*Sterna fuscata*
Southern emu wren	*Stipiturus malachurus*
Southern giant petrel	*Macronectes giganteus*
Sparkling violet-ear	*Colibri coruscans*
Spotted flycatcher	*Muscicapa striata*
Swainson's thrush	*Catharus ustulatus*
Sword-billed hummingbird	*Ensifera ensifera*
Tawny eagle	*Aquila rapax*
Tawny owl	*Strix aluco*
Thrush nightingale	*Luscinia luscinia*
Veery	*Catharus fuscescens*
Verdin	*Auriparus flaviceps*
Wandering albatross	*Diomedea exulans*
Western capercaillie	*Tetrao urogallus*
Western Grebe	*Aechmophorus occidentalis*
White-throated needletail	*Hirundapus caudacutus*
Willow grouse	*Lagopus lagopus*
Wilson's storm petrel	*Oceanites oceanicus*
Yellow-legged gull	*Larus cachinnans*
Zebra finch	*Taeniopygia guttata*

Appendix 2. Scientific names of birds

Accipiter gentiles	Northern goshawk
Accipiter nisus	Eurasian sparrow hawk
Aechmophorus occidentalis	Western Grebe
Alectoris rufa	Red-legged partridge
Amazilia amazilia	Amazilia hummingbird
Amazilia cyanifrons	Indigo-capped hummingbird
Amazilia fimbriata	Glittering-throated emerald
Amazilia tzacatl	Rufous-tailed hummingbird
Anas crecca	Common teal
Anas platyrhynchos	Mallard
Anas rubripes	American black duck
Anser brachyrhynchus	Pink-footed geese
Anser indicus	Bar-headed goose
Apus apus	Common swift
Apus pallidus	Pallid swift
Aquila rapax	Tawny eagle
Archilochus alexandri	Black-chinned hummingbird
Archilochus colubris	Ruby-throated hummingbird
Ardea cinerea	Grey heron
Ardea goliath	Goliath heron
Auriparus flaviceps	Verdin
Branta canadensis	Canada goose
Branta leucopsis	Barnacle goose
Buteo lagopus	Rough-legged buzzard
Calypte anna	Anna's hummingbird
Calypte costae	Costa's hummingbird
Carduelis carduelis	European goldfinch
Carduelis pinus	Pine siskin
Carduelis spinus	Eurasian siskin
Catharus fuscescens	Veery
Catharus guttatus	Hermit thrush
Catharus ustulatus	Swainson's thrush
Ceryle rudis	Pied kingfisher
Chelictinia riocourii	Scissor-tailed kite
Chlorostilbon aureoventris	Glittering-bellied emerald
Ciconia ciconia	European white stork
Cinnyris osea	Palestine sunbird
Colaptes auratus	Northern flicker
Colibri coruscans	Sparkling violet-ear
Colibri thalassinus	Green violet-ear
Colonia colonus	Long-tailed tyrant

Continued

Columba livia	Pigeon (rock dove)
Columba livia	Rock dove (pigeon)
Corvus corax	Common raven
Corvus corone cornix	Hooded crow
Corvus cryptoleucus	Chihuahuan raven
Corvus frugilegus	Rook
Corvus monedula	Eurasian jackdaw
Corvus ossifragus	Fish crow
Coturnix coturnix	Common quail
Cygnus olor	Mute swan
Delichon urbicum	Northern house martin
Dendragapus obscurus	Blue grouse
Dendroica striata	Blackpoll warbler
Diomedea exulans	Wandering albatross
Diomedea immutabilis	Laysan albatross
Dryocopus pileatus	Pileated woodpecker
Elanus caeruleus	Black-winged kite
Emberiza schoeniclus	Reed bunting
Ensifera ensifera	Sword-billed hummingbird
Erithacus rubecula	European robin
Eugenes fulgens	Magnificent hummingbird
Eulampis jugularis	Purple-throated carib
Falco eleonorae	Eleonora's falcon
Falco peregrinus	Peregrine falcon
Falco rupicoloides	Greater kestrel
Falco tinnunculus	Common kestrel
Florisuga fusca	Black Jacobin
Fregata magnificens	Magnificent frigatebirds
Fringilla montifringilla	Brambling
Fringilla coelebs	Chaffinch
Fulmarus glacialis	Northern fulmar
Gallus gallus	Chicken (red junglefowl)
Gyps africanus	African white-backed vulture
Gyps fulvus	Eurasion griffon
Hesperiphona vespertina	Evening grossbeak
Hirundapus caudacutus	White-throated needletail
Hirundapus giganteus	Brown-backed needletail
Hirundo rustica	Barn swallow
Hydrobates pelagicus	European storm petrel
Lagopus lagopus	Willow grouse
Lampornis clemenciae	Blue-throated hummingbird
Larus atricilla	Laughing gull
Larus audouinii	Audouin's gull
Larus cachinnans	Yellow-legged gull
Larus delawarensis	Ring-billed gull

Continued

Limosa lapponica	Bar-tailed godwit
Limosa limosa	Black-tailed godwit
Luscinia luscinia	Thrush nightingale
Lyrurus tetrix	Eurasian black grouse
Macronectes giganteus	Southern giant petrel
Macronectes halli	Northern giant petrel
Melanerpes lewis	Lewis' woodpecker
Melopsittacus undulatus	Budgerigar
Melospiza melodia	Song sparrow
Mergus merganser	Goosander
Monticola solitarius	Blue rock thrush
Morus bassanus	Northern gannet
Morus capensis	Cape gannet
Muscicapa striata	Spotted flycatcher
Nectarinia kilimensis	Bronzy sunbird
Nymphicus hollandicus	Cockatiel
Oceanites oceanicus	Wilson's storm petrel
Ocreatus underwoodii	Racquet-tailed puffleg
Oenanthe oenanthe	Northern wheatear
Pandion haliaetus	Osprey
Passer domesticus	House sparrow
Pelecanus onocrotalus	Great white pelicans
Perdix perdix	Grey partridge
Phalacrocorax aristotelis	European shag
Phalacrocorax brasilianus	Neotropical cormorant
Phalacrocorax penicillatus	Brandt's cormorant
Phasianus colchicus	Common pheasant
Phoebetria palpebrata	Light-mantled albatross
Pica pica	Common magpie
Picoides pubescens	Downy woodpecker
Picoides villosus	Hairy woodpecker
Progne subis	Purple martin
Prunella modularis	Dunnock
Ptyonoprogne rupestris	Crag martin
Puffinus puffinus	Manx shearwater
Pyrrhula pyrrhula	Eurasian bullfinch
Regulus regulus	Goldcrest
Riparia riparia	Collared sand martin
Sagittarius serpentarius	Secretary bird
Scolopax rusticola	Eurasian woodcock
Selasphorus platycercus	Broad-tailed hummingbird
Selasphorus rufus	Rufous hummingbird
Selasphorus sasin	Allen's hummingbird
Sphyrapicus nuchalis	Red-naped sapsucker
Stellula calliope	Calliope hummingbird

Continued

Stercorarius longicaudus	Long-tailed jaeger
Sterna fuscata	Sooty tern
Stipiturus malachurus	Southern emu wren
Streptopelia decaocto	Eurasian collared dove
Streptopelia risoria	Ringed turtle dove (Barbary dove?)
Strix aluco	Tawny owl
Struthio camelus	Ostrich
Sturnus roseus	Rosy starling
Sturnus vulgaris	European starling
Sula sula	Red-footed booby
Sylvia sarda	Marmora's warbler
Tachymarptis melba	Alpine swift
Taeniopygia guttata	Zebra finch
Tetrao urogallus	Western capercaillie
Thalassarche melanophrys	Black-browed albatross
Thalassarche chrysostoma	Grey-headed albatross
Tringa ochropus	Green sandpiper
Turdus merula	Eurasian blackbird
Turdus viscivorus	Mistle thrush
Tyto alba	Barn owl
Urotriorchis macrourus	Long-tailed hawk

Appendix 3. Explanations of scientific terms used in the text

Scientific term	English equivalent
Skeleton	
Caput humeri	Apex of the upper arm bone
Carina	Central keel on breast bone
Carpometacarpus	Fused carpals and metacarpals
Condyle	Rounded articulating end of a bone
Coracoid	Ventral breast bone (crow's beak bone)
Deltopectoral crest	Process on the proximal front part of the humerus
Furcula	Wishbone, united clavicles of birds
Glennoid cavity	Socket of a joint
Humerus	Upper arm bone
Olecranon	Large bony process on upper end of ulna
Pygostyle	Tail bone
Radiale	Wrist bone in line with the radius
Radialis	12th Spinal nerve
Radius	Outer forearm bone
Recticial bulbs	Fibro adipose structures in the tail
Scapula	The shoulder blade
Sesamoid bone	Bone developed within a tendon
Sternum	Breast bone
Synsacrum	Fused sacral vertebrae
Triosseal canal	Opening in the shoulder joint
Ulna	Inner forearm bone
Ulnare	Wrist bone in line with the ulna
Muscles	
Pectoralis	Major flight muscle (downstroke)
Supracoracoideus	Minor flight muscle (upstroke)
Biceps brachii Coracobrachialis caudalis Deltoideus major (cranial and caudal) Humerotriceps Scapulohumeralis caudalis Scapulotriceps Sternocoracoideus	Shoulder muscles: see Figs. 7.5 and 7.6

Continued

Scientific term	English equivalent
Bulbi recticium	Tail muscles: see Figs. 2.9 and 7.9
Caudofemoralis	
Depressor caudae	
Iliotrochantericus caudalis	
Iliotrochantericus caudalis	
Levator caudae	
Longissimus dorsi	
Subscapularis	
Pubocaudalis	
Feathers	
Alula	The bastard wing
Rectrices	Stiff tail feathers
Remiges	Primary wing feathers
Tectrices	Wing coverts
Feather structure	
Barbs	Interconnecting side branches of the shaft forming the vanes
Barbules	Interconnecting lateral projections from the barbs
Calamus	Proximal part of the shaft
Pennaceous	Barbs well structured and firmly interconnected by barbules
Pennulum	Delicate distal-most extensions of barbules
Plumulaceous	Barbs loosely or not at all interconnected
Rachis	Distal part of the shaft
Ramus	Lower part of the barbs
Tegmen	Plate-like extension of ventral ridges of barbs
Movements and directions	
Anterior	Towards the front
Caudal	Towards the tail end
Cranial	Towards the head end
Distal	Away from the body
Dorsal	Related to the upper side of an animal
Lateral	Situated at the side
Medial	Situated in the middle
Posterior	Towards the rear

Continued

Scientific term	English equivalent
Pronation	Forward rotation
Proximal	Close to the body
Supination	Rearward rotation
Ventral	Related to the underside of an animal

References

Jane's All the world's aircraft. 1989, 1–108.

Henderson's dictionary of biological terms. 1996, 11th edn, Longman, Singapore.

New perspectives on the origin and early evolution of birds. 2001, J. Gauthier and L. F. Gall, eds., Peabody Museum of Natural History, Yale University, New Haven, CT, pp. 1–613.

Mesozoic birds: above the heads of dinosaurs. 2002, L. M. Chiappe and L. M. Witmer, eds., The University of California Press, Berkeley and Los Angeles, CA.

The Howard and Moore complete checklist of the birds of the world. 2003, third edn, E. C. Dickinson ed., Christopher Helm, London.

Adams, N. J., Brown, C. R., and Nagy, K. A. 1986, 'Energy expenditure of free-ranging wandering albatrosses *Diomedea exulans*'. *Physiological Zoology*, 59(6), 583–591.

Adams, N. J., Abrams, R. W., Siegfried, W. R., Nagy, K. A., and Kaplan, I. R. 1991, 'Energy expenditure and food consumption by breeding Cape gannets *Morus capensis*'. *Marine Ecology Progress Series*, 70, 1–9.

Alerstam, T., Gudmundsson, G. A., and Larsson, B. 1993, 'Flight tracks and speeds of Antartic and Atlantic sea birds: radar and optical measurements'. *Philosophical Transactions of the Royal Society London B*, 340, 55–67.

Alexander, R. M. 1976, 'Estimates of speeds of dinosaurs'. *Nature*, 261, 129–130.

Alexander, R. M. 1992, *Exploring biomechanics: animals in motion*, Scientific American Library, New York.

Alexander, R. M. 1997, 'The U, J and L of bird flight'. *Nature*, 390, 13.

Anderson, J. D. jr. 1997, *A history of aerodynamics and its impact on flying machines*. Cambridge University Press, Cambridge.

Anderson, D. F. and Eberhardt, S. 2001, *Understanding flight*, McGraw-Hill, New York.

Arnould, J. P. Y., Briggs, D. R., Croxall, J. P., Prince, P. A., and Wood, A. G. 1996, 'The foraging behaviour and energetics of wandering albatrosses brooding chicks'. *Antarctic Science*, 8(3), 229–236.

Aschoff, J. and Pohl, H. 1970, 'Der Ruheumsatz von Vögeln als Funktion der Tageszeit und der Körpergrösze'. *Journal für Ornithologie*, 111(1), 38–47.

Ashill, P. R., Riddle, G. L., and Stanley, M. J. 1995, 'Separation control on highly-swept wings with fixed or variable camber'. *Aeronautical Journal*, October, 317–327.

Bäckman, J. and Alerstam, T. 2002, 'Harmonic oscillatory orientation relative to the wind in nocturnal roosting flights of the swift *Apus apus*'. *Journal of Experimental Biology*, 205, 905–910.

Balda, R. P., Caple, G., and Willis, W. R. 1985, 'Comparison of the gliding to flapping sequence with the flapping to gliding sequence.', in *The beginnings of birds*, M. K. Hecht, Ostrom, J. H., Viohl, G., and Wellnhofer, P., eds., Freunde des Jura-Museums, Eichstätt, pp. 267–277.

Ballance, L. T. 1995, 'Flight energetics of free-ranging red-footed boobies (Sula sula)'. *Physiological Zoology*, 68(5), 887–914.

Barnard, R. H. and Philpott, D. R. 1997, *Aircraft flight*, second edn, Longman, Harlow.

Barnes, J. 1991, *The complete works of Aristotle*, fourth edn, Princeton University Press, Princeton, NJ.

Bartholomew, G. A. and Lighton, J. R. B. 1986, 'Oxygen consumption during hover-feeding in free ranging Anna hummingbirds'. *Journal of Experimental Biology*, 123, 191–199.

Baumel, J. J. 1979, *Nomina anatomica avium*, Academic Press, London.

Baumel, J. J. 1988, 'Functional morphology of the tail apparatus of the pigeon (*Columbia livia*)'. *Advances in Anatomy, Embryology, and Cell Biology*, 110, 1–115.

Beebe, C. W. 1915, 'A tetrapteryx stage in the ancestry of birds'. *Zoologica*, 2, 39–52.

Beekman, J. H., Hollander, H. J. d., and Koffijberg, K. 1994, *Landsat satellite images for detection of submerged macrophytes: in search of potential stop-over sites for Bewick's Swans along their migratory route between arctic Russia and western Europe*, Ministry of Transport, Public Works and Water Management, Lelystad, RBA 1994-17.

Berg, C. v. d. and Rayner, J. M. V. 1995, 'The moment of inertia of bird wings and the inertial power requirement for flapping flight'. *Journal of Experimental Biology*, 198(8), 1655–1664.

Berg, C. v. d. and Ellington, C. P. 1997, 'The vortex wake of a 'hovering' model hawkmoth'. *Philosophical Transactions of the Royal Society London B*, 352, 317–328.

Berger, M. 1985, 'Sauerstoffverbrauch von Kolibris (*Colibri coruscans* und *Colibri Thalassinus*) beim Horizontalflug', in *Bird flight – Vogelflug*, BIONA-report, third edn, W. Nachtigall, ed., Gustav Fischer, Stuttgart and New York, pp. 307–314.

Berger, M. and Hart, J. S. 1972, 'Die Atmung beim Kolibri *Amazilia fimbriata* wärhend des Schwirrfluges bei verschiedenen Umgebungstemperaturen'. *Journal of Comparative Physiology*, 81, 363–380.

Berger, M., Hart, J. S., and Roy, O. Z. 1970, 'Respiration, oxygen consumption and heart rate in some birds during rest and flight'. *Zeitschrift für Vergleichende Physiolgie*, 66, 201–214.

Bernoulli, D. 1738, *Hydrodynamica*, J.R. Dulsecker, Strasbourg.

Bernstein, M. H., Thomas, S. P., and Schmidt-Nielsen, K. 1973, 'Power input during flight in the Fish Crow, *Corvus ossifragus*'. *Journal of Experimental Biology*, 58, 401–410.

Bevan, R. M., Woakes, A. J., Butler, P. J., and Boyd, I. L. 1994, 'The use of heart rate to estimate oxygen consumption of free-ranging black-browed albatrosses *Diomedea melanophrys*'. *Journal of Experimental Biology*, 193, 119–137.

Bevan, R. M., Butler, P. J., Woakes, A. J., and Prince, P. A. 1995, 'The energy expenditure of free-ranging black-browed albatrosses'. *Philosophical Transactions of the Royal Society London B*, 350, 119–131.

Biewener, A. A., Dial, K. P., and Goslow Jr, G. E. 1992, 'Pectoralis muscle force and power output during flight in the starling'. *Journal of Experimental Biology*, 164, 1–18.

Biewener, A. A., Corning, W. R., and Tobalske, B. W. 1998, 'In vivo pectoralis muscle force-length behavior during level flight in pigeons (*Columba livia*)'. *Journal of Experimental Biology*, 201, 3293–3307.

Bilo, D. 1971, 'Flugbiophysik von Kleinvögeln: I. Kinematik und Aerodynamik des Flügelabschlages beim Haussperling (*Passer domesticus* L.)'. *Zeitschrift für Vergleichende Physiologie*, 71, 382–454.

Bilo, D. 1972, 'Flugbiophysik von Kleinvögeln: II. Kinematik und Aerodynamik des Flügelaufschlages beim Haussperling (*Passer domesticus* L.)'. *Zeitschrift für Vergleichende Physiologie*, 76, 426–437.

Bilo, D. 1980, 'Kinematical peculiarities of the downstroke of a house sparrow's wing calling in question the applicability of steady state aerodynamics to the flapping flight of small Passeriformes', in *Instationäre Effekte an schwingenden Tierflügeln*, sixth edn, W. Nachtigall, ed., Akademie der Wissenschaften und der Literatur., Mainz, pp. 102–114.

Birt-Friesen, V. L., Montevecchi, W. A., Cairns, D. K., and Macko, S. A. 1989, 'Activity-specific metabolic rates of free-living Northern Gannets and other seabirds'. *Ecology*, 70, 357–367.

Blick, E. F., Watson, D., Belie, G., and Chu, H. 1975, 'Bird aerodynamic experiments', in *Swimming and flying in nature*, vol. 2, T. Y. T. Wu, C. J. Brokaw and C. Brennen, eds., Plenum Press, New York, pp. 939–952.

Boel, M. 1929, 'Scientific studies of natural flight'. *Transactions of the American Society of Mechanical Engineers*, 51, 217–242.

Boggs, D. F., Jenkins Jr, F. A., and Dial, K. P. 1997*a*, 'The effects of the wing beat cycle on respiration in black-billed magpies (*Pica pica*)'. *Journal of Experimental Biology*, 200(9), 1403–1412.

Boggs, D. F., Seveyka, J. J., Kilgore, D. L., and Dial, K. P. 1997*b*, 'Coordination of respiratory cycles with wing beat cycles in the black-billed magpie (*Pica pica*)'. *Journal of Experimental Biology*, 200(9), 1413–1420.

Bonser, R. H. C. 1995, 'Melanin and the abrasion resistance of feathers'. *The Condor*, 97, 590–591.

Bonser, R. H. C. and Purslow, P. P. 1995, 'The Young's modulus of feather keratin'. *Journal of Experimental Biology*, 198(4), 1029–1033.

Bonser, R. H. C. 1996, 'Comparative mechanics of bill, claw and feather keratin in the common starling, *Sturnus vulgaris*'. *Journal of Avian Biology*, 27(2), 175–177.

Bonser, R. H. C. and Rayner, J. M. V. 1996, 'Measuring leg thrust forces in the common starling'. *Journal of Experimental Biology*, 199, 435–439.

Bonser, R. H. C., Norman, A. P., and Rayner, J. M. V. 1999, 'Does substrate quality influence take-off decisions in common starlings?' *Functional ecology*, 13, 102–105.

Bonser, R. H. C., Saker, L., and Jeronimidis, G. 2004, 'Toughness anisotropy in feather keratin'. *Journal of Materials Science*, 39, 2895–2896.

Borelli, A. 1680, *De motu animalium*, Angeli Bernabo, Rome.

Brown, C. R. and Adams, N. J. 1984, 'Basal metabolic rate and energy expenditure during incubation in the wandering albatross (Diomedea exulans)'. *The Condor*, 86, 182–186.

Brown, R. E. and Fedde, M. R. 1993, 'Airflow sensors in the avian wing'. *Journal of Experimental Biology*, 179, 13–30.

Brown, R. E. and Cogley, A. C. 1996, 'Contributions of the propatagium to avian flight'. *Journal of Experimental Zoology*, 276, 112–124.

Bruderer, B. and Boldt, A. 2001, 'Flight characteristics of birds: I. Radar measurements of speeds'. *Ibis*, 143, 178–204.

Bryant, D. M. and Westerterp, K. R. 1980, The energy budget of the house martin *Delichon urbica*. *Ardea*, 68, 91–102.

Buisonjé, P. H. d. 1985, 'Climatological conditions during deposition of the Solnhofen limestones.', in *The beginnings of birds*, M. K. Hecht, Ostrom, J. H., Viohl, G., and Wellnhofer, P., eds., Freunde des Jura-Museums, Eichstätt, pp. 45–65.

Butler, M. and Johnson, A. S. 2004, 'Are melanized feather barbs stronger?' *Journal of Experimental Biology*, 207, 285–293.

Butler, P. J., West, N. H., and Jones, D. R. 1977, 'Respiratory and cardiovascular responses of the pigeon to sustained level flight in a windtunnel'. *Journal of Experimental Biology*, 71, 7–26.

Butler, P. J., Woakes, A. J., and Bishop, C. M. 1998, 'Behaviour and physiology of Svalbard barnacle geese, *Branta leucopsis*, during their autumn migration'. *Journal of Avian Biology*, 29, 536–545.

Cameron, G. J., Wess, T. J., and Bonser, R. H. C. 2003, 'Young's modulus varies with differential orientation of keratin in feathers'. *Journal of Structural Biology*, 143, 118–123.

Carpenter, R. E. 1975, 'Flight metabolism in flying foxes', in *Swimming and flying in nature*, T. Y. Wu, C. J. Brokaw, and C. Brennen, eds., Plenum, New York, pp. 883–889.

Carpenter, R. E. 1985, 'Flight physiology of flying foxes, *Pteropus poliocephalus*'. *Journal of Experimental Biology*, 114, 619–647.

Carpenter, R. E. 1986, 'Flight physiology of intermediate-sized fruit bats (Pteropodidae)'. *Journal of Experimental Biology*, 120, 79–103.

Cavé, A. J. 1968, 'The breeding of the Kestrel (*Falco tinnunculus* L.) in the reclaimed area Oostelijk Flevoland'. *Netherlands Journal of Zoology*, 18, 313–407.

Cayley, G. 1809, 'On aerial navigation part I'. *A Journal of Natural Philosophy, Chemistry, and the Arts*, 24(November), 164–174.

Cayley, G. 1810, 'On aerial navigation part II'. *A Journal of Natural Philosophy, Chemistry, and the Arts*, 25(February), 81–87.

Chai, P. and Dudley, R. 1995, 'Limits to vertebrate locomotor energetics suggested by hummingbirds hovering in heliox'. *Nature*, 377, 722–725.

Chai, P. and Millard, D. 1997, 'Flight and size constraints: hovering performance of large hummingbirds under maximal loading'. *Journal of Experimental Biology*, 200(21), 2757–2763.

Chai, P. and Dudley, R. 1999, 'Maximum flight performance of hummingbirds: capacities, constraints, and trade-offs'. *The American Naturalist*, 153(4), 398–411.

Chai, P., Chen, J. S. C., and Dudley, R. 1997, 'Transient hovering performance of hummingbirds under conditions of maximal loading'. *Journal of Experimental Biology*, 200(5), 921–929.

Chatterjee, S. 1997, *The rise of birds*, The Johns Hopkins University Press, Baltimore, MD.

Chen, P., Dong, Z., and Zhen, S. 1998, 'An exceptionally well-preserved theropod dinosaur from the Yixian Formation of China'. *Nature*, 391, 147–152.

Corning, W. R. and Biewener, A. A. 1998, '*In vivo* strains in pigeon flight feather shafts: implications for structural design'. *Journal of Experimental Biology*, 201(22), 3057–3065.

Costa, D. P. and Prince, P. A. 1987, 'Foraging energetics of Grey-headed Albatrosses *Diomedea chrysostoma* at Bird Island, South Georgia'. *Ibis*, 129, 149–158.

Cutts, C. J. and Speakman, J. R. 1994, 'Energy savings in formation flight of pink-footed geese'. *Journal of Experimental Biology*, 189, 251–261.

Daan, S., Masman, D., and Groenewold, A. 1990, 'Avian basal metabolic rates: their association with body composition and energy expenditure in nature'. *American Journal of Physiology*, 259, R333–R340.

Del Hoyo, J., Elliot, A. and Sargatal, J. eds. 1992, *Handbook of the brids of the world*, 1. Ostrich to ducks. Lynx Edicions, Barcelona.

Del Hoyo, J., Elliot, A. and Sargatal, J. eds. 1999, *Handbook of the birds of the world*, 5. Barn-owls to hummingbirds Lynx Edicions, Barcelona.

Deventer, R. W. v. 1983, '*Basiliscus basiliscus* (Chisbala, Garrobo, Basilisk, Jesus Lizard)', in *Costa Rican Natural History*, D. H. Janzen, ed., University of Chicago Press, Chicago, pp. 379–380.

Dial, K. P. 1992, 'Avian forelimb muscles and nonsteady flight: Can birds fly without using the muscles in their wings?' *The Auk*, 109(4), 874–885.

Dial, K. P. and Biewener, A. A. 1993, 'Pectoralis muscle force and power output during different modes of flight in pigeons (*Columba livia*)'. *Journal of Experimental Biology*, 176, 31–54.

Dial, K. P., Goslow Jr, G. E., and Jenkins Jr, F. A. 1991, 'The functional anatomy of the shoulder in the european starling *Sturnus vulgaris*'. *Journal of Morphology*, 207, 327–344.

Dial, K. P., Biewener, A. A., Tobalske, B. W., and Warrick, D. R. 1997, 'Mechanical power output of bird flight'. *Nature*, 390, 67–70.

Dickinson, M. H., Lehmann, F. O., and Sane, S. P. 1999, 'Wing rotation and the aerodynamic basis of insect flight'. *Science*, 284, 1954–1960.

Dingus, L. and Rowe, T. 1998, *The mistaken extinction: dinosaur evolution and the origin of birds*, W.H. Freeman, New York.

Dolnik, V. R. and Blyumental, T. I. 1967, 'Autumnal premigratory and migratory periods in the Chaffinch, *Fringilla coelebs* and some other temperate zone passerine birds'. *The Condor*, 69, 435–468.

Dolnik, V. R. and Gavrilov, V. M. 1973, 'Energy metabolism during flight of some passerines', in *Bird migrations: ecological and physiological factors*, B. E. Byikhovskii, ed., John Wiley & Sons, New York, pp. 288–296.

Duncker, H. R. and Güntert, M. 1985, 'The quantitative design of the avian respiratory system – from hummingbird to mute swan', in *Bird flight – Vogelflug*, BIONA-report, third edn, W. Nachtigall, ed., Gustav Fischer, Stuttgart and New York, pp. 361–378.

Dyck, J. 1985, 'The evolution of feathers'. *Zoologica Scripta*, 14(2), 137–154.

Ellington, C. P. 1984, 'The aerodynamics of hovering insect flight I. The quasi-steady analysis II. Morphological parameters III. Kinematics IV. Aerodynamic mechanisms V. A vortex theory VI. Lift and power requirements'. *Philosophical Transactions of the Royal Society London B*, 305(1122), 1–181.

Ellington, C. P., Machin, K. E., and Casey, T. M. 1990, 'Oxygen consumption of bumblebees in forward flight'. *Nature*, 347, 472–473.

Ellington, C. P., Berg, C. v. d., Willmott, A. P., and Thomas, A. L. R. 1996, 'Leading-edge vortices in insect flight'. *Nature*, 384, 626–630.

Ellis, H. I. 1984, 'Energetics of the free-ranging seabirds.', in *Seabird energetics*, G. C. Whittow and H. Rahn, eds., Plenum Press, New York, 203–234.

Elzanowski, A. 2002, 'Archaeopterygidae (upper Jurassic of Germany)', in *Mesozoic birds: above the heads of dinosaurs*, L. M. Chiappe and L. M. Witmer, eds., University of California Press, Berkeley and Los Angeles, CA, pp. 129–159.

Ennos, A. R., Hickson, J. R. E., and Roberts, A. 1995, 'Functional morphology of the vanes of the flight feathers of the pigeon *Columbia livia*'. *Journal of Experimental Biology*, 198(5), 1219–1228.

Epting, R. J. 1980, 'Functional dependence of the power for hovering on wing disc loading in hummingbirds'. *Physiological Zoology*, 53(4), 347–357.

Feduccia, A. 1993, 'Evidence from claw geometry indicating arboreal habits of Archaeopteryx'. *Science*, 259(5096), 790–793.

Feduccia, A. 1999, *The origin and evolution of birds*, second edn, Yale University Press, New Haven, CT.

Flint, E. N. and Nagy, K. A. 1984, 'Flight energetics of free-living sooty terns'. *The Auk*, 101, 288–294.

Gatesy, S. M. and Dial, K. P. 1993, 'Tail muscle activity patterns in walking and flying pigeons (*Columbia livia*)'. *Journal of Experimental Biology*, 176, 55–76.

Gavrilov, V. M. and Dolnik, V. R. 1985 'Basal metabolic rate, thermoregulation and existence energy in birds: World data'. Moscow, pp. 421–466.

George, J. C. and Berger, A. J. 1966, *Avian myology*, Academic Press, New York.

Gibbs-Smith, C. H. 1962, *Sir George Cayley's Aeronautics 1796–1855*, Her Majesty's Stationary Office, London.

Glasheen, J. W. and McMahon, T. A. 1996*a*, 'Size-dependence of water-running ability in basilisk lizards (*Basiliscus basiliscus*)'. *Journal of Experimental Biology*, 199, 2611–2618.

Glasheen, J. W. and McMahon, T. A. 1996*b*, 'A hydrodynamic model of locomotion in the Basilisk Lizard'. *Nature*, 380, 340–342.

Glasier, P. 1982, *Falconry and hawking*, Batsford, B.T. Ltd., London.

Goslow Jr, G. E. and Dial, K. P. 1990, 'Active stretch-shorten contractions of the m. pectoralis in the European starling (*Sturnus vulgaris*): evidence from electromyography and contractile properties'. *Netherlands Journal of Zoology*, 40(1–2), 106–114.

Goslow Jr, G. E., Dial, K. P., and Jenkins Jr, F. A. 1989, 'The avian shoulder: an experimental approach'. *American Zoologist*, 29, 287–301.

Graber, R. R. and Graber, J. W. 1962, 'Weight characteristics of birds killed in nocturnal migration'. *The Wilson Bulletin*, 74, 244–253.

Graham, R. R. 1931, 'Safety devices in wings of birds'. *British birds*, 24, 2–21.

Grant, G. S. and Whittow, G. C. 1983, 'Metabolic cost of incubation in the Laysan albatross and Bonin petrel.', *Comparative Biochemistry and Physiology*, 74A, 77–82.

Greenewalt, C. H. 1975, 'The flight of birds'. *Transactions of American Philosophical Society*, 65(4), 1–67.

Griffiths, P. J. 1996, 'The isolated *Archaeopteryx* feather'. *Archaeopteryx*, 14, 1–26.

Hails, C. J. 1979, 'A comparison of flight energetics in hirundines and other birds'. *Comparative Biochemistry and Physiology*, 63A, 581–585.

Hainsworth, F. R. and Wolf, L. L. 1969, 'Resting, torpid, and flight metabolism of the hummingbird, *Eulampis jugularis*'. *American Zoologist*, 9, 1100–1101.

Hambly, C., Harper, E. J., and Speakman, J. R. 2002, 'Cost of flight in the zebra finch (*Taenopygia guttata*): a novel approach based on elimination of ^{13}C labelled bicarbonate'. *Journal of Comparative Physiology B*, 172, 529–539.

Hambly, C., Harper, E. J., and Speakman, J. R. 2004, 'The energy cost of variations in wing span and wing asymmetry in the zebra finch *Taeniopygia guttata*'. *Journal of Experimental Biology*, 207, 3977–3984.

Hammond, K. A. and Diamond, J. 1997, 'Maximal sustained energy budgets in humans and animals.', *Nature*, 386, 457–462.

Hart, I. B. 1963, *The mechanical investigations of Leonardo da Vinci*, University of California Press, Berkeley, CA.

Hector, J. 1894, 'On the anatomy of flight of certain birds'. *Transations of the New Zealand Institute*, 27, 285–287.

Hedenström, A. and Rosén, M. 2003, 'Body frontal area in passerine birds'. *Journal of Avian Biology*, 34, 159–162.

Hedrick, T. L., Tobalske, B. W., and Biewener, A. A. 2003, 'How cockatiels (*Nymphicus hollandicus*) modulate pectoralis power output across flight speeds'. *Journal of Experimental Biology*, 206, 1363–1378.

Heppner, F. H. and Anderson, J. G. T. 1985, 'Leg thrust in flight take-off in the pigeon'. *Journal of Experimental Biology*, 114, 285–288.

Hertel, H. 1966, *Structure, Form and Movement*, Van Rostrand-Reinhold, New York.

Herzog, K. 1968, *Anatomie und Flugbiologie der Vögel*, Gustav Fischer Verlag, Stuttgart.

Hocking, B. 1953, 'On the intrinsic range and speed of flight of insects.', *Transactions of the Royal Entomological Society London*, 104, 223–345.

Hoerner, S. F. and Borst, H. V. 1975, *Fluid Dynamic Lift* Hoerner Fluid Dynamics, Brick Town, N. J.

Holmes, K. C. 1998, 'A powerful stroke'. *Nature Structural Biology*, 5(11), 940–942.

Holst, E. v. 1943, 'Über 'künstliche Vögel' als Mittel zum Studium des Vogelfluges'. *Journal für Ornithologie*, 91, 406–447.

Homberger, D. G. and Silva, de. K. N. 2000, 'Functional microanatomy of the feather-bearing integument: implications for the evolution of birds and avian flight'. *American Zoologist*, 40, 553–574.

Hou, L. 2001, *Mesozoic birds of China*, Phoenix Valley Provincial Aviary, Taiwan.

Hou, L., Zhou, Z., Martin, L. D., and Feduccia, A. 1995, 'A beaked bird from the Juassic of China'. *Nature*, 377, 616–618.

Hou, L., Martin, L. D., Zhou, Z., and Feduccia, A. 1996, 'Early adaptive radiation of birds: evidence from fossils from Notheastern China'. *Science*, 274, 1164–1167.

Hudson, D. M. and Bernstein, M. H. 1983, 'Gas exchange and energy cost of flight in the White-Necked Raven, *Corvus cryptoleucos*'. *Journal of Experimental Biology*, 103, 121–130.

Hummel, D. 1995, 'Formation flight as an energy-saving mechanism'. *Israel Journal of Zoology*, 41, 261–278.

Hummel, D. and Möllenstädt, W. 1977, 'On the calculation of the aerodynamic forces acting on a house sparrow (*Passer domesticus* L.) during downstroke by means of aerodynamic theory'. *Fortschritte der Zoologie*, 24(2/3), 235–256.

Hussel, D. J. T. 1969, 'Weight loss of birds during nocturnal migration'. *The Auk*, 86, 75–83.

Hussel, D. J. T. and Lambert, A. B. 1980, 'New estimates of weight loss in birds during migration'. *The Auk*, 97, 547–558.

Jenkins Jr, F. A., Dial, K. P., and Goslow Jr, G. E. 1988, 'A cineradiographic analysis of bird flight: the wishbone in starlings is a spring'. *Science*, 241, 1495–1498.

Josephson, R. K. 1985, 'Mechanical power output from striated muscle during cyclic contraction'. *Journal of Experimental Biology*, 114, 493–512.

Joudine, K. 1955, 'A propos du mécanisme fixant l'articulation du coude chez certains oiseaux (Tubinares)', *Proceedings of the 9th International Ornithological Congress*, pp. 279–283.

Kespaik, J. 1968, 'Heat production and heat loss of swallows and martins during flight'. *Eesti Nsv teaduste Akadeemia toimetised. XVII köide Biol.*, 12, 179–190.

Klaassen, M., Kvist, A., and Lindström, Å. 2000, 'Flight costs and fuel composition of a bird migrating in a wind tunnel'. *Condor*, 102(2), 444–451.

Kokshaysky, N. V. 1979, 'Tracing the wake of a flying bird'. *Nature*, 279, 146–148.

Kvist, A., Klaassen, M., and Lindström, Å. 1998, 'Energy expenditure in relation to flight speed: What is the power of mass loss estimates?' *Journal of Avian Biology*, 29, 485–498.

Kvist, A., Lindström, Å., Green, M., Piersma, T., and Visser, G. H. 2001, 'Carrying large fuel loads during sustained bird flight is cheaper than expected'. *Nature*, 413, 730–732.

Lack, D. 1956, *Swifts in a tower*, Chapman and Hall, London.

Laerm, J. 1974, 'A functional analysis of morphological variation and differential niche utilization in basilisk lizards'. *Ecology*, 55, 404–411.

Lasiewski, R. C. 1963, 'Oxygen consumption for torpid, resting, active and flying hummingbirds'. *Physiological Zoology*, 36, 122–140.

LeFebvre, E. A. 1964, 'The use of D_2O^{18} for measuring energy metabolism in *Columbia livia* at rest and in flight'. *The Auk*, 81, 403–416.

Lifson, N. and McClintock, R. 1966, 'Theory of use of the turnover rates of body water for measuring energy and material balance'. *Journal of Theoretical Biology*, 12, 46–74.

Lighthill, J. 1990, *An informal introduction to theoretical fluid mechanics*, Oxford University Press, Oxford.

Lilienthal, O. 1889, *Der Vogelflug als Grundlage der Fliegekunst*, third edn, R. Oldenbourg, Munchen und Berlin.

Lindström, Å. and Piersma, T. 1993, 'Mass changes in migrating birds: the evidence for fat and protein storage re-examined'. *Ibis*, 135, 70–78.

Lingham-Soliar, T. 2003, 'Evolution of birds: ichthyosaur integumental fibers conform to dromaeosaur protofeathers'. *Naturwissenschaften*, 90, 428–432.

Liu, H., Ellington, C. P., Kawachi, K., Berg, C. v. d., and Willmott, A. P. 1998, 'A computational fluid dynamic study of hawkmoth hovering'. *Journal of Experimental Biology*, 201, 461–477.

Lowson, M. V. and Riley, A. J. 1995, 'Vortex breakdown control by delta wing geometry'. *Journal of Aircraft*, 32, 832–838.

Lucas, A. M. and Stettenheim, P. R. 1972, *Avian Anatomy, Integument*, Agriculture handbook 362, U.S. Department of Agriculture, Washington, D.C.

Lyuleeva, D. S. 1970, 'Energy of flight in swallows and swifts'. *Doklady Akademii Nauk SSSR*, 190, 1467–1469.

Lyuleeva, D. S. 1973, 'Features of swallow biology during migration', in *Bird migrations, ecological and physiological factors*, B. E. Byikhovskii, ed., Halstead Press, New York, pp. 56–69.

Ma, M. S., Ma, W. K., Nieuwland, I., and Easley, R. R. 2002, 'Why *Archaeopteryx* did not run over water'. *Archaeopteryx*, 20, 51–56.

Machin, K. E. and Pringle, J. W. S. 1960, 'The physiology of insect fibrillar muscle: III. The effects of sinusoidal changes of length on a beetle flight muscle'. *Proceedings of the Royal Society of London B*, 152, 311–330.

Magnan, A., Perrilliat-Botonet, G., and Girerd, H. 1938, 'Essais d'enregistrements cinématographiques simultanées dans trois directions perpendiculaires deux à deux à l'écoulement de l'air autour d'un oiseau en', *Comptes Rendus Hebdomadaires Séances de l'Academie des Sciences, Paris*, 206, 374–377.

Maquet, P. 1989, *On the movement of animals*, Springer Verlag, Berlin.

Marey, E. J. 1890, *Le vol des oiseaux*, Masson, G., Paris.

Martin, L. D., Zhou, Z., Hou, L., and Feduccia, A. 1998, '*Confuciusornis sanctus* compared to *Archaeopteryx lithographica*.' *Naturwissenschaften*, 85, 286–289.

Mascha, E. 1904, 'Über die Schwungfedern'. *Zeitschrift für Wissenschaftliche Zoologie*, 77, 606–651.

Masman, D. and Klaassen, M. 1987, 'Energy expenditure during free flight in trained and free-living eurasian kestrels (*Falco tinnunculus*)'. *The Auk*, 104, 603–616.

Masman, D., Gordijn, M., Daan, S., and Dijkstra, C. 1986, 'Ecological energetics of the kestrel: field estimates of energy intake throughout the year'. *Ardea*, 74, 24–39.

Maxworthy, T. 1979, 'Experiments on the Weis-Fogh mechanism of lift generation by insects in hovering flight. Part 1. Dynamics of the "fling" '. *Journal of Fluid Mechanics*, 93(1), 47–63.

Mayr, E. 1963, *Animal species and evolution*, Harvard University Press, Cambridge, MA.

Mayr, G. Pohl, B. and Peters, D.S. 2005, 'A well-preserved Archaeopteryx specimen with theropod features'. *Science*, 310, 1483–1486.

Motte, A. 1729, *Sir Isaac Newton: the mathematical principles of natural philosophy*, B. Motte, London.

Müller, W. and Patone, G. 1998, 'Air transmissivity of feathers'. *Journal of Experimental Biology*, 201, 2591–2599.

Nachtigall, W. and Kempf, B. 1971, 'Vergleichende Untersuchungen zur flugbiologischen Funktion des Daumenfittichs (Alula spuria) bei Vögeln I. Der Daumenfittich als Hochauftriebserzeuger'. *Zeitschrift für Vergleichende Physiologie*, 71, 326–341.

Nachtigall, W., Rothe, U., Feller, P., and Jungmann, R. 1989, 'Flight of the honeybee III. Flight metabolic power calculated from gas analysis, thermoregulation and fuel consumption'. *Journal of Comparative Physiology*, 158, 729–737.

Nagy, K. A. 1980, 'CO_2 production in animals: analysis of potential errors in the doubly labeled water method'. *American Journal of Physiology*, 363(6119), R466–R473.

Nagy, K. A. and Costa, D. P. 1980, 'Water flux in animals: analysis of potential errors in the tritiated water method'. *Journal of Comparative Physiology*, 363(6119), R454–R465.

Necker, R. 1994, 'Sensorimotor aspects of flight control in birds: specialisations in the spinal cord'. *European Journal of Morphology*, 32, 207–211.

Necker, R. 2000, 'The somatosensory system', in *Sturkie's Avian Physiology*, fifth edn, G. C. Whittow, ed., Academic Press, London, pp. 57–69.

Newton, I. 1686, *Philosophiae naturalis principia mathematica*, S. Pepys, London.

Nisbet, I. C. T. 1963, 'Measurements with radar of the height of nocturnal migration over Cape Cod, Massachusetts'. *Bird-Banding*, 34(2), 57–67.

Norberg, U. M. 1990, *Vertebrate flight*, Springer, Berlin.

Novas, F. E. and Puerta, P. F. 1997, 'New evidence concerning avian origins from the late Cretaceous of Patagonia'. *Nature*, 387, 390–392.

Obst, B. S. and Nagy, K. A. 1992, 'Field energy expenditures of the southern giant-petrel'. *The Condor*, 94(4), 801–810.

Obst, B. S., Nagy, K. A., and Ricklefs, R. E. 1987, 'Energy utilization by Wilson's storm petrel (Oceanites oceanicus)'. *Physiological Zoology*, 60(2), 200–217.

Oehme, H. 1963, 'Flug und Flügel von Star und Amsel. Ein Beitrag zur Biophysik des Vogelfluges und zur vergleichenden Morphologie der Flugorgane. 2. Teil: Die Flugorgane'. *Biologisches Zentralblatt*, 5, 569–587.

Oehme, H. 1968, 'Der Flug des Mauerseglers (Apus apus)'. *Biologisches Zentralblatt*, 3, 288–311.

Oehme, H. and Kitzler, U. 1974, 'Über die Kinematik des Flügelschlages beim unbeschleunigten Horizontalflug. Untersuchungen zur Flugbiophysik und Flugphysiologie der Vögel I'. *Zoologische Jahrbücher Physiologie*, 78, 461–512.

Ostrom, J. H. 1979, 'Bird flight: How did it begin?' *American Scientist*, 67, 46–56.

Padian, K. and Chiappe, L. M. 1998, 'The origin and early evolution of birds'. *Biological Reviews*, 73, 1–42.

Park, K. J., Rosén, M., and Hedenström, A. 2001, 'Flight kinematics of the barn swallow (*Hirundo rustica*) over a wide range of speeds in a wind tunnel'. *Journal of Experimental Biology*, 204(15), 2741–2750.

Paul, G. S. 2002, *Dinosaurs of the air: the evolution and loss of flight in dinosaurs and birds*, The Johns Hopkins University Press, Baltimore.

Pearson, O. P. 1950, 'The metabolism of hummingbirds'. *The Condor*, 52(4), 145–152.

Pearson, O. P. 1964, 'Metabolism and heatloss during flight in pigeons'. *The Condor*, 66, 182–185.

Pennycuick, C. J. 1968, 'A wind-tunnel study of gliding flight in the pigeon *Columbia livia*'. *Journal of Experimental Biology*, 49, 509–526.

Pennycuick, C. J. 1971*a*, 'Gliding flight of the white-backed vulture *Gyps africanus*'. *Journal of Experimental Biology*, 55, 13–38.
Pennycuick, C. J. 1971*b*, 'Control of gliding angle in Rüppell's grifon vulture *Gyps rüppellii*'. *Journal of Experimental Biology*, 55, 39–46.
Pennycuick, C. J. 1972, 'Soaring behaviour and performance of some East African birds, observed from a motor-glider'. *Ibis*, 114, 178–218.
Pennycuick, C. J. 1973, 'The soaring flight of vultures'. *Scientific American*, 229(6), 102–109.
Pennycuick, C. J. 1974, *Handy matrices of unit conversion factors for biology and mechanics*, Edward Arnold, London.
Pennycuick, C. J. 1975, 'Mechanics of Flight', in *Avian Biology*, vol. 5, D. S. Farner and J. R. King, eds., Academic Press, London, pp. 1–75.
Pennycuick, C. J. 1982, 'The flight of petrels and albatrosses (procellariiformes), observed in South Georgia and its vicinity'. *Philosophical Transactions of the Royal Society London B*, 300, 75–106.
Pennycuick, C. J. 1989, *Bird flight performance*, Oxford University Press, Oxford.
Pennycuick, C. J. 1995, 'The use and misuse of mathematical flight models'. *Israel Journal of Zoology*, 41, 307–319.
Pennycuick, C. J. 2002, 'Gust soaring as a basis for the flight of petrels and albatrosses (Procellariiformes)'. *Avian Science*, 2(1), 1–12.
Pennycuick, C. J., Alerstam, T., and Hedenström, A. 1997, 'A new low-turbulence wind tunnel for bird flight experiments at Lund University, Sweden'. *Journal of Experimental Biology*, 200, 1441–1449.
Pennycuick, C. J., Klaassen, M., Kvist, A., and Lindström, Å. 1996, 'Wingbeat frequency and the body drag anomaly: wind-tunnel observations on a thrush nightingale (*Luscinia luscinia*) and a teal (*Anas crecca*)'. *Journal of Experimental Biology*, 199, 2757–2765.
Peters, D. S. and Görgner, E. 'A comparative study on the claws of *Archaeopteryx*', *Papers in avian paleontology. Honoring Pierce Brodkorb*. K. E. Campbell, ed., Natural History Museum Los Angeles Science Series, 36, 29–37.
Peter, D. and Kestenholz, M. 1998, 'Sturtzflüge von Wanderfalke *Falco peregrinus* und Wüstenfalke *F. pelegrinoides*'. *Ornithologische Beobachter*, 95, 107–112.
Pettigrew, J. B. 1873, *Animal locomotion, or walking, swimming and flying*, Henry S. King and Co., London.
Pettit, T. N., Nagy, K. A., Ellis, H. I., and Whittow, G. C. 1988, 'Incubation energetics of the Laysan Albatross'. *Oecologia*, 74, 546–550.
Piersma, T. and Jukema, J. 1990, 'Budgeting the flight of a long-distance migrant: changes in nutrient reserve levels of bar-tailed godwits at successive spring staging sites'. *Ardea*, 78, 315–337.
Polhamus, E. C. 1971, "Predictions of vortex-lift characteristics by a leading-edge suction analogy", *Journal of aircraft*, vol. 8, no. 4, pp. 193–199.
Polus, M. 1985, 'Quantitative and qualitative respiratory measurements on unrestrained free-flying pigeons by AMACS (Airborne Measuring and

Control Systems)', in *Bird flight – Vogelflug*, BIONA-report, third edn, W. Nachtigall, ed., Gustav Fischer, Stuttgart and New York, pp. 297–305.

Poore, S. O., Sánchez-Haiman, A., and Goslow, G. E. 1997*a* , 'Wing upstroke and the evolution of flapping flight'. *Nature*, 387, 799–802.

Poore, S. O., Ashcroft, A., Sánchez-Haiman, A., and Goslow Jr, G. E. 1997*b*, 'The contractile properties of the M.supracoracoideus in the pigeon and starling: a case for long-axis rotation of the humerus'. *The Journal of Experimental Biology*, 200, 2987–3002.

Prince, P. A. and Morgan, R. A. 1987, 'Diet and feeding ecology of Procellariiformes', in *Seabirds, feeding ecology and role in marine ecosystems*, J. P. Croxall, ed., Cambridge University Press, Cambridge, pp. 135–171.

Proctor, N. S. and Lynch, P. J. 1993, *Manual of ornithology: avian structure and function*, Yale University Press, New Haven, CT and London.

Purslow, P. P. and Vincent, J. F. V. 1978, 'Mechanical properties of primary feathers from the pigeon'. *Journal of Experimental Biology*, 72, 251–260.

Rand, A. S. and Marx, H. 1967, 'Running speed of the lizard *Basiliscus basiliscus* on water'. *Copeia*, 1, 230–233.

Rayleigh, L. 1883, 'The soaring of birds'. *Nature*, 27, 534–535.

Rayner, J. 1977, 'The intermittent flight of birds', in *Scale effects in animal locomotion*, T. J. Pedley, ed., Academic Press, London, pp. 437–444.

Rayner, J. M. V. 1979*a*, 'A vortex theory of animal flight. Part 1. The vortex wake of a hovering animal'. *Journal of Fluid Mechanics*, 91(4), 697–730.

Rayner, J. M. V. 1979*b*, 'A vortex theory of animal flight. Part 2. The forward flight of birds'. *Journal of Fluid Mechanics*, 91(4), 731–763.

Rayner, J. M. V. 1979*c*, 'A new approach to animal flight mechanics'. *Journal of Experimental Biology*, 80, 17–54.

Rayner, J. M. V. 1985, 'Mechanical and ecological constraints on flight evolution', in *The beginnings of birds*, M. K. Hecht, Ostrom, J. H., Viohl, G., and Wellnhofer, P., eds., Freunde des Jura-Museums, Eichstätt, 279–288.

Rayner, J. M. V. 1985, 'Cursorial gliding in Proto-birds, an expanded version of a discussion contribution', in *The beginnings of birds*, M. K. Hecht, Ostrom, J. H., Viohl, G., and Wellnhofer, P., eds., Freunde des Jura-Museums, Eichstätt, pp. 289–292.

Rayner, J. M. V. 1985*c*, 'Bounding and undulating flight in birds'. *Journal of Theoretical Biology*, 117, 47–77.

Rayner, J. M. V. 1988, 'Form and function in avian flight' in *Current ornithology*, vol. 5, R. F. Johnston, ed., Plenum Press, New York, pp. 1–66.

Rayner, J. M. V. 1995, 'Dynamics of the vortex wakes of flying and swimming vertebrates', in *Biological fluid dynamics*, C. P. Ellington and T. J. Pedley, eds., The Company of Biologists Limited, Cambridge, pp. 131–155.

Reynolds, O. 1883, 'An experimental investigation of the circumstances which determine whether the motion of water shall be direct or sinuous, and of the law of resistance in parallel channels'. *Philosophical Transactions of the Royal Society of London*, 174, 935–982.

Richet, C., Richet, C., and Richet, A. 1909, 'Observations relatives au vol des oiseaux'. *Archivio di fisiologia*, 7, 301–321.

Rietschel, S. 1985, 'Feathers and wings of *Archaeopteryx*, and the question of her flight ability', in *The beginnings of birds*, M. K. Hecht, Ostrom, J. H., Viohl, G., and Wellnhofer, P., eds., Freunde des Jura-Museums, Eichstätt, pp. 251–260.

Rosén, M., Spedding, G. R., and Hedenström, A. 2003, 'The relationship between wingbeat kinematics and vortex wake of a thrush nightingale', in *Birds in the flow: mechanics, wake dynamics and flight performance,* PhD thesis M. Rosén, Department of Animal Ecology, Lund University, Sweden, 111–124.

Rothe, H. J., Biesel, W., and Nachtigall, W. 1987, 'Pigeon flight in a wind tunnel: II. Gas exchange and power requirements'. *Journal of Comparative Physiology*, 157B, 99–109.

Sambursky, S. 1987, *The physical world of late antiquity*, first paperback edn, Princeton University Press, Princeton, NJ.

Sane, S. P. 2003, 'The aerodynamics of insect flight'. *Journal of Experimental Biology*, 206, 4191–4208.

Sanz, J. L., Chiappe, L. M., Pérez-Moreno, B. P., Buscalioni, A. D., Moratalla, J. J., Ortega, F., and Poyato-Ariza, F. J. 1996, 'An early Cretaceous bird from Spain and its implications for the evolution of avian flight'. *Nature*, 382, 442–445.

Schuchmann, K. L. 1979*a*, 'Metabolism of flying hummingbirds'. *Ibis*, 121, 85–86.

Schuchmann, K. L. 1979*b*, 'Energieumsatz in Abhängigkeit von der Umgebungstemperatur beim Kolibri *Ocreatus u. underwoodii*'. *Journal für Ornithologie*, 120, 311–315.

Shipman, P. 1998, *Taking wing, Archaeopteryx and the evolution of bird flight*, Simon and Schuster, New York.

Sick, H. 1937, 'Morphologisch-funktionelle Untersuchungen über die Feinstruktur der Vogelfeder'. *Journal für Ornithologie*, 85(2), 206–372.

Slijper, E. J. 1950, *De vliegkunst in het dierenrijk*, Brill, E.J., Leiden.

Snow, D. W. and Perrins, C. M. 1998, *The birds of the Western Palearctic*, Oxford University Press, Oxford.

Speakman, J. R. and Racey, P. A. 1991, 'No cost of echolocation for bats in flight'. *Nature*, 350, 421–423.

Spedding, G. R. 1986, 'The wake of a jackdaw (*Corvus monedula*) in slow flight'. *Journal of Experimental Biology*, 125, 287–307.

Spedding, G. R. 1987, 'The wake of a kestrel (*Falco tinnunculus*) in flapping flight'. *Journal of Experimental Biology*, 127, 59–78.

Spedding, G. R., Hedenström, A., and Rosén, M. 2003, 'Quantitative studies of the wakes of freely flying birds in a low-turbulence wind tunnel'. *Experiments in Fluids*, 34, 291–303.

Spedding, G. R., Rayner, J. M. V., and Pennycuick, C. J. 1984, 'Momentum and energy in the wake of pigeon (Columbia livia) in slow flight'. *The Journal of Experimental Biology*, 111, 81–102.

Srygley, R. B. and Thomas, A. L. R. 2002, 'Unconventional lift-generating mechanisms in free-flying butterflies'. *Nature*, 420, 660–664.

Stamhuis, E. J. and Videler, J. J. 1995, 'Quantitative flow analysis around aquatic animals using laser sheet particle image velocimetry'. *Journal of Experimental Biology*, 198(2), 283–294.

Stamhuis, E. J., Videler, J. J., Duren, L. A. v., and Müller, U. K. 2002, 'Applying digital particle image velocimetry to animal-generated flows: traps, hurdles and cures in mapping steady and unsteady flows in *Re* regimes between 10^{-2} and 10^{5}'. *Experiments in Fluids*, 33, 801–813.

Stanfield, R. I. 1967, *Flying manual and pilot's guide*, Ziff-Davis, New York.

Steinbeck, J. 1958, *The log from the Sea of Cortez*. William Heinemann Ltd., London.

Stolpe, M. and Zimmer, K. 1939, 'Der Schwirrflug des Kolibri im Zeitlupenfilm'. *Journal für Ornithologie*, 87(1), 136–155.

Sun, M. and Tang, J. 2002, 'Unsteady aerodynamic force generation by a model fruit fly wing in flapping motion'. *Journal of Experimental Biology*, 205, 55–70.

Sunada, S. and Ellington, C. P. 2000, 'Approximate added-mass method for estimating induced power for flapping flight'. *AIAA Journal*, 38(8), 1313–1321.

Sutton, O. G. 1953, *Micrometeorology*, McGraw-Hill, New York.

Sy, M. 1936, 'Funktionell-anatomische Untersuchungen am Vogelflügel'. *Journal für Ornithologie*, 84(2), 199–296.

Tatner, P. and Bryant, D. M. 1986, 'Flight cost of a small passerine measured using doubly labelled water: implications for energetics studies'. *The Auk*, 103, 169–180.

Taylor, J. W. R. 1968, *Jane's all the world's aircraft*, McGraw-Hill, New York.

Taylor, W. R. and Van Dyke, G. C. 1985, 'Revised procedures for staining and clearing small fishes and other vertebrates for bone and cartilage'. *Cybium*, 9(2), 107–119.

Teal, J. M. 1969, 'Direct measurement of CO_2 production during flight in small birds'. *Zoologica*, 54(1), 17–23.

Thomas, A. L. R. 1995, *On the tail of birds*, PhD thesis Department of Animal Ecology, Lund University, Sweden.

Thomas, A. L. R. 2003, 'Insect flight: lift generating mechanisms and unsteady aerodynamics'. *Comparative Biochemistry and Physiology*, 134(3), S39.

Thomas, S. P. 1975, 'Metabolism during flight in two species of bats, *Phyllostomus hastatus* and *Pteropus gouldii*'. *Journal of Experimental Biology*, 63, 273–293.

Thomas, S. P. 1981, 'Ventilation and oxygen extraction in the bat *Pteropus gouldii* during rest and steady flight'. *Journal of Experimental Biology*, 94, 231–250.

Thulborn, R. A. and Hamley, T. L. 1985, 'A new palaeontological role for *Archaeopteryx*', in *The beginnings of birds*, M. K. Hecht, Ostrom, J. H.,

Viohl, G., and Wellnhofer, P., eds., Freunde des Jura-Museums, Eichstätt, pp. 81–90.

Tobalske, B. W. and Dial, K. P. 1996, 'Flight kinematics of black-billed magpies and pigeons over a wide range of speeds'. *Journal of Experimental Biology*, 199, 263–280.

Tobalske, B. W. 1996, 'Scaling of muscle composition, wing morphology and intermittent flight behavior in woodpeckers'. *The Auk*, 113(1), 151–177.

Tobalske, B. W., Peacock, W. L., and Dial, K. P. 1999, 'Kinematics of flapping flight in the zebra finch over a wide range of speeds'. *Journal of Experimental Biology*, 202(13), 1725–1739.

Tobalske, B. W., Altshuler, D. L., and Powers, D. R. 2004, 'Take-off mechanics in hummingbirds (Trochilidae)'. *Journal of Experimental Biology*, 207, 1345–1352.

Tobalske, B. W., Hedrick, T. L., Dial, K. P., and Biewener, A. A. 2003, 'Comparative power curves in bird flight'. *Nature*, 421, 363–366.

Torre-Bueno, J. R. and LaRochelle, J. 1978, 'The metabolic cost of flight in unrestrained birds'. *Journal of Experimental Biology*, 75, 223–229.

Tubaro, P. L. 2003, 'A comparative study of aerodynamic function and flexural stiffness of outer tail feathers in birds'. *Journal of Avian Biology*, 34, 243–250.

Tucker, V. A. 1968, 'Respiratory exchange and evaporative water loss in the flying Budgerigar'. *Journal of Experimental Biology*, 48, 67–87.

Tucker, V. A. 1972, 'Metabolism during flight in the laughing gull, *Larus atricilla*'. *American Journal of Physiology*, 222(2), 237–245.

Tucker, V. A. 1974, 'Energetics of Natural Avian Flight', in *Avian Energetics*, R. Paynter, ed., Cambridge University Press, Cambridge, MA, pp. 298–333.

Tucker, V. A. 1975, 'Aerodynamics and energetics of vertebrate fliers', in *Swimming and flying in nature*, vol. 2, T. Y. T. Wu, C. J. Brokaw, and C. Brennen, eds., Plenum, New York, pp. 845–868.

Turner, A. K. 1982*a*, 'Timing of laying by Swallows (*Hirundo rustica*) and Sand Martins (*Riparia riparia*)'. *Journal of Animal Ecology*, 51, 29–46.

Turner, A. K. 1982*b*, 'Optimal foraging by the Swallow (*Hirundo rustica*): prey size selection'. *Animal Behaviour*, 30, 862–872.

Usherwood, J. R. and Ellington, C. P. 2002*a*, 'The aerodynamics of revolving wings: I. Model hawkmoth wings'. *Journal of Experimental Biology*, 205, 1547–1564.

Usherwood, J. R. and Ellington, C. P. 2002*b*, 'The aerodynamics of revolving wings: II. Propeller force coefficients from mayfly to quail'. *Journal of Experimental Biology*, 205, 1565–1576.

Utter, J. M. and LeFebvre, E. A. 1970, 'Energy expenditure for free flight by the purple martin (*Progne subis*)'. *Comparative Biochemistry and Physiology*, 35, 713–719.

Van Tyne, J. and Berger, A. J. 1976, *Fundamentals of ornithology*, second edn, John Wiley, New York.

Vanden Berge, J. C. 1979, 'Myologia', in *Nomina anatomica avium*, J. J. Baumel, ed., Academic Press, London, pp. 175–219.

Vazquez, R. J. 1992, 'Functional osteology of the avian wrist and the evolution of flapping flight'. *Journal of Morphology*, 211, 259–268.

Vazquez, R. J. 1994, 'The automating skeletal and muscular mechanisms of the avian wing (Aves)'. *Zoomorphology*, 114, 59–71.

Videler, J. J. 1993, *Fish swimming*, Chapman and Hall, London.

Videler, J. J. 1997, *Bidden voor de kost*, Backhuys, Leiden.

Videler, J. J. 2000, '*Archaeopteryx*: A dinosaur running over water?' *Archaeopteryx*, 18, 27–34.

Videler, J. and Groenewold, A. 1991, 'Field measurements of hanging flight aerodynamics in the kestrel *Falco tinnunculus*'. *Journal of Experimental Biology*, 155, 519–530.

Videler, J. J., Weihs, D., and Daan, S. 1983, 'Intermittent gliding in the hunting flight of the kestrel, *Falco tinnunculus* L'. *Journal of Experimental Biology*, 102, 1–12.

Videler, J. J., Stamhuis, E. J., and Povel, G. D. E. 2004, 'Leading-edge vortex lifts swifts'. *Science*, 306, 1960–1962.

Videler, J. J., Vossebelt, G., Gnodde, M., and Groenewegen, A. 1988*a*, 'Indoor flight experiments with trained kestrels: I. Flight strategies in still air with and without added weight'. *Journal of Experimental Biology*, 134, 173–183.

Videler, J. J., Groenewegen, A., Gnodde, M., and Vossebelt, G. 1988*b*, 'Indoor flight experiments with trained kestrels: II. The effect of added weight on flapping flight kinematics'. *Journal of Experimental Biology*, 134, 185–199.

Viohl, G. 1985, 'Geology of the Solnhofen lithographic limestones and the habitat of *Archaeopteryx*', in *The beginnings of birds*, M. K. Hecht, Ostrom, J. H., Viohl, G., and Wellnhofer, P., eds., Freunde des Jura-Museums, Eichstätt, pp. 31–44.

Visser, G. H., Dekinga, A., Achterkamp, B., and Piersma, T. 2000, 'Ingested water equilibrates isotopically with the body water pool of a shorebird with unrivaled water fluxes'. *Am. J. Physiol. Regulatory Integrative Comp Physiol*, 279, R1795–R1804.

Voigt, C. C. and Winter, Y. 1999, 'Energetic cost of hovering flight in nectar-feeding bats (Phyllostomidae: Glossophaginae) and its scaling in moths, birds and bats'. *Journal of Comparative Physiology B*, 169, 38–48.

Wagner, H. 1925, 'Über die Entstehung des dynamischen Auftriebes von Tragflügeln.', *Zeitschrift für Angewandte Mathematik und Mechanik*, 5(1), 17–35.

Walsberg, G. E. and Wolf, B. O. 1995, 'Variation in the respiratory quotient of birds and implications for indirect calorimetry using measurements of carbon dioxide production'. *Journal of Experimental Biology*, 198(1), 213–219.

Ward, S., Möller, U., Rayner, J. M. V., Jackson, D. M., Bilo, D., Nachtigall, W., and Speakman, J. R. 2001, 'Metabolic power, mechanical power and efficiency during wind tunnel flight by European starlings *Sturnus vulgaris*.', *Journal of Experimental Biology*, 204, 3311–3322.

Ward, S., Bishop, C. M., Woakes, A. J., and Butler, P. J. 2002, 'Heart rate and the rate of oxygen consumption of flying and walking barnacle geese (*Branta leucopsis*) and bar-headed geese (*Anser indicus*)'. *Journal of Experimental Biology*, 205, 3347–3356.

Warrick, D. R., Dial, K. P., and Biewener, A. A. 1998, 'Asymmetrical force production in the maneuvring flight of pigeons'. *The Auk*, 115(4), 916–928.

Weimerskirch, H., Martin, J., Clerquin, Y., Alexandre, P., and Jiraskova, S. 2001, 'Energy saving in flight formation'. *Nature*, 413, 697–698.

Weis-Fogh, T. 1952, 'Fat combustion and metabolic rate of flying locusts (*Schistocerca gregaria Forskål*)'. *Philosophical Transactions of the Royal Society London B*, 237B, 1–36.

Weis-Fogh, T. 1973, 'Quick estimates of flight fitness in hovering animals, including novel mechanisms for lift production'. *Journal of Experimental Biology*, 59, 169–230.

Welham, C. V. J. 1994, 'Flight speeds of migrating birds: a test of maximum range speed predictions from three aerodynamic equations'. *Behavioural Ecology*, 5(1), 1–8.

Wellnhofer, P. 1985, 'Remarks on the digit and pubis problems of *Archaeopteryx*.', in *The beginnings of birds*, M. K. Hecht, Ostrom, J. H., Viohl, G., and Wellnhofer, P., eds., Freunde des Jura-Museums, Eichstätt, pp. 113–122.

Wellnhofer, P. 1993, 'Das siebte Exemplar von *Archaeopteryx* aus den Solnhofer Schichten'. *Archaeopteryx*, 11, 1–47.

Wells, D. J. 1993, 'Ecological correlates of hovering flight of hummingbirds'. *Journal of Experimental Biology*, 178, 59–70.

Westerterp, K. R. and Bryant, D. M. 1984, 'Energetics of free existence in swallows and martins (Hirundinidae) during breeding: a comparative study using doubly labeled water'. *Oecologia*, 62, 376–381.

Westerterp, K. R. and Drent, R. H. 'Energetic costs and energy-saving mechanisms in parental care of free-living passerine birds as determined by the $D_2^{18}O$ method.', *Proceedings of the 18th International Ortnithological Congress*. V. D. Ilyichev and V. M. Gavrilov, eds., Moscow, 392–398.

Westerterp, K. R., Saris, W. H. M., Vanes, M., and Tenhoor, F. 1986, 'Use of the doubly labelled water technique in humans during heavy sustained exercise'. *Journal of Applied Physiology*, 61(6), 2162–2167.

Wikelski, M., Tarlow, E. M., Raim, A., Diehl, R. H., Larkin, R. P., and Visser, G. H. 2003, 'Costs of migration in free-flying songbirds'. *Nature*, 423, 704.

Williamson, M. R., Dial, K. P., and Biewener, A. A. 2001, 'Pectoralis muscle performance during ascending and slow level flight in mallards (*Anas platyrhynchos*)'. *Journal of Experimental Biology*, 204, 495–507.

Wilson, J. A. 1975, 'Sweeping flight and soaring by albatrosses'. *Nature*, 257, 307–308.

Winter, Y. 1999, 'Flight speed and body mass of nectar-feeding bats (Glossophaginae) during foraging'. *Journal of Experimental Biology*, 202, 1917–1930.

Winter, Y. and Helversen, O. v. 1998, 'The energy cost of flight: Do small bats fly more cheaply than birds?' *Journal of Comparative Physiology B*, 168, 105–111.

Withers, P. C. 1979, 'Aerodynamics and hydrodynamics of the 'hovering' flight of Wilson's storm petrel'. *Journal of Experimental Biology*, 80, 83–91.

Witmer, L. M. 2002, 'The debate on avian ancestry: phylogeny, function, and fossils', in *Mesozoic birds: above the heads of dinosaurs*, L. M. Chiappe and L. M. Witmer, eds., University of California Press, Berkeley and Los Angeles, CA, pp. 3–30.

Woledge, R. C., Curtin, N. A., and Homsher, E. 1985, *Energetic aspects of muscle contraction*, Academic Press, London.

Wolf, L. L., Hainsworth, F. R., and Gill, F. B. 1975, 'Foraging efficiencies and time budgets in nectar feeding birds'. *Ecology*, 56, 117–128.

Wolf, T. J., Schmid-Hempel, P., Ellington, C. P., and Stevenson, R. D. 1989, 'Physiological correlates of foraging efforts in honey-bees: oxygen consumption and nectar load'. *Functional Ecology*, 3, 417–424.

Worcester, S. E. 1996, 'The scaling of the size and stiffness of primary flight feathers'. *Journal of Zoology, London.*, 239(3), 609–624.

Xu, X., Zhou, Z., Wang, X., Kuang, X., Zhang, F., and Du, X. 2003, 'Four-winged dinosaurs from China'. *Nature*, 421, 335–340.

Yalden, D. W. 1984, 'What size was Archaeopteryx?' *Zoological Journal of the Linnean Society*, 82, 177–188.

Yudin, K. A. 1957, 'O nyekotoryikh prisposobityel'nyikh osobyennostyakh kryila trubkonosyikh ptits (otryad Tubinares). [On certain adaptive properties of the wing in birds of the order Tubinares.]'. *Zoologiceskij Zurnal*, 36, 1859–1873.

Zhang, F. and Zhou, Z. 2000, 'A primitive Enantiornithine bird and the origine of feathers'. *Science*, 290, 1955–1959.

Zhou, Z. and Zhang, F. 2002, 'Largest bird from the early Cretaceous and its implications for the earliest avian ecological diversification'. *Naturwissenschaften*, 89, 34–38.

Zhou, Z. and Hou, L. 2002, 'The discovery and study of Mesozoic birds in China', in *Mesozoic birds: above the heads of dinosaurs*, L. M. Chiappe and L. M. Witmer, eds., University of California Press, Berkeley and Los Angeles, CA, pp. 129–159.

Index

Page numbers in *italics* refer to information in boxes.